Lecture Notes in Computer Science 11268

Commenced Publication in 1973
Founding and Former Series Editors:
Gerhard Goos, Juris Hartmanis, and Jan van Leeuwen

More information about this series at http://www.springer.com/series/7409

Leong Hou U · Haoran Xie (Eds.)

Web and Big Data

APWeb-WAIM 2018 International Workshops:
MWDA, BAH, KGMA, DMMOOC, DS
Macau, China, July 23–25, 2018
Revised Selected Papers

 Springer

Editors
Leong Hou U
University of Macau
Macao, China

Haoran Xie
Education University of Hong Kong
Hong Kong, China

ISSN 0302-9743 ISSN 1611-3349 (electronic)
Lecture Notes in Computer Science
ISBN 978-3-030-01297-7 ISBN 978-3-030-01298-4 (eBook)
https://doi.org/10.1007/978-3-030-01298-4

Library of Congress Control Number: 2018948814

LNCS Sublibrary: SL3 – Information Systems and Applications, incl. Internet/Web, and HCI

This Springer imprint is published by the registered company Springer Nature Switzerland AG
The registered company address is: Gewerbestrasse 11, 6330 Cham, Switzerland

Preface

The Asia Pacific Web (APWeb) and Web-Age Information Management (WAIM) Joint Conference on Web and Big Data is a leading international conference for researchers, practitioners, developers, and users to share and exchange their cutting-edge ideas, results, experiences, techniques, and tools in connection with all aspects of Web data management. As the first joint event, APWeb-WAIM 2018 was held in Macau SAR, China, during July 23–25, 2018, and it attracted participants from all over the world.

Along with the main conference, APWeb-WAIM workshops intend to provide an international forum for researchers to discuss and share research results. After reviewing the workshop proposals, we were able to accept five workshops, which covered topics in Web data, health-care data, massive open online course data, knowledge graph data, data mining, and data science. The diversity of topics in these workshops contributed to the main themes of the APWeb-WAIM conference. For these workshops, we accepted 31 full papers that were carefully reviewed from 44 submissions. The five workshops were as follows:

- Mobile Web Data Analytics 2018 (MWDA 2018)
- Big Data Analytics for Health Care 2018 (BAH 2018)
- The First International Workshop on Knowledge Graph Management and Analysis (KGMA 2018)
- Data Management and Mining on MOOCs 2018 (DMMOOC 2018)
- APWeb-WAIM Data Science Workshop 2018 (DS 2018)

August 2018

Leong Hou U
Haoran Xie

Organization

Workshop Co-chairs

Leong Hou U	University of Macau, SAR China
Haoran Xie	Education University of Hong Kong, SAR China

Mobile Web Data Analytics 2018 (MWDA 2018)

General Co-chairs

Li Li	Southwest University, China
Li Liu	Chongqing University, China
Xiangliang Zhang	King Abdullah University of Science and Technology, Saudi Arabia

Program Committee

Shiping Chen	CSIRO, Australia
Jiong Jin	Swinburne University of Technology, Australia
Ming Liu	Southwest University, China
Guoxin Su	National University of Singapore, Singapore
Min Gao	Chongqing University, China
Basma Alharbi	King Abdullah University of Science and Technology, Saudi Arabia
Zehui Qu	Southwest University, China
Yonggang Lu	Lanzhou University, China
Zhen Dong	National University of Singapore, Singapore
Ye Liu	National University of Singapore, Singapore
Jun Zeng	Chongqing University, China
Aiguo Wang	Hefei University of Technology, China
Xuejun Li	Anhui University, China
Chenren Xu	Peking University, China
Danni Wang	Chongqing University, China
Kunxia Wang	Texas A&M University, USA

Big Data Analytics for Health Care 2018 (BAH 2018)

General Co-chairs

Genlang Chen	Ningbo Institute of Technology, Zhejiang University
Shiting Wen	Ningbo Institute of Technology, Zhejiang University

Program Committee

An Liu	Soochow University, China
Chaoyang Zhu	Zhejiang University, China
Lihua Yue	University of Science and Technology of China, China
Chaoyi Pang	CSIRO, Australia
Yang Yang	Jiangsu University, China
Detian Zhang	Jiangnan University, China
Peiquan Jin	University of Science and Technology of China, China
Jinqiu Yang	NIT, Zhejiang University, China

The First International Workshop on Knowledge Graph Management and Analysis (KGMA 2018)

Honorary Co-chairs

Junhu Wang	Griffith University, Australia
Xiaojie Yuan	Nankai University, China

General Co-chairs

Xin Wang	Tianjin University, China
Jianxin Li	University of Western Australia, Australia

Program Committee

Gao Cong	Nanyang Technological University, Singapore
Huajun Chen	Zhejiang University, China
Tim French	University of Western Australia, Australia
Jun Gao	Peking University, China
Md Saiful Islam	Griffith University, Australia
Ronghua Li	Beijing Institute of Technology, China
Yuanfang Li	Monash University, Australia
Jiaheng Lu	University of Helsinki, Finland
Jeff Z. Pan	The University of Aberdeen, UK
Chuitian Rong	Tianjin Polytechnic University, China
Jijun Tang	University of South Carolina, USA
Haofen Wang	Shenzhen Gowild Robotics Co. Ltd., China
Hongzhi Wang	Harbin University of Industry, China
Kewen Wang	Griffith University, Australia
Guohui Xiao	Free University of Bozen-Bolzano, Italy
Yajun Yang	Tianjin University, China
Haiwei Zhang	Nankai University, China
Qingpeng Zhang	City University of Hong Kong, SAR China
Wei Zhang	Alibaba, China
Xiaowang Zhang	Tianjin University, China
Xuyun Zhang	The University of Auckland, New Zealand

| W. Jim Zheng | The University of Texas, Health Science Center at Houston, USA |
| Yuanyuan Zhu | Wuhan University, China |

Data Management and Mining on MOOCs 2018 (DMMOOC 2018)

Workshop Organizers

Wenjun Wu	Beihang University, China
Yan Zhang	Peking University, China
Yongxin Tong	Beihang University, China

Program Committee

Dawei Gao	Beihang University, China
Hailong Liu	Northwestern Polytechnical University, China
Jun Liu	Xian Jiaotong University, China
Xinjun Mao	National University of Defense Technology, China
Longfei Shangguan	Princeton University, USA
Tianshu Song	Beihang University, China
Zhiyang Su	Microsoft, China
Jie Tang	Tsinghua University, China
Qian Tao	Beihang University, China
Libin Wang	Beihang University, China
Qiong Wang	Peking University, China
Ting Wang	National University of Defense Technology, China
Jia Xu	Guangxi University, China
Wei Xu	Tsinghua University, China
Gang Yin	National University of Defense Technology, China
Yuxiang Zeng	Hong Kong University of Science and Technology, SAR China
Zimu Zhou	ETH Zurich, Switzerland

APWeb-WAIM Data Science Workshop 2018 (DS 2018)

Honorary Chair

| Zhiguo Gong | University of Macau, SAR China |

Workshop Co-chairs

| Leong Hou U | University of Macau, SAR China |
| Man Lung YIU | Hong Kong Polytechnic University, SAR China |

Program Committee Members

Gao Cong	Nanyang Technological University, Singapore
Haibo Hu	Hong Kong Polytechnic University, SAR China
Reynold Cheng	The University of Hong Kong, SAR China
Fai Wong	University of Macau, SAR China

Contents

APWeb-WAIM Data Science Workshop

Mobile Web Data Analytics

Hybrid Decision Based Chinese News Headline Classification

Yukun Cao, Xiaofei Xu, Ye Du, Jun He, and Li Li$^{(\boxtimes)}$

College of Computer and Information Science, Southwest University,
Chongqing 400715, China
yukun.cao@outlook.com , lily@swu.edu.cn

Abstract. In recent years, short text classification is attracting more attention. With the development of social platforms such as micro blogging and wechatting, Chinese short text classification has great impact on public opinion analysis and sentiment mining. Among social media texts, news headline classification has substantial influence on both academia and Internet economy. The issues such as semantic sparsity caused by the limited length of texts, and the grammatical nonstandard of the text, have prevented the performance of classification. In the paper, a Chinese news headline classification method based on multi model decision is proposed. First, an effective Convolutional Neural Network (CNN) is applied as one of text classifiers, at the same time, a Long Short-Term Memory (LSTM) is used as another text classifier as well. The aim is to obtain both abstract semantics of news headlines through CNN and context information between word sequences through LSTM. Second, an efficient text categorization tool - fastText (Facebook) is introduced to get the most excellent and balanced results. Finally, a decision model is proposed to favor the best performance of classification. A simple but very effective voting system is proposed and the result is very promising. Experiments based on the dataset from nlpcc 2017 Task2 has proved the efficiency of our method. Our method achieves much higher performance (F_1 of 79%) than the baseline provided by nlpcc 2017.

Keywords: CNN · LSTM · fastText · Short text classification
Hybrid decision

1 Introduction

This paper focuses on classification of Chinese news headlines. The classification of Chinese headlines is one of tasks in short text classification [1]. A news headline can be categorized into a precise classification from the semantic point of view. Compared to other text-based classification tasks, news headline classification faces with a couple of challenges. The following examples are from nlpcc 2017 Evaluation Task 2[1].

---

[1] https://github.com/FudanNLP/nlpcc2017_news_headline_categorization.

© Springer Nature Switzerland AG 2018
L. H. U and H. Xie (Eds.): APWeb-WAIM 2018, LNCS 11268, pp. 3–12, 2018.
https://doi.org/10.1007/978-3-030-01298-4_1

(Challenge 1) (The English translation: The tragedy will never happen again.) 这样的悲剧请别再发生.. (The sentence is short with little or no context. No significant semantic feature is available in this case.)

(Challenge 2)汉王刘邦为什么得天下？什么原因？ (The English translation: Why Liu became the founding emperor of Han dynasty? What was the reason behind it?) (The sentence fails to follow standard grammatical structure.)

(1) due to limited length of the text, only sparse semantics will be available. In other words, it is difficult for traditional statistical machine learning methods to abstract semantic features on a significant level. (2) the style of news headlines (coming from different networks, newspapers and magazines) might be totally different. Additionally, headlines usually are not completely compliant with standard grammar rules [2].

We first apply the Convolutional Neural Network(CNN) to deal with Challenge 1, the semantic sparseness of news headlines. The word embedding in the convolution calculation [3] is regarded as the abstract-level representation of a sentence. Unlike CNN which depicts textual features in neighboring spaces, the Recurrent Neural Network (RNN) is able to abstract textual features in time series. It takes into account the semantic relationship information with more concise sequences [4]. For Challenge 2, RNN will attempt to learn fine information when a sentence rule is confusing and the front and back components are not dependent. The RNN variant model, the Long Short-Term Memory (LSTM) is commonly used in the field of textual text classification. In order to improve the accuracy and stability of classification, the fastText classifier is introduced[2]. It provides an easy to manage but effective method for text representation and classification.

In our work, the above three models (CNN, LSTM and fastText classifier) are combined as a hybrid system. Internally, a voting mechanism is available to guide the best selection of results from three models. In the experiment, heuristic rules are provided which work well to cope with above two challenges. This is evidenced by the enhancement of classification accuracy of some ambiguous categories. We then apply our method Task2 of nlpcc 2017 Evaluation. In the experiment, F_1 measure reaches 79%, which is much higher than the baseline provided by nlpcc Evaluation and the average F_1 measure of all participating teams of the task.

The remaining of the paper is organized as follows: Sect. 2 will briefly summarize the related work. Section 3 will introduce the details of our method. Section 4 details the experiment and discussions. Finally, Sect. 5 is the conclusion.

2   Related Work

Short text classification has been widely studied and concerned in the text categorization task. News headline categorization is a very typical short text classification task. Due to short text and irregular syntax, the short text features

---

[2] https://fasttext.cc.

are sparse. It is difficult to obtain good results from the " feature engineering + classifier" method based on traditional machine learning.

Some studies use external corpus to enrich the semantic information of short texts. For example, Banerjee et al. [5] use network resources to expand short texts. However, it is difficult to obtain and organize high quality corpus from a large number of network resources. At the same time, adding external corpus makes the text representation more complicated. Other studies have classified short texts by constructing Latent Dirichlet Allocation(LDA) models [6] to mine hidden topics from external large-scale corpus.

In recent years, due to the development of deep learning. The distributed representation of words is obtained through a neural probabilistic language model [7], namely the word embedding process, which provides a powerful feature representation of words. Mikolov et al. [8] proposed a simpler and more effective word2vec word embedding method. It has been well validated in the semantic dimension and has greatly advanced the process of text analysis. Combined word embedding, a large number of studies have used CNN/RNN and other deep learning networks and their variants to obtain a text classification model. For example, Xu et al. [9] proposes a CNN model for short text classification. Young et al. [10] combines RNN and CNN for efficient short text classification.

Deep learning methods have achieved good results in text categorization. However, there are also problems with excessive training time and costs. The fastText developed by facebook, which provides a simple and effective method for text representation and classification [11], making text classification faster and more efficient.

Referring to the previous methods, this paper constructs simple but effective CNN short text classifier and LSTM short text classifier combined with the classification results of fastText as a baseline. In order to enhance the stability and accuracy of classification, this paper proposes a hybrid decision voting model based on the experimental results.

3   Methodology

3.1   CNN Short Text Categorization Model

In short text categorization, because of its limited length and compact structure, it can express semantics independently. So it possible for CNN to deal with this type of issue. Yoon Kim [12] proposed a short text classification model based on CNN. It mainly consists of 4 parts:

Input layer. The input layer is the n × k matrix of the word vectors corresponding to the words in the sentences (assuming there are n words in the sentence, the dimension of the vector is k), let $X_i \in \mathbb{R}^k$ be the k-dimensional word vector corresponding to the i-th word in the sentence. So $X_{i:i+j}$ refer to the concatenation of words X_i, X_{i+1}, ..., X_{i+j}, where the sentence with degree n is represented as

$$X_{1:n} = X_1 \oplus X_2 \oplus \ldots X_n. \tag{1}$$

where $\oplus$ is the concatenation operator.

Convolutional layer. In the convolutional layer, which obtains several feature maps through a convolution operation. The convolution window size is h × k (h denotes the number of words, and k denotes the dimension of the word vector amount). Through this large convolution window, we will eventually get a number of feature maps of 1 column. For example, feature e_i is obtained from a window of words $X_{i:i+h-1}$ by

$$e_i = f(W \cdot X_{i:i+h-1} + b). \tag{2}$$

Here $b \in \mathbb{R}$ is a bias term, f is a non-linear function such as hyperbolic tangent and a *filter* $W \in \mathbb{R}^{hk}$. Finally, a feature map will be generated from each possible window of words in the sentence $\{X_{1:h}, X_{2:h+1}, \ldots, X_{n-h+1:n}\}$:

$$\mathbf{e} = [e_1, e_2, \ldots, e_{n-h+1}]. \tag{3}$$

with $\mathbf{e} \in \mathbb{R}^{n-h+1}$.

Pooling layer. The model uses a "max-over-timepooling" method [13] to simply obtain the maximum value from a one-dimensional feature map, which is interpreted as representing the most important signal, ie:

$$\widehat{e} = max\{\mathbf{e}\}. \tag{4}$$

Fully connected layer. The one-dimensional vector output by the pooling layer is connected to a softmax layer by a full connection. Eventually, model can get and output the probability distribution on the different categories.

This paper learn from the above model to propose a simple CNN text classification model. The model is shown in Fig. 2. Firstly, an **Embedding layer** is constructed to process text into a vector.The main process is shown in Fig. 1. secondly, the dimension of the vector is reduced by a convolutional layer and a pooling layer, and then a flatten layer is added to compress the vector dimension to 1D. finally, model get the output through two fully connected layers.

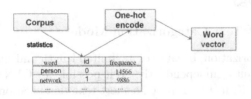

Fig. 1. Embedding process

3.2   LSTM Short Text Categorization Model

RNN can use sequence data and capture context information very well, so it can also be applied to text classification. As a variant of the RNN. LSTM(Long Short

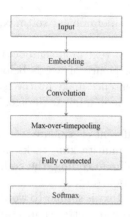

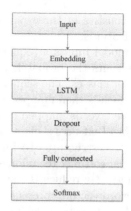

Fig. 2. CNN categorization model   **Fig. 3.** LSTM categorization model

Term Memory), which is a sequence model based on recurrent neural networks to memory each node of the input sequence [14]. This method has a feature which is very important in processing and predicting long interval and delay events. Therefore, this method performs very well in processing sequenced text data.

LSTM has many different variations, the typical model is used for illustration. A cell consists of three gates (input, forget, output) and a cell unit. A gate uses a sigmoid activation function, while input and cell state usually use tanh to convert. LSTM's cell can be defined using the following equation:

Gates:

$$i_t = g\left(W_{xi}x_t + W_{hi}h_{t-1} + b_i\right) \tag{5}$$

$$f_t = g\left(W_{xf}x_t + W_{hf}h_{t-1} + b_f\right) \tag{6}$$

$$o_t = g\left(W_{xo}x_t + W_{ho}h_{t-1} + b_o\right) \tag{7}$$

Input Transformation:

$$c\_in_t = tanh\left(W_{xc}x_t + W_{hc}h_{t-1} + b_{c\_in}\right) \tag{8}$$

State Update:

$$c_t = f_t * c_{t-1} + i_t * c\_in_t \tag{9}$$

$$h_t = o_t * tanh\left(c_t\right) \tag{10}$$

This paper presents a simple LSTM short text classification model. Model shown in Fig. 3.

3.3   fastText

The fastText classifier is developed by Facebook that provides simple and efficient text categorization [11]. It is a method of characterizing learning [15]. Its method consists of three parts: model architecture, hierarchical softmax, and N-gram features.

The model inputs a sequence of words (a piece of text or a sentence), and finally outputs the probability distribution of the sequence of words belonging to different classes. The sequences of words in the sentence make up the feature vector, which is then mapped to the middle layer through a linear change. Then the middle layer is mapped to the label.

The fastText uses a hierarchical classifier and different classes are integrated into the tree structure. To improve the runtime of multitasking tasks, it uses hierarchical softmax techniques. On the basis of Huffman coding, the coding of tags can greatly reduce the number of model prediction targets.

The word vector representation method of the fastText is similar to word2vec, but it adds character-level N-gram features, so it performs better than word2vec in classification problems [15].

This paper selects the fastText classifier as part of the hybrid model. It can provide good baseline classification results with little additional training overhead.

3.4   Multi-model Hybrid Decision Voting Model

This paper ultimately uses results voting decisions for CNN, LSTM, and the fastText. We presented the hybrid decision model based on the performance of the individual classifiers obtained from the experiments and the voting system from Zhu et al. [2] on nlpcc 2017 (see Fig. 4).

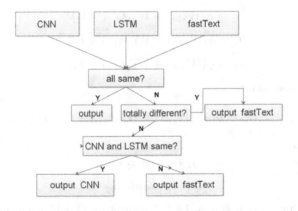

Fig. 4. The hybrid decision model

Rule1: If the three models have the same judgments, we directly output the result.

Rule2: If the three models have totally different judgments, we will eventually be biased towards the results of the fastText.

Rule3: In the remaining case. We focuses on whether the CNN and LSTM results are same. If they are the same, the results of fastText will be ignored, biased the results of deep learning networks. If not, we still retains the result of fastText.

For **Rule1**, the idea is obvious. From the experiment, the correct intersection of the three models accounts for 70% of the test cases, and we list these parts as trustworthy items. This also reflects the possibility that we use the voting rules to increase the classification accuracy in the remaining 30%.

Rule2 is based on the individual performance of the three models. Through experimental testing, the fastText provides a more stable classification effect. So when the three models have different judgments, we are more inclined to the fastText.

Rule3 is based on the experimental analysis of [2], we believe that when the deep learning method has the same judgment, it has a very high degree of confidence. However, when their judgements are different, CNN and LSTM are inconsistent in different categories, and it is difficult to have a good judgment mechanism. At this point we completely ignore their judgments and still choose the result of fastText.

Finally, through this hybrid decision-making mechanism, all indicators on the training data have been significantly improved.

4   Experiments and Results

4.1   Datasets

The datasets used in this paper is from nlpcc 2017 task2. The goal of the task is to categorize a Chinese news headline into one of the predefined 18 categories. The following table lists the details of the data set (Table 1):

Table 1. The statistic of datasets

Category	Train	Dev	Test	Category	Train	Dev	Test
Entertainment	10000	2000	2000	Military	10000	2000	2000
Sports	10000	2000	2000	History	10000	2000	2000
Car	10000	2000	2000	Baby	10000	2000	2000
Society	10000	2000	2000	Fashion	10000	2000	2000
Tech	10000	2000	2000	Food	10000	2000	2000
World	10000	2000	2000	Discovery	4000	2000	2000
Finance	10000	2000	2000	Story	4000	2000	2000
Game	10000	2000	2000	Regimen	4000	2000	2000
Travel	10000	2000	2000	Essay	4000	2000	2000

This paper will use 156,000 samples of data as a training set and 36,000 samples of data as a development set. Finally, the model will be evaluated with 36,000 test data.

4.2 Experiment Setup

Data preprocessing. In the CNN and RNN models, the same data preprocessing process is used in this paper. Firstly, the Chinese word segmentation is performed using the jieba[3]. Later, when the sentence is processed into a sequence of word indexes, 35 is determined as the maximum sentence input length after a number of experimental verifications. We padding zero-length sequences with zeros,and the label is finally processed as an 18-dimensional ont-hot vector.

CNN and LSTM. In this work, we choose kreas[4] to build the deep learning model. It is worth noting that at the word embedding layer, we set the dimensions of the sentence vector to 300 dimensions, which proved to be effective.

Fasttext. This paper uses the fastText toolkit[5] provided by facebook.

4.3 Results and Discussion

We use macro-averaged precision, recall, F1 and accuracy to evaluate the performance of the hybrid decision model. Nlpcc 2017 task2 uses the results of LSTM, CNN, and NBOW as baselines, this paper introduces them and compares them with the model. In addition, we count all the evaluation results of participating in nlpcc 2017 task 2.

The final experimental results are shown in Table 2.

Table 2. Experimental results

Model	Precision	Recall	F1-Score	Accuracy
LSTM(nlpcc 2017 baseline)	0.760	0.747	0.750	0.747
CNN(nlpcc 2017 baselines)	0.769	0.763	0.764	0.763
NBOW(nlpcc 2017 baseline)	0.791	0.783	0.784	0.783
CNN(our model)	0.780	0.773	0.774	0.773
LSTM(our model)	0.738	0.728	0.730	0.768
Fasttext	0.792	0.785	0.787	0.786
Hybrid decision model	**0.796**	**0.789**	**0.790**	**0.789**

We can see that in our three basic approaches, the performance have clearly outperformed the baseline provided by nplcc 2017. The fastText is the best and the LSTM is general. The possible reason is that the amount of training data is small. LSTM is difficult to play its due effect.We will regard this as a direction for further in-depth study, and explore a LSTM short text classification model with better performance. The performance of the CNN model and the fastText model is relatively stable. This shows that for short text classification

---

[3] http://github.com/fxsjy/jieba.

[4] https://github.com/keras-team/keras.

[5] https://fasttext.cc.

problems, they can provide a good baseline result. However, in the embedding we do not consider much optimization. Further research can incorporate word vector models pre-trained in corpus of news fields, or enrich the semantic features of news headlines to achieve better results. The hybrid decision model achieved the best performance on all evaluation indicators, which also indicated the validity of the voting rules. Based on the voting model of this paper, the results of the three basic models can be combined in the most stable manner. Compared to the baseline results of nlpcc, the voting model undoubtedly shows great advantages. This is due in large to the good results of the three basic models. However, compared to the three basic models, the voting model appears to be more stable and reasonable, balancing the results of the large differences between the models.

Table 3 shows a comparison of our results with all the evaluation results of participating nlpcc 2017 task2.

Table 3. The performances of all evaluation results and our model

Model	Precision	Recall	F1-Score	Accuracy
Best in nlpcc 2017	0.831	0.829	0.830	0.830
Median in nlpcc 2017	0.789	0.785	0.787	0.785
Average in nlpcc 2017	0.733	0.722	0.727	0.722
Hybrid decision model	**0.796**	**0.789**	**0.790**	**0.789**

Our results are upstream in all assessments. We noticed that in the [2], the results of CNN, GRU, and SVM were combined with similar voting models and they also showed better performance. Therefore, it is possible to combine traditional machine learning methods with deep learning models. However, simply balancing and selecting the classification results of different models is not an innovative method. Method with the highest evaluation result, we can see that the better extraction and expansion of the semantic features of short texts is an extremely important aspect as shown in [16].

5   Conclusion

In the paper, a multi-model decision voting mechanism is proposed to improve performance of the short text classification. The aim is to group Chinese news headlines into the predefined categories automatically. We incorporate two classifiers and use the fastText classifier to tune the difference between deep learning classifiers. A more stable result is achieved in the experiment. Experimental results show that our hybrid method is effective and efficient. We are planning to consider more semantic features of sentence vectors, for instance, introducing a large amount of corpus of current and domestic news to enrich word embedding models and expand semantic representations of the sentences.

Acknowledgement. This work was supported by National Undergraduate Training Program for Innovation and Entrepreneurship(NO.201810635003)

References

1. Song, G., Ye, Y., Du, X., Huang, X., Bie, S.: Short text classification: a survey. J. Multimed. **9**(5) (2014)
2. Zhu, F., Dong, X., Song, R., Hong, Y., Zhu, Q.: A multiple learning model based voting system for news headline classification. In: Huang, X., Jiang, J., Zhao, D., Feng, Y., Hong, Y. (eds.) NLPCC 2017. LNCS (LNAI), vol. 10619, pp. 797–806. Springer, Cham (2018). https://doi.org/10.1007/978-3-319-73618-1_69
3. Mikolov, T., Sutskever, I., Chen, K., Corrado, G.S., Dean, J.: Distributed representations of words and phrases and their compositionality. In: Advances in Neural Information Processing Systems, pp. 3111–3119 (2013)
4. Liu, P., Qiu, X., Huang, X.: Recurrent neural network for text classification with multi-task learning. arXiv preprint arXiv:1605.05101 (2016)
5. Banerjee, S., Ramanathan, K., Gupta, A.: Clustering short texts using wikipedia. In: Proceedings of the 30th Annual International ACM SIGIR Conference on Research and Development in Information Retrieval, pp. 787–788. ACM (2007)
6. Blei, D.M., Ng, A.Y., Jordan, M.I.: Latent Dirichlet allocation. J. Mach. Learn. Res. **3**(Jan), 993–1022 (2003)
7. Bengio, Y., Ducharme, R., Vincent, P., Jauvin, C.: A neural probabilistic language model. J. Mach. Learn. Res. **3**(Feb), 1137–1155 (2003)
8. Mikolov, T., Chen, K., Corrado, G., Dean, J.: Efficient estimation of word representations in vector space. arXiv preprint arXiv:1301.3781 (2013)
9. Wang, P., Xu, B., Xu, J., Tian, G., Liu, C.-L., Hao, H.: Semantic expansion using word embedding clustering and convolutional neural network for improving short text classification. Neurocomputing **174**, 806–814 (2016)
10. Lee, J.Y., Dernoncourt, F.: Sequential short-text classification with recurrent and convolutional neural networks. arXiv preprint arXiv:1603.03827 (2016)
11. Joulin, A., Grave, E., Bojanowski, P., Mikolov, T.: Bag of tricks for efficient text classification. arXiv preprint arXiv:1607.01759 (2016)
12. Kim, Y.: Convolutional neural networks for sentence classification. arXiv preprint arXiv:1408.5882 (2014)
13. Collobert, R., Weston, J., Bottou, L., Karlen, M., Kavukcuoglu, K., Kuksa, P.: Natural language processing (almost) from scratch. J. Mach. Learn. Res. **12**(Aug), 2493–2537 (2011)
14. Hochreiter, S., Schmidhuber, J.: Long short-term memory. Neural Comput. **9**(8), 1735–1780 (1997)
15. Bojanowski, P., Grave, E., Joulin, A., Mikolov, T.: Enriching word vectors with subword information. arXiv preprint arXiv:1607.04606 (2016)
16. Yin, Z., Tang, J., Ru, C., Luo, W., Luo, Z., Ma, X.: A semantic representation enhancement method for Chinese news headline classification. In: Huang, X., Jiang, J., Zhao, D., Feng, Y., Hong, Y. (eds.) NLPCC 2017. LNCS (LNAI), vol. 10619, pp. 318–328. Springer, Cham (2018). https://doi.org/10.1007/978-3-319-73618-1_27

A Self-representation Model for Robust Clustering of Categorical Sequences

Kunpeng Xu[1,2], Lifei Chen[1,2(✉)], Shengrui Wang[3], and Beizhan Wang[4]

[1] School of Mathematics and Informatics, Fujian Normal University,
Fuzhou 350117, China
[2] Digit Fujian Internet-of-Things Laboratory of Environmental Monitoring,
Fujian Normal University, Fuzhou 350117, China
`chrisxkp1994@163.com, clfei@fjnu.edu.cn`
[3] Department of Computer Science, University of Sherbrooke,
Sherbrooke, QC J1K2R1, Canada
[4] School of Software, Xiamen University, Xiamen 361005, China

Abstract. Robust clustering on categorical sequences remains an open and challenging task due to the noise data and lack of an inherently meaningful measure of pairwise similarity between sequences. In this paper, a self-representation model is proposed as a representation of categorical sequences. Based on the model, we transform the robust clustering to a subspace clustering problem. Furthermore, an efficient algorithm for robust clustering of categorical sequences is also proposed, which provides the new measure with high-quality clustering results and the elimination of noise sequences using the subspace method. The experimental results on the synthetic and real world data demonstrate the promising performance of the proposed method.

Keywords: Categorical sequences · Self-representation
Robust clustering

1 Introduction

The past few years have been a rapid increase in the amount of sequence data in mobile web domain, such as speech sequences, text documents, web usage data, behavior or event sequences, etc. With the upsurge in the amount of sequence data, its automatic analysis has become increasingly important, yet still remains a challenging task in the data mining domain [1]. In this paper, we focus on the challenging problem of clustering categorical sequences, in which the instances are ordered list of nominal categories (such as the click-through events of web usage sequences, the 20 possible amino acids making up protein sequences, and so on).

Supported by the National Natural Science Foundation of China under Grant 61672157; Innovative Research Team of Probability and Statistics: Theory and Application (IRTL1704).

L. H. U and H. Xie (Eds.): APWeb-WAIM 2018, LNCS 11268, pp. 13–23, 2018.
https://doi.org/10.1007/978-3-030-01298-4_2

Unlike vectorial data, categorical sequences may be of different length and have inherent structure that a fully nonparametric method cannot capture. Various algorithms have been developed for categorical sequences. Disregarding their difference, the key point behind them is to transform raw categorical sequences into the form of vectorial data where traditional clustering algorithms can be directly applied. These approaches defined for categorical sequences in the literature [2] can be roughly divided into two groups: directly approach and indirectly approach, respectively, in the following representation.

The former approach, such as n-gram [3], SCS [4], etc., directly measures the similarity between sequences. These methods vary in the definition of a pairwise similarity measure: n-gram finds the concurrence of the common subsequences to measure the similarity of two sequences; SCS suggests a different approach to find the significant subsequence, which is not the simple n-gram. In fact, SCS constructs a natural language processing method using symbol strings as "words" and sequences as texts to measure the similarity of the two sequences, the method of finding "words" is similar to the method of biological sequences searching for matching fragments. However, such methods cannot capture the hidden properties underlying the categorical sequences such as global structural natures in the sequences, which is important for pattern extractions.

Such an approach is able to capture the global structural natures hidden in sequences, which we also called model-based approaches, first project sequences into a new space, and then define the similarity between the sequences in the new space. Model-based approaches aim at exploring a best model to fit the observation sequences. In this type of methods, HMM has attracted increasing attention over the last decade. Blasiak [5] makes use of the transfer matrix of the HMM model to represent a sequence. Smyth [6] presented a probabilistic model-based approach to clustering sequences using HMM. This approach first devises a pairwise distance matrix between observation sequences by computing a symmetrized similarity. This similarity is obtained by training an HMM for each sequence, so that the log-likelihood of each sequence, given each model, can be computed. While model-based approaches provide a general probabilistic framework for clustering categorical sequences, the measure between two sequences only uses the model parameters generated by the two sequences. Obviously, these methods do not take into account overall statistics and if the sequence is very small over dataset, the effectiveness of this statistically-based method will be reduced.

Another challenging issue is noise in categorical sequences. In fact, most categorical sequences are infected with significant quantities of noise [7], which confuses the identification of cluster structures. In such cases, the performance of clustering could be disturbed by noise data [8]. Figure 1 gives an example. The sequence set shown in Fig. 1 consists of six sequences in two groups c_1 and c_2 and the sequence data O_3 and O_4 differ greatly from the rest of sequences, which we called noise data. If the cluster analysis is performed directly, this leads to seriously affected the performance of clustering due to the noise data disturbing the partition of the rest of the data.

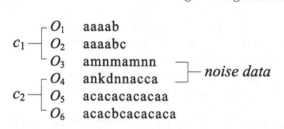

Fig. 1. The noise data in clusters;

In this paper, we propose a self-representation model to solve these problem. Our model is derived from the representational learning perspective, based on the discrete probability distribution using Hidden Markov Model(HMM) and kernel embedding of distribution(KED). In this model, the sequence is represented by all other sequences and to construct a uniform-length feature vector to capture the overall statistical information. Because of each feature of vector represents a object in dataset, the feature selection is equivalent to sample selection, which means that we can transform the robust clustering into a subspace clustering problem to deal with noise data. To provide robust clustering results for the clustering process, we also propose a soft subspace clustering algorithm for categorical sequences, named SSCS. A series of experiments on synthetic and real-world categorical sequences are conducted to examine the performance of the proposed clustering algorithm, and the experimental results show their effectiveness.

The remainder of this paper is organized as follows. Section 2 discusses the background knowledge on HMM and KED. Section 3 describes the proposed self-representation model in detail. In Sect. 4, we present the design of SSCS algorithm. Section 5 experimentally evaluates the proposed method. Section 6 gives our conclusion and discusses the directions of future work.

2 Background

2.1 Hidden Markov Model

HMM is a statistical generative model. It has been popularly used in temporal sequence analysis for bioinformatics, speech recognition, motion pattern detection, text classification and sign language identification. HMM possess many advantages that make it practical both in research and applications.

A discrete-time hidden Markov model λ can be viewed as a Markov model whose states are not directly observed: instead, each state is characterized by a probability distribution function, modelling the observations corresponding to that state. More formally, an HMM is defined by the following entities:

- $S=\{S_1,S_2,...,S_D\}$ the finite set of possible (hidden) states;
- The transition matrix $\boldsymbol{A}=\{a_{ij}, 1 \leqslant j \leqslant D\}$ representing the probability of moving from state S_i to state S_j,
$$a_{ij}=P[q_{t+1}=S_j|q_t=S_i], 1 \leqslant i,j \leqslant D,$$
with $a_{ij} \geqslant 0, \sum_{j=1}^{D} a_{ij} = 1$, and where q_t denotes the state occupied by the model at time t.
- The emission matrix $\boldsymbol{B} = \{b(o|S_j)\}$, indicating the probability of emission of symbol $o \in V$ when system state is S_j; V can be a discrete alphabet or of continuous set (e.g. $V = \mathbb{R}$), in which case $b(o|S_j)$ is a probability density function.
- $\boldsymbol{\pi} = \{\pi_i\}$, the initial state probability distribution,
$$\pi_i = P[q_1 = S_i], 1 \leqslant i \leqslant D,$$
with $\pi_i \geqslant 0$ and $\sum_{i=1}^{D} \pi_i = 1$.

For convenience, we represent an HMM by a triplet $\boldsymbol{\lambda} = (\boldsymbol{A}, \boldsymbol{B}, \boldsymbol{\pi})$.

Let O_i be an observation sequence, and let $\boldsymbol{\lambda}$ be the parameters of an HMM. The following are the main tasks of an HMM learning algorithm [9]:

- Compute the posterior probability of observation sequence given the model, i.e., $P(O_i|\boldsymbol{\lambda})$ (the forward-backward algorithm).
- Find an optimal sequence of states that maximizes the probability of the observation sequence O_i (the Viterbi algorithm).
- Learn the parameters $\boldsymbol{\lambda}$ that maximize the probability of the observation sequence $P(O_i|\boldsymbol{\lambda})$ (the Baum-Welch algorithm).

2.2   Kernel Embedding of Distribution

Here, we briefly review the kernel embedding of distribution approach [10]. The key idea of this line of work is to implicitly map distribution into infinite dimensional feature spaces using kernel, such that subsequent comparisons and manipulations of distributions can be achieved via feature space operations. This can be thought of as a generalization of the feature mapping of individual points, as used in classical kernel methods. By mapping probabilities into infinite-dimensional feature spaces, we can ultimately capture all the statistical features of arbitrary distributions [11].

In the following we denote by $\mathcal{X}$ the space, and let $k : \mathcal{X} \times \mathcal{X} \to \mathbb{R}$ be a positive definite kernel. For any positive definite kernel k, there exists a Hilbert space $\mathcal{H}$ uniquely associated with the kernel, called the reproducing kernel Hilbert space (RKHS), which consists of functions on $\mathcal{X}$ [12]. It is known that the reproducing property holds for $\mathcal{H}$: for any $x \in \mathcal{X}$ and $f \in \mathcal{H}$, we have $< k(\cdot, x), f >_{\mathcal{H}}= f(x)$, where $< \cdot, \cdot >_{\mathcal{H}}$ denotes the inner product of $\mathcal{H}$. Here, $k(\cdot, x) =: \phi(x)$ is the (possibly infinite-dimensional) feature vector of x. Then we can compute the inner product between feature vectors by $< \phi(x), \phi(x') >_{\mathcal{H}}= k(x, x')$ using the reproducing property.

Let $\mathcal{P}$ be the set of all probability distributions on $\mathcal{X}$. Then we represent any probability distribution $P \in \mathcal{P}$ as an embedding into $\mathcal{H}$ defined by [10]:

$$m_P := \boldsymbol{E}_{\mathcal{X} \sim P}[\phi(\mathcal{X})] \tag{1}$$

Namely, we represent the distribution P by the expectation of the feature vector $\phi(X)$. We refer to $m_P \in \mathcal{H}$ as the kernel embedding of distribution P. We have a reproducing property for m_P in terms of expectation: for any $f \in \mathcal{H}$, we have $< f, m_P >_{\mathcal{H}} = E_{X \sim P}[f(X)]$ [10].

If the map $\mathcal{P} \to \mathcal{H} : P \to m_P$ is injective, i.e. $m_P = m_Q$, then the kernel k used for embedding is called characteristic [13]. In other words, if we use a characteristic kernel, m_P uniquely identifies the distribution P. Thus, if our objective is to estimate the difference between probability distributions P and Q, it suffices to estimate the difference of m_P and m_Q, see details in Fig. 2. Note that, the discrete probability distribution is just one of the special cases.

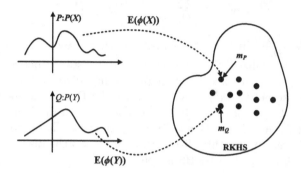

Fig. 2. Geometric interpretation of kernel embedding of distribution, P, Q represent the probability distribution of random variables X and Y, respectively.

3   A Self-representation Model

In this section, we present a detailed description of our self-representation model for categorical sequences set, in which all the sequences are composed from the finite alphabet Ω.

The traditional sample self-representation method is to represent every sample as the linear combination of other samples, such that make good use of the correlation between samples, e.g. let $\mathcal{O} = < O_1, O_2, ..., O_N >$ be the sample space, and the self-representation of O_i in [14] is defined as $\hat{O}_i = \mathcal{O} Z_i$, where Z_i is a column vector of $N \times 1$. Note that, this approach cannot be directly used for categorical sequences due to its implicit assumption that the dimensions of the elements in $\mathcal{O}$ should be consistent.

To represent sequence by self-representation, we adopt HMM to modeling each sequence and the posterior probability of each sequence given each model can be computed, we represent O_i by an N-dimensional vector composed of posterior probabilities. Formally, we define the self-representation of O_i as:

$$\hat{O}_i = < P(O_i|\lambda_1), P(O_i|\lambda_2), ..., P(O_i|\lambda_N) > \qquad (2)$$

Where $P(O_i|\lambda_j)(1 \leqslant i,j \leqslant N)$ represents the probability of model λ_j to generate O_i by using the standard forward-backward algorithm [9] and λ_j is the HMM parameters of observation sequence O_j by applying the Baum-Welch algorithm [9]. In fact, the probability of model λ_j to generate O_i represents the correlation between samples O_i and O_j, which is essentially the same as the traditional self-representation model.

In HMM, if the length of categorical sequence is large, the posterior probability will approaches to 0. From a statistical perspective, such a probability typically is meaningless. Thus, we enlarge the difference in these probabilities by normalization. The normalized vector $\widehat{O}_i$ is defined as:

$$\widehat{O}_{i\_norm} = \frac{\widehat{O}_i}{\sum \widehat{O}_i} = < P_{i1}, P_{i2}, ..., P_{iN} > \tag{3}$$

Where $P_{ij} = \frac{P(O_i)|\lambda_j}{\sum \widehat{O}_i}$. In fact, the sequence O_i is represented as a probability distribution by normalized. This means that the distance between sequences is equivalent to the difference between probability distributions. We use the Kronecker delta kernel $k(x,x') = \delta(x,x')$ [11] in KED, the feature map $\phi(x)$ is then the standard basis of e_x in $\mathbb{R}^N$. We denote by X a random variable with distribution $P(X)$ and define $P(x = j) = P_{ij}$, it is reasonable to say that the $P(X)$ is the discrete probability distribution of O_i. So the discrete probability distribution of O_i mapped to RKHS is defined as:

$$\begin{pmatrix} P_{i1} \\ \cdot \\ \cdot \\ \cdot \\ P_{iN} \end{pmatrix} = \begin{pmatrix} P(x=1) \\ \cdot \\ \cdot \\ \cdot \\ P(x=N) \end{pmatrix} = E_X(\phi(X)) = \sum_{x=1}^{N} P(x)e_x = m_{\widehat{O}_i} \tag{4}$$

For convenience, we use m_i to denote $m_{\widehat{O}_i}$. Note that, $m_i = < m_{i1}, m_{i2}, .., m_{iN} >$ is an N-dimensional vector in RKHS and the self-representation model of categorical sequences is outlined in Algorithm 1.

Algorithm 1 The self-representation model of categorical sequences

---

Input: N categorical sequences $\{O_1, O_2, .., O_N\}$
Output: N elements in RKHS $\{m_1, m_2, ..., m_N\}$

1. For each sequence O_i, set the transition matrix uniformly and set the number of hidden states
2. Fit one HMM for each O_i using Baum-Welch algorithm to obtain HMM models parameters $\lambda = \{\lambda_1, \lambda_2, ..., \lambda_N\}$
3. Calculate posterior probability of sequences using forward-backward algorithm, the form of each sequence in Eq. (2) can be obtained, and normalize Eq. (2) according to Eq. (3).
4. According to Eq. (4), obtain $\{m_1, m_2, ..., m_N\}$

---

4 Robust Clustering of Categorical Sequences

In the subsection, we propose a soft subspace clustering algorithm for categorical sequences. Suppose that we are given a set of N sequences $\{O_1, O_2, ..., O_N\}$ which is equivalent to $\{m_1, m_2, ..., m_N\}$ from which k clusters are searched for. We denote the kth cluster by c_k with n_k being its number of objects, such that $\bigcup_{k=1}^{k} c_k = DB$ and $\sum_{k=1}^{K} n_k = N$. The set of K subsets is denoted by $C = \{c_k | k = 1, 2, ..., K\}$. Based on the above, the clustering criterion to be minimized can be defined as:

$$J(C, W) = \sum_{k=1}^{K} \sum_{m_i \in c_k} \sum_{d=1}^{N} w_{kd}^{\theta} (m_{id} - \mu_{kd})^2 + \sum_{k=1}^{K} \zeta_k (1 - \sum_{d=1}^{N} w_{kd}) \qquad (5)$$

where $\mu_{kd} = \frac{1}{n_k} \sum_{m_i \in c_k} m_{id}$, $W = \{w_{kd}\}_{K \times N}$ is the weight matrix, the w_{kd} indicates the degree of contribution of the dth dimension to the cluster c_k and subjects to the constraint that $\sum_{d=1}^{N} w_{kd} = 1, w_{kd} \geqslant 0$. Here, weighting exponent θ controlling the strength of the incentive for clustering on more attributes and ζ_k for $\forall k$ the Lagrange multipliers enforcing the constraints for the weights.

Given DB to be clustered into K clusters of C, the goal is to look for a saddle point by minimizing the weighted within-cluster scatters with respect to C and W, and maximizing with respect to the Lagrange multipliers. The usual method of achieving this is to use partial optimization for each parameter. Following this method, minimization of $J(C, W)$ can be performed by optimizing C and W in a sequential structure analogous to the mathematics of the EM algorithm [15]. In each iteration, we first set $W = \widehat{W}$, and solve C as $\widehat{C}$ to minimize $J(C, \widehat{W})$. Next, $C = \widehat{C}$ is set and the optimal W, say $\widehat{W}$ is solved to minimize $J(\widehat{C}, W)$.

The first problem can be solved by assigning each input m_i to its most similar center, by comparing the object-to-center distances of m_i to the K cluster centers. Formally, we assign m_i to cluster k according to

$$k' = argmin_{\forall k} \sum_{d=1}^{N} \widehat{w}_{kd}^{\theta} (m_{id} - \mu_{kd})^2 \qquad (6)$$

The second optimization problem is solved by setting the gradients of $J(\widehat{C}, \widehat{W})$ with respect to each $\widehat{w}_{kd}$ and $\widehat{\zeta}_k$ to zero, derived

$$\widehat{w}_{kd} = \left[\sum_{m_i \in c_k} (m_{id} - \mu_{kd})^2 \right]^{\frac{1}{1-\theta}} \sum_{d=1}^{N} \left[\sum_{m_i \in c_k} (m_{id} - \mu_{kd})^2 \right]^{-\frac{1}{1-\theta}} \qquad (7)$$

It can be seen that SSCS is an extension to the EM algorithm [16]. Therefore, we refer the reader to that paper for the discussions of convergence. The time complexity of SSCS is $O(KND)$.

5 Experimental Evaluation

In this section, we evaluate our proposed algorithm on both synthetic data and real-world data. We set the initial hidden Markov model parameters and choose the value of ? that makes the result the best. For comparison, three different clustering approaches were used as the baselines. The first one is called KHMM which adopt the transfer matrix of the HMM model to represent a sequence and map each row of the transition matrix to RKHS with KED, and then it is the same as SSCS; The second one is referred to as Kmeans-SSCS which differs with SSCS in that it performs standard k-means algorithm in RKHS instead of subspace method; The last one is n-gram which is the mainstream sequence similarity measurement method.

We intentionally select above three algorithms as baselines in order to study three aspects of proposed algorithm: 1. How the overall statistics information of sequence affects the clustering accuracy (KHMM); 2. The sensitivity to noisy data (Kmeans-SSCS); 3. Contrast with mainstream methods (n-gram).

5.1 Synthetic Data

We generated 100 sequences ranging from 100–200 in length from 2-component HMM mixture ($K = 2$, 50 sequences from each), with 3 hidden states. Specifically, we set the transfer matrix and emission matrix of two Hidden Markov Models to be different. We also generated 10 sequences ranging from 100–200 in length from other 2-component HMM mixture as the noise data (5 sequences from each). We use α to denote the average degree of deviation between the noise model and the sequence model and larger α means noise data are more deviated from the clusters.

Each algorithm will run 20 times under different α value and we average clustering accuracy for all trials (see Fig. 3). It can be seen form Fig. 3 that with the α increasing (means the noise level increases), the performance of all the algorithms decreases. It can be clearly seen that SSCS is less sensitive to the variation of α than other algorithms. This is because, SSCS makes use of subspace method to identify the noise data. While KHMM also adopt subspace method, it repre-

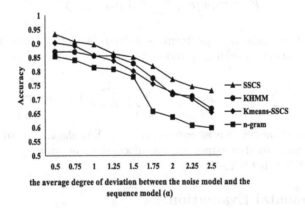

Fig. 3. Clustering accuracy of algorithm under intrinsic noise

sents sequences as transfer matrix rather than self-representation, so KHMM is actually also affected by noise.

5.2   Real-World Data

This set of experiments was designed to examine the performance of SSCS in real-world applications, these algorithms are tested on three sequence data sets from different domains. The first one consists of sequences used for speech recognition [9,17], the second one is from the SWISS PROT protein sequence data bank and the third one is homologous vertebrate gene database(HOVERGEN) which is from PBIL. The detailed parameters of each data set are listed in Table 1.

Table 1. Summary of the parameters for the real-world datasets

Dataset	The number of sequences	The average length of the sequences	The number of sequence classes
Speech sequences	50	1898	5
Protein sequences	33	871	5
Homologous vertebrate genes	285	1307	6

We set the number of hidden states is 4, and each algorithm will run 50 times. The average performances are reported in Table 2 with the format average ± 1 standard deviation and the best result are marked in bold typeface. Table 2 shows our proposed algorithm outperforms the other four algorithms by a large margin in terms of both clustering quality and stability. The two algorithms based on the posterior probability(SSCS, Kmeans-SSCS) perform better than KHMM, which demonstrates that adopt the transfer matrix of the HMM model to represent a sequence is not useful approach, due to the lack of overall statistical significance. Because the hidden properties underlying the categorical sequence are not captured, the n-gram has lower accuracy than other algorithms.

Table 2. Comparison of Accuracy by different algorithms, on the real-world datasets

	Dataset	SSCS	KHMM	Kmeans-SSCS	n-gram
Accuracy	Speech sequences	**0.9817 ± 0.02**	0.8813±0.10	0.9126±0.21	0.7049±0.24
	Protein sequences	**0.9126 ± 0.12**	0.8669±0.09	0.8873±0.16	0.8710±0.15
	Homologous vertebrate genes	**0.8743 ± 0.06**	0.8498±0.14	0.8574±0.17	0.8010±0.12

6 Conclusion

In this paper, we present a self-representation model for categorical sequences and a robust clustering algorithm is proposed which is relatively insensitive to noise and has improved accuracy on categorical sequences clustering. We first derived the form of self-representation model of categorical sequence based on HMM and KED. Based the new model, we transform the robust clustering to a subspace clustering problem. The experimental results on both synthetic and real-world data show its outstanding effectiveness.

For future work, we may consider the self-representation model with representative sample instead of entire datasets. Another effort will be directed toward multidimensional sequences.

References

1. Xing, Z., Pei, J., Keogh, E.: A brief survey on sequence classification. ACM SIGKDD Explor. Newsl. **12**, 40–48 (2010)
2. Xiong, T., Wang, S., Jiang, Q., Huang, J.Z: A new Markov model for clustering categorical sequences. In: IEEE International Conference on Data Mining, pp. 854–863 (2011)
3. Kondrak, G.: N-Gram similarity and distance. In: International Conference on String Processing and Information Retrieval, pp. 115–126 (2005)
4. Kelil, A., Wang, S.: SCS: a new similarity measure for categorical sequences. In: Eighth IEEE International Conference on Data Mining, pp. 343–352 (2008)
5. Blasiak, S., Rangwala, H.: A hidden Markov model variant for sequence classification. In: IJCAI 2011, Proceedings of the International Joint Conference on Artificial Intelligence, pp. 1192–1197. Barcelona, Catalonia, Spain (July 2012)
6. Smyth, P.: Clustering sequences with hidden Markov models. Nips **9**, 648–654 (1997)
7. Guo, G., Chen, L., Ye, Y., Jiang, Q.: Cluster validation method for determining the number of clusters in categorical sequences. IEEE Trans. Neural Netw. Learn. Syst. **28**, 2936–2948 (2017)
8. Yao, S.: A robust hidden Markov model based clustering algorithm. In: Information Technology and Artificial Intelligence Conference, pp. 259–264 (2011)
9. Rabiner, L.R.: A tutorial on hidden Markov models and selected applications in speech recognition. Read. Speech Recognit. **77**, 267–296 (1990)
10. Smola, A., Gretton, A., Song, L.: A Hilbert space embedding for distributions. In: International Conference on Algorithmic Learning Theory, pp. 13–31 (2007)
11. Song, L.: Kernel embeddings of conditional distributions. IEEE Signal Process. Mag. **30**, 98–111 (2013)
12. Schlkopf, B., Smola, A.: Learning with Kernels: support vector machines. Regularization, optimization, and beyond, publications of the American statistical association **98**, 489–489 (2002)
13. Fukumizu, K.: Kernel measures of conditional dependence. In: Conference on Neural Information Processing Systems, pp. 167–204. Vancouver, British Columbia, Canada (December 2007)
14. Elhamifar, E., Vidal, R.: Sparse subspace clustering: algorithm, theory, and applications. IEEE Trans. Pattern Anal. Mach. Intell. **35**, 2765–2781 (2012)

15. Bezdek, J.C.: A convergence theorem for the fuzzy ISODATA clustering algorithms. IEEE Trans. Pattern Anal. Mach. Intell. **2**, 1–8 (1980)
16. Jordan, M., Xu, L.: On convergence properties of the EM algorithm for Gaussian mixtures. Neural Comput. **8**, 129–151 (1995)
17. Loiselle, S., Rouat, J., Pressnitzer, D., Thorpe, S.: Exploration of rank order coding with spiking neural networks for speech recognition. In: 2005 IEEE International Joint Conference on Neural Networks, IJCNN 2005. Proceedings, vol. 2074, pp. 2076–2080 (2005)

Diversified Spatial Keyword Query
on Topic Coverage

Zhihu Qian[1], Ling Zhang[2(✉)], Haifeng Zhu[1], and Jiajie Xu[1]

[1] School of Computer Science and Technology, Soochow University, Suzhou, China
[2] National Earthquake Response Support Service, Beijing, China
zhangling903@163.com

Abstract. Spatial keyword queries are widely used in location based service systems nowadays to find the spatial web object people need. The returned objects usually are relevant to the query but not diversified each other. Motivated by this, we study the problem of diversified spatial keyword query on topic coverage, which returns k relevant objects close to the query, and they together can cover a certain number of topics for the purpose of diversification. We devise two novel algorithms, one aims to iteratively include the objects with minimum marginal penalty on top of a carefully designed indexing structure; the other adopts a hierarchy based selection policy, and its effectiveness can be confirmed by an error bound derived through the theoretical analysis. Empirical study based on real check-in dataset demonstrate the good effectiveness and efficiency of our proposed algorithms.

Keywords: Spatial keyword query · Diversification · Topic coverage
Error bound

1 Introduction

With the widespread use of smart phones and mobile internet nowadays, increasing volume of spatial web objects are becoming available on the web that may represent Point of Interests (PoIs) such as restaurants, hotels and shops. Querying on these data has been considered as one of the most frequently used queries today, especially in map services or POI recommendation [16–18]. This calls for techniques to support efficient processing of spatial keyword queries, which take a geo-location and a set of keywords from user as input, and return most relevant PoIs that match the inputs. The study of spatial keyword search has attracted great deal of attention from both academic and industrial communities, with numerous effective indexing structures and query processing methodologies [9,10,13,19,20] designed for recommending users suitable spatial web objects efficiently. However, the ranking approaches of existing methods mainly focus on the similarity (regarding to spatial and textual similarity) between the objects and query only, without considering their diversification among returned objects.

© Springer Nature Switzerland AG 2018
L. H. U and H. Xie (Eds.): APWeb-WAIM 2018, LNCS 11268, pp. 24–34, 2018.
https://doi.org/10.1007/978-3-030-01298-4_3

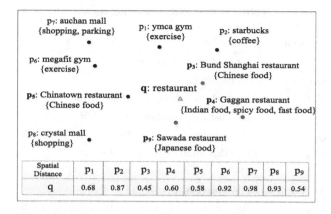

Fig. 1. An example of map search services

Example 1. Consider the example in Fig. 1, a user issues a query q with a location and a keyword 'restaurant' to find three PoIs for recommendation. Conventional spatial keyword query approach tends to select the most close objects that match the given keywords, i.e. $\{p_3, p_5, p_9\}$. But we find out p_3 and p_5 have the same tags, which means the two objects are similar. A more rational result is $\{p_3, p_4, p_9\}$ since p_4 is as close as p_5 to q and has more uncovered tags.

For the example in Fig. 1, every tag of the objects corresponds to a topic that represents the user's possible intention. The query result should cover more topics to meet the user's various needs. Some objects in the dataset may have few tags but a lot of user comments, so that we can use these comments to derive a set of latent topics by the techniques of *word embedding*. Every latent topic is a tag with statistical meanings. Motivated by the above example, in this paper we study the problem of spatial keyword querying with proportional topic coverage. Given a set of spatial objects, we aim to find k objects that satisfy the following conditions: (1) each of them contains the keywords in the query; (2) are close to the query location; (3) cover at least a threshold of topics.

To sum up, the main contributions of this paper can be briefly summarized as follows:

- We formally define the problem of diversified spatial keyword querying on topic coverage.
- We propose a greedy heuristic algorithm to efficiently retrieve objects in the increasing order of *marginal penalty*. To further optimize the query, we adopt novel indices, which prune the search space effectively.
- We propose an algorithm with provable approximation bound, which divides the objects that satisfy the keyword constraint into levels and selects the ones with most uncovered topics.
- Comprehensive experiments on real and synthetic dataset demonstrate the effectiveness and efficiency of our proposed methods.

Table 1. The summary of notations

Notation	Definition
$\mathcal{O}$	A set of objects in the dataset
o, o^* (q)	An object (query)
$\mathcal{R}, \mathcal{R}^*$	A set of objects returned as output
$f(q, \mathcal{R})$	The distance function of q and $\mathcal{R}$
ξ_{max}	The maximum spatial distance in the search
Γ_{max}	The threshold of topics to be covered
$\mathcal{D}(q, o)$	The spatial distance between q and o
$MPen(o, \mathcal{R})$	The marginal penalty of o
$\mathcal{D}_{opt}$	Sum of spatial distance of the optimal result
$\mathcal{T}_{opt}$	Number of all topics of the optimal result
$\mathcal{D}$	The estimator of $\mathcal{D}_{opt}$

2   Problem Definition

In this section, we give some basic definitions and then formalize the problem. Table 1 summarizes the mathematical notations used in this paper.

Definition 1 *(Spatial Web Object). A spatial web object is described by a spatial point in 2-dimensional space, a set of terms (keywords) from a vocabulary $\mathcal{V}$, and a few topics it belongs to. It is formalized as $o = (o.loc, o.T, o.\zeta)$, where o.loc is the coordinate of o, o.T denotes the terms to describe this object, and o.ζ is a set of topics it covers. For presentation simplicity, we use* spatial object *to represent it in short.*

Definition 2 *(Candidate Object Set). Given a set $\mathcal{O}$ of spatial objects, a query $q = (q.loc, q.T)$ and a threshold Γ_{max}, we say a subset $\mathcal{S}$, $\mathcal{S} \subseteq \mathcal{O}$, to be a candidate object set if and only if it satisfies*

- *keyword constraint. Every object o in $\mathcal{S}$ contains all the query keywords, i.e., $\forall o \in \mathcal{S}, q.T \subseteq o.T$. This guarantees all the returned objects to be textually related to q.*
- *diversification requirement. All the objects in $\mathcal{S}$ cover at least Γ_{max} topics together, i.e., $|\cup_{o \in \mathcal{S}} o.\zeta| \geq \Gamma_{max}$, which indicates that the user can be informed of diversified information subject to the query.*

Definition 3 *(Distance Function). Given a set $\mathcal{O}$ of spatial objects and a query q, we define a distance function $f(q, \mathcal{R})$ for a subset $\mathcal{R} \subseteq \mathcal{O}$ with $|\mathcal{R}| = k$, which is formalized as follows.*

$$f(q, \mathcal{R}) = \sum_{o \in \mathcal{R}} \frac{\mathcal{D}(q, o)}{\xi_{max}} \qquad (1)$$

where $\mathcal{D}(q, o)$ denotes the Euclidean distance between q and o in spatial, and ξ_{max} is the maximum spatial distance among all objects in the given dataset.

Problem Statement. In this paper, we investigate the problem of the *diversified spatial keyword query on topic coverage* (*DSK query*). Given a set of spatial objects $\mathcal{O}$, a query q, a distance function f, an integer k and a threshold Γ_{max} (Γ_{max} can be set as k by default), we aim to find a candidate object set $\mathcal{R}$ of size k, i.e., $\mathcal{R} \subseteq \mathcal{O}$, $|\mathcal{R}| = k$, $\forall o \in \mathcal{R}$, $q.T \subseteq o.T$, and $|\cup_{o \in \mathcal{R}} o.\zeta| \geq \Gamma_{max}$, such that $f(q, \mathcal{R})$ is minimized.

3 Marginal Penalty Based (MP) Algorithm

This section proposes a heuristic method, called MP based algorithm, to find rational result that can find a good balance between objective function and diversification constraint.

The basic idea of the MP algorithm is to select the k objects one by one according to certain criterion. The whole searching process is divided into two stages: (1) When the diversification is not satisfied yet, we aim to include those objects (containing all keywords in query) that are not only spatially close to query, but also to diversify the result. To find a proper balance between them, we use the concept of *marginal penalty* to rank the unselected objects. Suppose that we have already picked several objects into the result set $\mathcal{R}$, then the marginal penalty of an object o can be calculated as

$$MPen(o, \mathcal{R}) = \frac{\mathcal{D}(q, o)}{|o.\zeta \setminus \cup_{o^* \in \mathcal{R}} o^*.\zeta|} \tag{2}$$

where $\mathcal{D}(q, o)$ is the spatial distance from o to q, and $|o.\zeta \setminus \cup_{o^* \in \mathcal{R}} o^*.\zeta|$ is the number of topics covered by o but not by all the objects in $\mathcal{R}$. Obviously, the marginal penalty help us to select those object. (2) If the diversification requirement has been satisfied, we select the closest objects that contains all keywords in query to minimize the distance function $f(q, \mathcal{R})$.

To speed up the query processing of MP algorithm, we propose some indexing structures that can integrate spatial, keywords and topic information seamlessly.

IQ* -tree base method. We carefully design a variant of IR-tree, i.e. IQ*-tree, to avoid large dead space for *DSK query*. For spatial, we utilize the Quad-tree instead of R-tree to organize the objects. The Quad-tree partitions the spatial space by recursively subdividing it into four quadrants or regions. Each quad-tree node has (1) a set of inverted lists to denote the keyword information; (2) a bitmap to denote the topics it covers, i.e. the i-th bit is set as 1 if there exists an object in this node covers the i-th topic, otherwise set as 0. In this way all the required information in query processing can be incorporated in the tree structure.

S2I based method. S2I [13] is a text-first index that utilizes keywords constraint to prune the search space preferentially, which coincides to our problem seamlessly. As shown in Fig. 2, the structure of S2I is composed of two parts: the term inverted list and the aggregated R-tree (aR-tree) or block. The term inverted list is the structure to store the information for every term emerging in the objects, which includes the term frequency, the type and the pointer to a block or an aR-tree. The index maps each term into an aR-tree or a block. When the frequency of a term exceeds a given threshold, we store the objects that contain it in an aR-tree, otherwise, the objects are put into blocks in a file.

The block stores a set of objects whose frequency of certain keywords is relatively less. For every object in the block, we store its location and a bitmap to denote if each topic can be covered by the objects in the block. The objects in the same block are disordered.

The aggregated R-tree (aR-tree) is based on the traditional R-tree with the covered topics embedded in its node. The leaf node stores the information of the objects contained in it, while a non-leaf node utilizes a bitmap to store the topics covered by the objects that belong to this node, similar to the IQ*-tree.

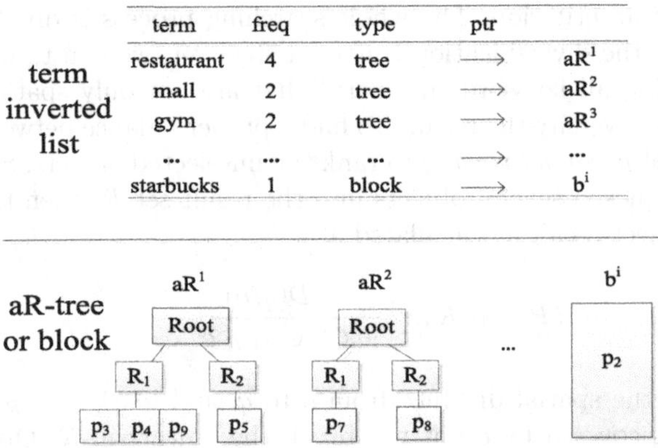

	term	freq	type	ptr	
	restaurant	4	tree	$\longrightarrow$	aR^1
term inverted list	mall	2	tree	$\longrightarrow$	aR^2
	gym	2	tree	$\longrightarrow$	aR^3
	...	...	...	$\longrightarrow$	...
	starbucks	1	block	$\longrightarrow$	b^i

Fig. 2. An example of *S2I* on objects in Fig. 1

In addition, we use two extra lists to store the number of nodes and objects in the aR-trees and blocks respectively.

Since the S2I is a text-first index, we just need to consider the aR-trees or blocks pointed at by all the keywords (terms) in q. Recall that every object returned must satisfy the keyword constraint, i.e., contain all the keyword in the query, so we only have to process one aR-tree and single block to get the objects with the minimum marginal penalties according to Lemma 1.

Lemma 1. *Given an index S2I, we only have to traverse one aR-tree and single block in the S2I to get the objects with the minimum marginal penalties.*

Proof. For the query q, we do not need to traverse all the aR-trees and blocks in the S2I to get the desired result, because the objects that contain all the query keywords emerge in all the aR-trees or blocks that contain one of the keyword in the query. So the one with minimum marginal penalty in one aR-tree or a block also has the minimum value in the others.

In the query processing, we visit the aR-tree and the block with the minimum number of nodes and objects respectively on behalf of all the aR-trees and blocks that contain one single query keyword. During the search in an aR-tree, we visit its node N in increasing order of the *best match marginal penalty*, which is defined as follows.

$$MPen_{bm}(N, \mathcal{R}) = \frac{D(q, N.mbr)}{|Occu_{i=1}|} \qquad (3)$$

where $D(q, N_{mbr})$ is spatial distance of the MBR of N to the query q and $|Occu_{i=1}|$ is the number of occurrences of 1 in the bitmap of N, i.e., the number of covered topics by the objects in the node. We keep finding the next object that satisfies the keyword constraint with the minimum marginal penalty and maintain the top-k best objects. For a block, we traverse the objects in it in the increasing order of the marginal penalty and keep at most k best ones. After that, we store all the objects found in the aR-tree and block, then iteratively choose the one with the minimum marginal penalty and add it to the result. The search algorithm terminates when we have already found k objects or all the desired number of topics have been covered.

4   Hierarchical Greedy (HG) Algorithm

The MP algorithm retrieves k objects by a local greedy way, but it may not guarantee both diversification constraint and distance function accuracy. With an aim to improve the quality of the result, we introduce another solution called hierarchical greedy (HG) algorithm. It adopts the method to partition all objects into different hierarchical levels, and then selects a specific number of objects from each level.

The basic idea of HG algorithm is to divide the objects into levels based on their spatial distance to the query and then select the objects with the most uncovered topics in each level.

Suppose that the user wants to retrieve k objects, then we divide the objects within the radius of $\mathcal{D}$ centered at the query location into $1 + \lceil log_2k \rceil$ levels according to their spatial distance to q using the following formula:

$$L_i = \begin{cases} \{o \in \mathcal{C}, \mathcal{D}(q, o) \in (\frac{\mathcal{D}}{2^i}, \frac{\mathcal{D}}{2^{i-1}}]\}; & i = 1, 2, ..., \lfloor log_2k \rfloor \\ \{o \in \mathcal{C}, \mathcal{D}(q, o) \in (\frac{\mathcal{D}}{k}, \frac{\mathcal{D}}{2^{i-1}}]\}; & i = \lceil log_2k \rceil \\ \{o \in \mathcal{C}, \mathcal{D}(q, o) \in (0, \frac{\mathcal{D}}{k}]\}; & i = 1 + \lceil log_2k \rceil \end{cases} \qquad (4)$$

where $\mathcal{C}$ is the set of objects that satisfy the keyword constraint. The objects in the first level L_1 are the most distant from the query q with spatial distance between $\frac{\mathcal{D}}{2}$ and $\mathcal{D}$; the second level L_2 contains objects whose distances to q are between $\frac{\mathcal{D}}{4}$ and $\frac{\mathcal{D}}{2}$, and so on; in the last level $L_{1+\lceil log_2k \rceil}$, the spatial distance of objects is at most $\frac{\mathcal{D}}{k}$.

After the division, we select different number of objects in different levels. For the lower levels, we choose less objects since their spatial distances to q are larger than those in the higher ones; on the contrary, we choose more from the higher levels. Provided that the objects in the search region of radius $\mathcal{D}$ centered at the query location are divided into $1 + \lceil log_2 k \rceil$ levels, in the first $\lceil log_2 k \rceil$ levels, we select up to $2^i (1 \leq i \leq \lceil log_2 k \rceil)$ objects and we pick k ones from the last level. At each level, we choose the objects in increasing order of the number of uncovered topics.

When $\mathcal{D}$ is small, the objects found in the region closer to the query may not cover the desired number of topics. Then $\mathcal{D}$ is increased at a rate so that we can find a combination that covers more topics. By redividing all the objects within the radius of the enlarged $\mathcal{D}$, we select limited number of objects in each level to constitute a new result.

The HG algorithm terminates when all the desired number of topics is covered or $\mathcal{D}$ exceeds the sum of k maximum spatial distances.

If there exists an optimal set of k objects that minimize the distance function with total spatial distance of $\mathcal{D}_{opt}$ and covers T_{opt} topics, the HG algorithm returns a collection of objects with size up to $5k$ and total spatial distance at most $(1 + b) * (2\lceil log_2 k \rceil + 1) * \mathcal{D}_{opt}$ that covers at least $(1 - \frac{1}{e}) * T_{opt}$ topics [7].

5 Experiments

5.1 Experiment Settings

We use a real dataset by extracting the PoIs from the check-in records of Foursquare within the area of Los Angles. The dataset contains 294338 PoIs and we tag them with some random tags. According to the statistical results, each object in the dataset has 3 topics on average. In the experiments, the default values of the parameters are summarized in Table 2.

Table 2. Default parameter values

Parameter	Default value	Description
k	10	No. of returned objects
Γ_{max}	15	No. of topics to be covered
$\mathcal{O}$	100 K	No. of objects
b	1	Increasing rate of $\mathcal{D}$

5.2 Performance Evaluation

We investigate the efficiency of the proposed algorithms when varying the values of the parameters shown in Table 2. In the following part of this section, we use **MP-NIX** to represent the MP algorithm without an index; **MP-IQ*** and

MP-S2I on behalf of the MP algorithm using the indices IQ*-tree and S2I respectively. The HG algorithm is represented as **HG**.

Effect of k. The number of returned objects, k, has interesting impact on the performance of all algorithms as shown in Fig. 3. For all MP based algorithms, their running time all increases as k goes up since it takes more iterations to compute the marginal penalty of the remaining objects and choose ones with the minimum values. For the HG algorithm, the running time decreases as k increases. This is due to the difficulty of finding a set of objects to cover Γ_{max} topics if k is relatively small, which incurs the HG algorithm to keep searching and enlarging the radius until the subspace is large enough, and it is harder to terminate the search in advance by using bounds. The number of visited objects of **MP-NIX** and **HG** are the same (i.e., the default value of dataset size).

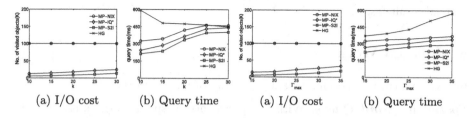

(a) I/O cost   (b) Query time   (a) I/O cost   (b) Query time

Fig. 3. Effect of k **Fig. 4.** Effect of Γ_{max}

Effect of Γ_{max}. Figure 4 plots the query time and number of visited objects of all methods when varying the value of parameter Γ_{max}. In overall, the HG algorithm runs slower than the others as it tends to find more accurate result. We can easily observe that the querying time of MP algorithms are quite stable without surprise, as the number of iterations in query processing are subject to k. While the HG algorithm requires more querying time when Γ_{max} increases, because it incurs additional cost to find an object to cover Γ_{max} topics with relatively small spatial distance cost.

Effect of $|\mathcal{O}|$. In Fig. 5, we can see the significant influence of the size of dataset $\mathcal{O}$ on the efficiency of the algorithms. The larger the dataset is, the greater I/O and query time are consumed to process the query by all proposed algorithms as expected. The **MP-S2I** approach is the most efficient algorithm on all values of $|\mathcal{O}|$ on the dataset. From the figures we notice the linear growth of the number of objects of **MP-NIX** and **HG** on the dataset, but that of **MP-S2I** and **HG** are relatively stable due to the use of indexing structure and query optimization strategies.

Effect of b on HG algorithm. Figure 6 explains the effect of the parameters b. We can infer the number of visited objects of **HG** stays constant (i.e., the dataset size) whenever k and b vary, because the algorithm calls for the scan over all objects to derive the final result. Moreover, when the value of b remains

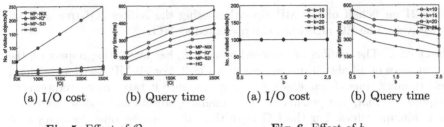

(a) I/O cost   (b) Query time   (a) I/O cost   (b) Query time

Fig. 5. Effect of $\mathcal{O}$ Fig. 6. Effect of b

unchanged, the running time of the HG algorithm decreases when k increases, which is not beyond expectation because the algorithm keeps searching and enlarging the radius until the subspace is large enough. Note that, in Fig. 8(b), the query time decreases as b increases since larger b indicates $\mathcal{D}$ will meet the optimal result faster.

6 Related Work

Below, we introduce two important categories of related work.

Spatial Keyword Query. With the prevalence of spatial objects associated with textual information on the Internet, spatial keyword queries that exploit both location and textual description are gaining in prominence. A prototypical spatial keyword query takes a user location and user-supplied keywords as arguments and returns web objects that are spatially and textually relevant to these arguments. A range of contributions are already made in the literature that study different aspects of spatio-textual querying [4,9–14,19,20]. Many novel indexing structures are proposed to support efficient processing on *SKBQ* and *SKAQ*, such as S2I [13]. Numerous work studies the problem of spatial keyword query on *why-not questions* [5], *interactive querying* [20], etc.

Search Result Diversification. Diversification is an aspect to measure the search result quality. In recent years, search result diversification has been extensively studied [2,3,6,8,15] to satisfy different intentions (e.g., content, novelty and coverage) in different applications. The approaches used for search result diversification have been surveyed in [1]. The search result diversification problem can be classified into the following two categories: implicit an explicit. Our problem belongs to the former category. Most work on search result diversification aims to find a set of k items based on a scoring function that considers both relevance and diversity.

However, the problems studied in the above work substantially differ from our diversified spatial keyword query on topic coverage and their approaches cannot directly applied in our work.

7   Conclusion and Future Work

This paper studies the problem of retrieving spatial objects close to the query location and diversified to cover sufficient topics. We propose two novel algorithms, one iteratively includes the objects in increasing order of the marginal penalty and novel indices are designed to speed up the query processing; the other adopts a hierarchy based selection policy, so as to guarantee the result not only to be diversified, but also has an error bound regarding distance function. In the future, it will be interesting to incorporate the problem of diversified spatial keyword query on topic coverage without road network, to recommend users more informative results in real applications.

References

1. Abid, A., Hussain, N., Abid, K., et al.: A survey on search results diversification techniques. Neural Comput. Appl. **27**(5), 1207–1229 (2015)
2. Agrawal, R., Gollapudi, S., Halverson, A., et al. Diversifying search results. In: WSDM (2009)
3. Carbonell, J., Goldstein, J.: The use of mmr, diversity-based reranking for reordering documents and producing summaries. In: SIGIR (1998)
4. Chen, J., Xu, J., Liu, C., Li, Z., Liu, A., Ding, Z.: Multi-objective spatial keyword query with semantics. In: DASFAA, pp. 34–48 (2017)
5. Chen, L., Cong, G.: Diversity-aware top-k publish/subscribe for text stream. In: SIGMOD (2015)
6. Fraternali, P., Martinenghi, D., Tagliasacchi, M.: Top-k bounded diversification. In: SIGMOD (2012)
7. Golab, L., Korn, F., Li, F., Saha, B., Srivastava, D.: Size-constrained weighted set cover. In: ICDE (2015)
8. Gollapudi, S., Sharma, A.: An axiomatic approach for result diversification. In: IW3C2 (2009)
9. Guo, L., Shao, J., Aung, H.H., Tan, K.-L.: Efficient continuous top-k spatial keyword queries on road networks. GeoInformatica **19**(1), 29–60 (2015)
10. Li, G., Feng, J., Jing, X.: Desks: direction-aware spatial keyword query. In: ICDE (2012)
11. Liu, H., Xu, J., Zheng, K., Liu, C., Du, L., Wu, X.: Semantic-aware query processing for activity trajectories. In: WSDM, pp. 283–292 (2017)
12. Qian, Z., Xu, J., Zheng, K., Zhao, P., Zhou, X.: Semantic-aware top-k spatial keyword queries. In: World Wide Web, pp. 1–22 (2017)
13. Rocha-Junior, J.B., Gkorgkas, O., Jonassen, S., Nørvåg, K.: Efficient processing of top-k spatial keyword queries. In: Pfoser, D., et al. (eds.) SSTD 2011. LNCS, vol. 6849, pp. 205–222. Springer, Heidelberg (2011). https://doi.org/10.1007/978-3-642-22922-0_13
14. Sun, J., Xu, J., Zheng, K., Liu, C.: Interactive spatial keyword querying with semantics. In: CIKM, pp. 1727–1736 (2017)
15. Vallet, D., Castells, P.: On diversifying and personalizing web search. In: SIGIR (2011)
16. Yin, H., Cui, B., Sun, Y., Hu, Z., Chen, L.: Lcars: a spatial item recommender system. In: TOIS, p. 11 (2014)

17. Yin, H., Wang, W., Wang, H., Chen, L., et al.: Spatial-aware hierarchical collaborative deep learning for poi recommendation. TKDE **29**, 2537–2551 (2017)
18. Yin, H., Zhou, X., Cui, B., Wang, H., Zheng, K., et al.: Adapting to user interest drift for poi recommendation. TKDE **28**, 2566–2581 (2016)
19. Zhang, C.: Inverted linear quadtree: efficient top k spatial keyword search. In: ICDE (2013)
20. Zheng, K., et al.: Interactive top-k spatial keyword queries. In: ICDE (2015)

Sequence-As-Feature Representation for Subspace Classification of Multivariate Time Series

Liang Yuan[1], Lifei Chen[2,3]([✉]), Rong Xie[4], and Huihuang Hsu[5]

[1] Network Operation Maintenance Center, University of Electronic Science and
Technology of China, Chengdu, China
[2] College of Mathematics and Informatics, Fujian Normal University, Fuzhou, China
[3] Digital Fujian Internet-of-Things Laboratory of Environmental Monitoring,
Fuzhou, China
clfei@fjnu.edu.cn
[4] School of Computer Science, Wuhan University, Wuhan, China
[5] Department of Computer Science and Information Engineering, Tamkang
University, New Taipei City, China

Abstract. Subspace classification of multivariate time series (MTS)
data is currently a challenging problem, due to the difficulties in defining a meaningful similarity measure for distinguishing between the series
features. In this paper, a new representation model called Sequence-As-
Feature (SAF) is proposed, where each MTS in the new representation
is a vector of sequential attributes, each being an univariate time series,
such that the common similarity measures can be readily adapted. An
attribute-weighted measure is then defined using a conditional probability distribution (CPD) modeling method. Based on these, a prototype
classifier is derived to classify the test MTS in a linear time complexity, with the MTS prototype optimized according to the CPD model
of sequential attributes. Experimental results on MTS data sets from
three real-world domains are given to demonstrate the performance of
the proposed methods.

Keywords: Multivariate time series · Representation model ·
Conditional probability model · Attribute weighting · Subspace
prototype classification

1 Introduction

A time series is a series of numerical data in order, generally, taken at succes-
sive equally spaced points in time. Well-known examples include the temporal

Supported by the National Natural Science Foundation of China under Grant
No. 61672157, and the Program of Probability and Statistics: Theory and Application (No. IRTL1704).

L. H. U and H. Xie (Eds.): APWeb-WAIM 2018, LNCS 11268, pp. 35–45, 2018.
https://doi.org/10.1007/978-3-030-01298-4_4

measurements of environmental sensors in a ubiquitous computing system, and the electroencephalogram (EEG) signal measured on a dynamical system that presents brain activity. In practice, such signals could be collected through multiple channels, composing a multivariate time series for the sample. Currently, multivariate time series analysis has received wide interest in the broad communities including machine learning and data mining, owing to its capacity in extracting meaningful statistics and discovering interesting patterns from the complex data [1,2].

Classification of time series, however, is a nontrivial task. This is due to the fact that, compared to the common attribute-value data, where dissimilarity between samples can be calculated directly on the attributes, the mathematical structure is much more complex in the domain of time series. As a result, popular dissimilarity measures such as the Euclidean distance are inappropriate, because the chronological dependencies inherent in the series would be ignored. To bypass this limitation, space transformation methods aimed at embedding time series in vector spaces are commonly used. Examples include the feature-extraction methods [3–5] and deep learning methods, which extract subsequences of interest (for example, the shapelets [6]) or learn the deep representation (for example, the DCNN networks [7]) to feature the series. Such methods prevent the structural information in the series from being lost, but the transformation would be inefficient and highly domain-dependent.

Another challenging issue is attribute weighting, which entails distinguishing the contribution of the data features for class prediction. Technically, this is related to the *subspace classification* problem [8], aimed at learning appropriate sets of attributes that span projected subspaces where the classes exist. Such a technique is essential for many machine learning applications, especially with high-dimensional data. For example, in the multivariate EEG data discussed in the beginning of this section, the dimensionality of the data collected at each time point reaches 64. In such high-dimensional data, there would be many noisy features that do not contribute to class prediction [9]. Classification with attribute weighting for such data becomes crucial as it effectively classifies the new sample in the space spanned solely by some relevant features. However, few attempts have been made to apply such techniques to multivariate time series classification. The main obstacle is that the existing methods are mainly based on the hypothesis that one multivariate time series is precisely a *sequence of vectors* [10]. As such, distinguishing between different data features becomes unrealistic because here each multi-dimensional vector is treated as an indivisible whole for dissimilarity computation.

In this paper, a new representation called *sequence-as-feature* (abbreviated SAF) representation, is proposed for multivariate time series. Each multivariate time series in the new representation is a vector of sequential attributes, such that the common attribute-weighting technique can be readily adapted. Based on the new representation, an efficient classifier is proposed for subspace classification on multivariate time series, with the prototypes and the attribute-weighted measures defined on the probability model of the sequential attributes.

A series of experiments on real-world multivariate time series data from various domains are conducted and the experimental results show their effectiveness.

2 Subspace Classification of Multivariate Time Series

The prototype-based classifier for subspace classification on multivariate time series (MTS), abbreviated *PCMT*, is presented in this section. We begin by describing some preliminaries and a few related work, followed by the new representation model for MTS.

2.1 Preliminaries

In what follows, the training set is denoted by $Tr = \{(\tilde{M}_i, C_i) | i = 1, 2, \ldots, N\}$, where $\tilde{M}_i$ is the ith multivariate time series of length T with $C_i \in [1, K]$ being its class label. Here, T is the number of time points at which the data of time series are taken. The kth training class is denoted by $\pi_k = \{\tilde{M}_i | (\tilde{M}_i, k) \in Tr\}$, where $k = 1, 2, \ldots, K$ and K is the number of pre-defined classes. The number of samples in π_k is denoted by $|\pi_k|$. Suppose that there are D observations in the t-th time point of each $\tilde{M}_i$ and denote the observations on the dth channel by s_{id}^t, where $t = 1, 2, \ldots, T$ and $d = 1, 2, \ldots, D$. As discussed in Sect. 1, each $\tilde{M}_i$ is traditionally viewed as a series of D-dimensional vectors, i.e.,

$$\langle s_{i1}^1, s_{i2}^1, \ldots, s_{iD}^1 \rangle \langle s_{i1}^2, s_{i2}^2, \ldots, s_{iD}^2 \rangle \ldots \langle s_{i1}^t, s_{i2}^t, \ldots, s_{iD}^t \rangle \ldots \langle s_{i1}^T, s_{i2}^T, \ldots, s_{iD}^T \rangle.$$

The similarity of $\tilde{M}_i$ and $\tilde{M}_j$ could thus be computed using the simple Euclidean distance, based on the assumption that $\tilde{M}_i$ and $\tilde{M}_j$ have the same length T.

To capture the structure information in MTS data, a number of feature-extraction methods have been suggested. In such methods, a set of sequential patterns is extracted and used to convert each MTS into a vector. Such patterns include the Shapelets [6], principle components [11] and the Local features at Thinned-out Keypoints (LTK) [10], etc. Model-based methods have also been used for the purpose, for instance, by learning a hidden Markov model for each time series and then representing it as a vector of the model parameters [12]. Generally, they have a high time complexity as extraction for the desired patterns is typically a time-consuming process. Recently, LSTM network (i.e., a recurrent neural network composed of long short-term memory units) has been used to learn a deep representation model for time series [7]. However, the problem becomes difficult when the time series are in multi-dimensionality, as it is typically designed for univariate time series.

Different from such indirect methods based on space transformation, a few *direct* measures have been proposed to compute the dissimilarity on the original MTS. For example, in Dynamic Time Warping (DTW) [13], the time series are stretched or shrunk nonlinearly along the time points with certain restrictions (typically, within a window of a fixed length), in order to compute an optimal match between them. The distances between the time points in the warping path are finally aggregated to measure the dissimilarity. Like the indirect methods discussed before, DTW also does not distinguish between the time series data from different channels in the multivariate case.

2.2 The Sequence-As-Feature Representation

In the traditional sequence-of-vectors representation model, each MTS $\tilde{M}_i$ is precisely the $D \times T$ matrix where each column corresponds to one D-dimensional vector, $\langle s_{i1}^t, s_{i2}^t, \ldots, s_{iD}^t \rangle$ for $t = 1, 2, \ldots, T$. Modeling such matrix variate distributions, however, is currently a challenging problem, as most of the established methods focus on the multivariate case. We thus suggest the sequence-as-feature (SAF) representation, by considering each $\tilde{M}_i$ as *a vector of sequential attributes*, i.e., representing $\tilde{M}_i$ by

$$\langle \tilde{S}_{i1}, \tilde{S}_{i2}, \ldots, \tilde{S}_{id}, \ldots, \tilde{S}_{iD} \rangle,$$

where $\tilde{S}_{id} = s_{id}^1 s_{id}^2 \ldots s_{id}^t \ldots$ is the univariate time series consisting of the dth observations at all the time points.

This new representation allows the time series collected from different channels having different lengths, since they are represented by vectors of the same length D. Thus, in the following pages, we will use $L_{id} = |\tilde{S}_{id}|$ to denote the length of $\tilde{S}_{id}$ instead of the previous constant T. Furthermore, we transform each univariate time series into a categorical sequence, by discretizing its numeric values into symbols. By the transformation [4,14], the time series $\tilde{S}_{id} = s_{id}^1 s_{id}^2 \ldots$ is discretized into the categorical sequence $\tilde{X}_{id} = x_{id}^1 x_{id}^2 \ldots$, which is a string of L_{id} symbols each being a category discretized from the original numeric value at the time point. This finally results in a new representation for multivariate time series, as shown in Definition 1.

Definition 1. *The multivariate time series $\tilde{M}_i$ in the SAF representation is denoted by*

$$\tilde{M}_i = \langle \tilde{X}_{i1}, \tilde{X}_{i2}, \ldots, \tilde{X}_{id}, \ldots, \tilde{X}_{iD} \rangle$$

where $\tilde{X}_{id} = x_{id}^1 x_{id}^2 \ldots x_{id}^t \ldots x_{id}^{L_{id}}$ with $x_{id}^t \in \mathcal{X}_d$ is a 1-dimensional categorical sequence, called the d-th sequential attribute of $\tilde{M}_i$. Here, $\mathcal{X}_d$ is the set of symbols collected from the sequences on the d-th sequential attribute.

The SAF representation thus allows the similarity between $\tilde{M}_i$ and $\tilde{M}_j$ to be measured by summarizing the individual similarity on each sequential attribute, using the traditional measures defined for attribute-value data. Such measures include the popular class-dependent weighting definition [9], given by

$$Sim_{\pi_k}(\tilde{M}_i, \tilde{M}_j) = \sum\nolimits_{d=1}^{D} w_{kd}^{\beta} \times sim(\tilde{X}_{id}, \tilde{X}_{jd}) \tag{1}$$

subject to

$$\forall k : \sum\nolimits_{d=1}^{D} w_{kd} = 1 \text{ and } \forall k, d : w_{kd} \geq 0, \tag{2}$$

based on the assumption that the univariate time series in one MTS from different channels are conditional independent. Here, $sim(\cdot, \cdot)$ measures the structural similarity of $\tilde{M}_i$ and $\tilde{M}_j$ on the dth sequential attribute. The exponential weight $w_{kd}{}^{\beta}$ measures the contribution of that attribute for class prediction, with the weighting exponent $\beta \neq 0$ and $\neq 1$ controlling the strength of the incentive for classification on more attributes. The greater the contribution, the higher the attribute weight.

2.3   Prototype-Based Classification

The dissimilarity measures such as that defined in Eq. (1) can be easily applied to derive a distance-based classifier (for example, the k-NN classifier [1,2]) for MTS classification. In the work described here, we are interested in the prototype-based classifier [8,15], because of its inherent simplicity, clear geometrical interpretations and its effectiveness reported in the literature. Roughly speaking, in the prototype classifier, one prototype or representative is built for each class from the training samples; the class of a new sample is then predicted as the class of its most similar prototype. However, the existing prototype methods cannot be directly applied to MTS data, due to the difficulties in formulating the class prototypes for the data (obviously, the concept of class centroid as frequently defined for numeric data is meaningless in this case). We thus turn to the probability model-based method, as follows:

The main idea is to learn a multivariate probability model for each training class based on the SAF representation, such that the class and consequently its prototype can be represented by the model parameters. Here, the conditional probability distribution (CPD) model [12] is used, which is a special Markov chain model based on the hypothesis that occurrence of each symbol x_{id}^t in the sequence $\tilde{X}_{id} = x_{id}^1 x_{id}^2 \ldots x_{id}^t \ldots$ is closely related to its preceding subsequence $y_{id}^t = x_{id}^1 \ldots x_{id}^{t-1}$. Hence, the length-normalized probability of $\tilde{M}_i$ with regard to the class π_k can be estimated, given by

$$p_k(\tilde{M}_i) = \prod_{d=1}^{D} p_k(\tilde{X}_{id}) = \prod_{d=1}^{D} \left(\prod_{t=1}^{L_{id}} p_k(x_{id}^t|y_{id}^t) \right)^{\frac{1}{L_{id}}} \tag{3}$$

where $p_k(x_{id}^t|y_{id}^t)$ with $p_k(x_{id}^1|y_{id}^1) = p_k(x_{id}^1)$ is the conditional probability measuring the sequential dependency of x_{id}^t on its preceding subsequence with regard to π_k. In practice, we use the CPD model of order n, where each preceding subsequence is truncated to a length that does not exceed n.

Based on the CPD modeling method, the class prototype for each training class π_k ($k = 1, 2, \ldots, K$) can be formulated as a set of the model parameters, i.e., $\theta_k = \{p_k(x)|x \in \mathcal{X}_d, d \in [1, D]\} \bigcup \{p_k(x|y)|x \in \mathcal{X}_d, y \in \mathcal{Y}_d, d \in [1, D]\}$, when the similarities over all the samples in π_k reaches the maximum in the sense of Eq. (3). Here, $\mathcal{Y}_d$ denotes the set of preceding subsequences collected from the discretized time series on the dth sequential attribute. This suggests the following Definition 2 (in the logarithm domain).

Definition 2. *Denote the class prototype of π_k by $\tilde{A}_k$, which is the average series parameterized by the set θ_k that maximizes*

$$J(\theta_k) = \sum_{\tilde{M}_i \in \pi_k} \ln p_k(\tilde{M}_i)$$

subject to

$$\forall d \text{ and } y \in \mathcal{Y}_d : \sum_{x \in \mathcal{X}_d} p_k(x|y) = 1.$$

Using the Lagrangian multiplier technique, the problem of Definition 2 can be transformed into an unconstrained optimization problem, and the optimal definitions for the variables can then be solved by setting the gradients to zero, yielding $p_k(x|y) = f_{\pi_k,d}(yx)/f_{\pi_k,d}(y)$ and $p_k(x) = f_{\pi_k,d}(x)/\sum_{\tilde{M}_i \in \pi_k} L_{id}$. To surmount the zero-probability problem, we then use the add-one smoothing method and obtain

$$p_k(x|y) = \frac{f_{\pi_k,d}(yx) + 1}{f_{\pi_k,d}(y) + |\mathcal{X}_d|} \tag{4}$$

and

$$p_k(x) = \frac{f_{\pi_k,d}(x) + 1}{\sum_{\tilde{M}_i \in \pi_k} L_{id} + |\mathcal{X}_d|} \tag{5}$$

for $\forall x \in \mathcal{X}_d$ and $\forall y \in \mathcal{Y}_d$. Here, $f_{\pi_k,d}(\cdot)$ is the times a subsequence occurs in the sequences of π_k on the dth sequential attribute.

The above derivation also allows the sample-to-prototype similarity measure to be defined. In the sense of Eq. (1), the similarity is given by $Sim_{\pi_k}(\tilde{M}_i, \tilde{A}_k)$ for the sample $\tilde{M}_i$ and the prototype $\tilde{A}_k$ of π_k. Since the prototype has been formulated as an optimized CPD model according to Definition 2, the similarity can be computed by the log-likelihood of $\tilde{M}_i$ to the model. The larger the likelihood, the more similar the sample to the prototype. Replacing $sim(\cdot, \cdot)$ in Eq. (1) with the log-likelihood results in

$$Sim_{\pi_k}(\tilde{M}_i, \tilde{A}_k) = \sum_{d=1}^{D} w_{kd}^{\beta} \times \ln p_k(\tilde{X}_{id}) \tag{6}$$

where $p_k(\tilde{M}_i)$ is computed using Eq. (3) with the parameters set to those in the optimized θ_k, given by Eqs. (4) and (5).

Now, the training algorithm of our *PCMT* can be defined. The goal is to learn the classification model $V = (\Theta, W)$, with $\Theta = \{\theta_k | k = 1, 2, \ldots, K\}$ being the set of class prototypes (learned according to Definition 2) and $W = \{w_{kd} | k = 1, 2, \ldots, K; d = 1, 2, \ldots, D\}$ the set of attribute-weights. To learn the weight w_{kd} for the sequential attribute d of π_k, we define the following objective function that needs to be maximized: $J(W) = \sum_{k=1}^{K} \sum_{\tilde{M}_i \in \pi_k} \sum_{d=1}^{D} Sim_{\pi_k}(\tilde{M}_i, \tilde{A}_k)$ subject to the constraints defined in Eq. (2). Again, using the Lagrangian multiplier technique and setting the gradients with respect to the variables to 0, the resulting weight can be obtained as

$$w_{kd} = \alpha_{kd}^{\frac{1}{1-\beta}} \times \left(\sum_{d'=1}^{D} \alpha_{kd'}^{\frac{1}{1-\beta}}\right)^{-1} \tag{7}$$

with

$$\alpha_{kd} = - \sum_{\tilde{M}_i \in \pi_k} \ln p_k(\tilde{X}_{id}).$$

Using the learned model V, the prediction algorithm of $PCMT$ is called to classify each test sample $\tilde{M}_i$ First, K sample-to-prototype similarities between $\tilde{M}_i$ and the class prototypes are computed, using Eq. (6) and the parameters in V. Then, the class label of $\tilde{M}_i$ is predicted as the most similar prototype k' according to

$$k' = \operatorname{argmax}_{k=1,2,\ldots,K} Sim_{\pi_k}(\tilde{M}_i, \tilde{A}_k). \tag{8}$$

Table 1. Details of the real-world multivariate time series.

Data set	# Classes (K)	# Samples (N)	# Time points (T)	# Attributes/Channels (D)
EEG-S1	2	200	256	64
EEG-S2	2	192	256	64
Robot-LP4	3	117	15	6
JapVowels	9	640	7~29	12

By applying the classification algorithms, in practice, we first collect the preceding subsequences by scanning each sequence, and save them in a prefix tree where each path corresponds to a subsequence. The height of the tree is thus $n + 1$ (including the root). Since the number of preceding subsequences for a time series of length T would reach T in the worst case, the time complexity for building the trees is $O(nTD|\mathcal{X}|)$, where $|\mathcal{X}|$ denotes the average number of symbols appearing on the sequential attributes. Considering N samples of length T in average, the time complexity of the $PCMT$ training algorithm can thus be given as $O(nTDN|\mathcal{X}|)$, in the worst case. The time complexities of the prediction algorithm is $O(nTDK|\mathcal{X}|)$, which is independent of the size of the training set.

3   Experimental Evaluation

In this section, we evaluate the performance of $PCMT$ on real-world data sets. Four multivariate time series sets from the UCI Machine Learning Repository were used, as Table 1 shows. The data sets belong to three real-world domains. We obtained two EEG series sets, EEG-S1 and EEG-S2, from the domain of genetic predisposition studies. The difference is that EEG-S1 consists of 200 subjects exposed to a single stimulus (S1), while for the EEG-S2 set the subjects were exposed to two stimuli (S1 and S2) where S2 differed from S1. The time series in the fourth set, named Robot-LP4, were obtained from the failure detection domain. The set named JapVowels is the Japanese Vowels database, belonging to the speech recognition domain. The set contains time series of 12 LPC cepstrum coefficients taken from 9 speakers, with the length ranging from 7 to 29 (this means that JapVowels contains time series having different lengths).

3.1   Attribute-Weighting Results

This set of experiments was designed to examine the effectiveness of *PCMT* in terms of attribute weighting. In applying *PCMT*, the weighting exponent β and the order n of the CPD model should be specified in advance. In the experiments, each data set was partitioned into one training set (66.7%) and one test set (33.3%), and the performance was evaluated with respect to the parameters, in terms of classification accuracy. The results show that *PCMT* is robust when $\beta < 0$, and the classification accuracy tends to drop with a large β. We also observe that the accuracy drops when $n > 3$ with a fixed β. Below, the classification results are reported by setting $\beta = -0.5$ with the order of the CPD model used by *PCMT* is fixed at 3 based on these observations.

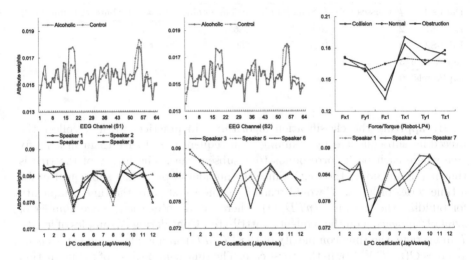

Fig. 1. The attribute weights learned by *PCMT* for the individual classes in the four real-world multivariate time series sets.

Different from the existing classifiers, *PCMT* is able to produce a set of attribute weights, indicating the individual contributions of the time series from different channels based on the SAF representation. Figure 1 shows the change in the attribute weights for each data set, from which one can see different contributions of the same attribute for the classes in the same data set. The figures also show obvious changes in the weights for the sequential attributes. This sort of weights distribution allows us to extract a reduced attribute subset for class prediction. According to the figures, for example, the electrode time series from the channels falling in $\{1, 2, 5, 6, 7, 12, 16, 22, 25, 32, 39, 48, 57, 62, 63, 64\}$ for EEG-S1 and $\{1, 2, 5, 7, 12, 13, 16, 21, 22, 25, 32, 39, 57, 62, 63, 64\}$ for EEG-S2 can be removed, with 25% dimensionality reduction, because of their low contribution in identifying the correlation of genetic predisposition to alcoholism. The performance of the reduced attribute subset will be examined in Sect. 3.2.

3.2   Performance Comparisons

The second set of experiments was designed to evaluate *PCMT* with comparison to other mainstream methods for multivariate time series classification. Four representative similarity measures were used. The first one is the simple Euclidean distance as discussed in Sect. 2.1, which is abbreviated EU in the experiments. We chose the well-known DTW [13] as the second competing measure. Two indirect measures: the PCA-based measure [11] and the recently published LTK [10], were also chosen.

Table 2. Average accuracy of different methods on the real-world time series sets.

Data set	EU	PCA	DTW	LTK	*PCMT*
EEG-S1	0.625 ± 0.056	0.565 ± 0.100	0.660 ± 0.080	0.555 ± 0.079	**0.850 ± 0.074**
EEG-S2	0.630 ± 0.124	0.636 ± 0.108	0.599 ± 0.081	0.578 ± 0.084	**0.848 ± 0.073**
Robot-LP4	0.775 ± 0.077	0.785 ± 0.081	**0.843 ± 0.092**	0.783 ± 0.113	0.812 ± 0.048
JanVowels	-	-	0.925 ± 0.038	0.892 ± 0.049	**0.953 ± 0.034**

Table 3. Average classification accuracy of the methods on the reduced data sets with the removal of nonsignificant attributes identified by *PCMT*.

Data set	D	*PCMT*+EU	*PCMT*+PCA	*PCMT*+DTW	*PCMT*+LTK
EEG-S1	48	0.640 ± 0.083 ↑	0.563 ± 0.073	0.670 ± 0.071 ↑	0.590 ± 0.136 ↑
EEG-S2	48	0.578 ± 0.079 ↓	0.542 ± 0.098 ↓	0.604 ± 0.086 ↑	0.583 ± 0.087 ↑
Robot-LP4	5	0.853 ± 0.097 ↑	0.818 ± 0.090 ↑	0.836 ± 0.066 ↓	0.801 ± 0.109 ↑
JanVowels	9	-	-	0.928 ± 0.023 ↑	0.905 ± 0.036 ↑

Table 2 shows the average classification accuracy of different methods on the four data sets. In the experiments, each data set is classified by each classifier using ten-fold cross-validation, and the average accuracy is reported in the format *average ± 1 standard deviation*. From the results in the table, we see that *PCMT* is able to obtain high-quality results, especially on the EEG sets, where they outperform the competing methods by more than 20% except the EEG-S1 case of DTW. LTK performs the worst, while PCA and EU fail in classifying the JapVowels set because of the varying length of the times series in the set.

To understand the strength of the attribute-weighting method used in *PCMT*, we created four reduced data sets each for one series set by removing those nonsignificant attributes, according to the global weights learned by *PCMT* as shown in Fig. 1. In details, the 3rd sequential attribute associated with the force Fz1 was removed from the Robot-LP4 set; for the JapVowels set, the attributes corresponding to the 4th, 8th and 12th LPC coefficient were deleted. The EEG sets were reduced based on the discussion in Sect. 3.1. Table 3 shows the change in average accuracy of the competing methods on the reduced sets. It can be seen that, interestingly, the four methods obtain more accurate results on the reduced data than on the original data in the most cases. This indicates that the attribute-weighting method used in our *PCMT* is able to capture significant structural information hidden in the sequential attributes, from which the existing methods can also benefit.

4   Conclusions

In this paper, we defined a new classifier for subspace classification on multivariate time series data. Unlike the existing methods, which are mainly based on the sequence-of-vectors representation, we proposed to represent each multivariate time series as a vector of sequential attributes (called sequence-as-feature representation), such that the similarity can be computed by an attribute-weighted measure. We also proposed a prototype-based classifier, called *PCMT*, by learning the class prototype and the attribute weights assigned to the sequential attributes, based on the conditional probability distribution modeling on the new representation. Experiments are conducted on four real-world multivariate time series data sets from three domains, and the experimental results show the effectiveness of *PCMT* compared with the existing methods. For future work, we would like to extend the similarity measure to a nonlinear formulation without the independence assumption.

References

1. Bagnall, A., Lines, J., Bostrom, A., Large, J., Keogh, E.: The great time series classification bake off: a review and experimental evaluation of recent algorithmic advances. Data Min. Knowl. Discov. **31**(32), 606–660 (2017)
2. Spiegel, S., Gaebler, J., Lommatzsch, A., Luca, E., Albayrak, S.: Pattern recognition and classification for multivariate time series. In: Proceedings of the Fifth International Workshop on Knowledge Discovery from Sensor Data, pp. 34–42 (2011)
3. Baydogan, M.G., Runger, G., Tuv, E.: A bag-of-features framework to classify time series. IEEE Trans. Pattern Anal. Mach. Intell. **35**(11), 2796–2802 (2013)
4. Baydogan, M.G., Runger, G.: Learning a symbolic representation for multivariate time series classification. Data Min. Knowl. Discov. **29**, 400–422 (2015)
5. Weng, X., Shen, J.: Classification of multivariate time series using locality preserving projections. Knowl.-Based Syst. **21**(7), 581–587 (2008)
6. Ye, L., Keogh, E.J.: Time series shapelets: a novel technique that allows accurate, interpretable and fast classification. Data Min. Knowl. Discov. **22**(1–2), 149–182 (2011)
7. Zheng, Y., Liu, Q., Chen, E., Ge, Y., Zhao, J.L.: Exploiting multi-channels deep convolutional neural networks for multivariate time series classification. Front. Comput. Sci. **10**(1), 96–112 (2016)
8. Zhang, J., Chen, L., Guo, G.: Projected-prototype-based classifier for text categorization. Knowl. Based Syst. **49**, 179–189 (2013)

9. Chen, L., Wang, S., Wang, K., Zhu, J.: Soft subspace clustering of categorical data with probabilistic distance. Pattern Recognit. **51**, 322–332 (2016)
10. Fang, Y., Huang, H.H., Kawagoe, K., Modified A-LTK: improvement of a multi-dimensional time series classification method. In: Proceedings of the International Conference on Computer Science, pp. 212–216 (2015)
11. Singhal, A., Seborg, D.E.: Pattern matching in historical batch data using PCA. IEEE Control. Syst. Mag. **22**(5), 53–63 (2002)
12. Xiong, T., Wang, S., Mayers, A., Monga, E.: DHCC: divisive hierarchical clustering of categorical data. Data Min. Knowl. Discov. **24**(1), 103–135 (2012)
13. Keogh, E., Ratanamahatana, C.A.: Exact indexing of dynamic time warping. Knowl. Inf. Syst. **7**, 358–386 (2005)
14. Moskovitch, R., Shahar, Y.: Classification of multivariate time series via temporal abstraction and time intervals mining. Knowl. Inf. Syst. **45**(1), 35–74 (2015)
15. Han, E.-H.S., Karypis, G.: Centroid-based document classification: analysis and experimental results. In: Zighed, D.A., Komorowski, J., Żytkow, J. (eds.) PKDD 2000. LNCS (LNAI), vol. 1910, pp. 424–431. Springer, Heidelberg (2000). https:// doi.org/10.1007/3-540-45372-5_46

Discovering Congestion Propagation Patterns by Co-location Pattern Mining

Ying He, Lizhen Wang[✉], Yuan Fang, and Yurui Li

School of Information Science and Engineering, Yunnan University, Kunming
650091, China
heyingzy@163.com, lzhwang@ynu.edu.cn

Abstract. Traffic congestion has been an important problem all over the world. It is necessary to discover meaningful traffic patterns such as congestion propagation patterns from the massive historical dataset. Existed methods focusing on discovering congestion propagation patterns can't mine transitivity of time and space very well. The spatio-temporal co-location pattern mining discovers the subsets of features which are located together in adjacent time periods frequently. So we propose using the spatio-temporal co-location pattern mining to discover congestion propagation patterns. Firstly, we propose the concepts of Spatio-Temporal Congestion Co-location Pattern (STCCP). Secondly, we give a framework and an algorithm for mining STCCPs. Finally, we validate our algorithm on real data sets. The results show that our method can effectively discover congestion propagation patterns.

Keywords: Co-location pattern · Traffic congestion pattern · Spatio-temporal congestion co-location

1 Introduction

In the field of transportation, due to the rapid development and expansion of the city, the number of cars has increased dramatically in the cities. The inconvenience caused by urban congestion has become an urgent problem to be solved in the current social development [5]. Hence, it is essential to discover interesting patterns of congestion propagations in the traffic networks [6].

Existed works focusing on discovering propagation relationships among spatio-temporal congestions mainly include Region-based method [2–4], Road-based method [1], etc. The Region-based method modeled the traffic networks by partitioning the urban area into several regions [2, 3] or junctions [4], a regional graph was built with nodes as regions and traffic flows were represented by links between them. However, since continuous road network was partitioned into isolated regions, the Region-based method is failed to reveal the complete spatial transitivity. For example, Fig. 1 shows 4 snapshots of a crossroad in a day. If road C and road D are partitioned into one region then we can't discover the propagation relationships between road C and road D. For solving this shortcoming, a Road-based congestion propagation identification method from sensor data was proposed in [1], this work discovers traffic congestion from sensor data and constructs causality trees from traffic congestions to estimate their

L. H. U and H. Xie (Eds.): APWeb-WAIM 2018, LNCS 11268, pp. 46–55, 2018.
https://doi.org/10.1007/978-3-030-01298-4_5

propagation probabilities. However, there still exist several limitations. Firstly, the robustness of traditional sensor data is not guaranteed. Once a sensor is damaged, a large number of data will be lost, which is not conducive to the mining of traffic congestion propagation patterns. Secondly, the flexibility also not perform well on traditional sensor data. Sensor data is easy to be limited to the interval of the monitor and the sensor data is discontinuous. So, it is difficult to find a short time congestion propagation pattern. For example, as shown in Fig. 1(c), the sensor on road C can't detect the actual congestion of road B.

Motivated by above issues, we designed a new road-based congestion propagation identification method on floating car data (FCD) [7], by introducing the spatio-temporal

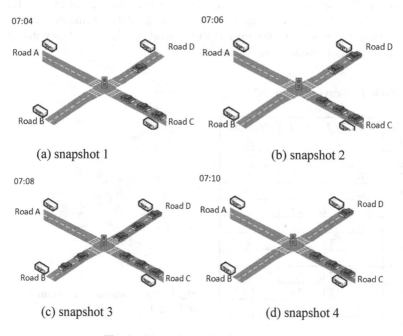

(a) snapshot 1   (b) snapshot 2

(c) snapshot 3   (d) snapshot 4

Fig. 1. 4 snapshots of traffic on one day

co-location congestion pattern mining. The traditional spatio-temporal co-location pattern mining discovers the subsets of features which are located together in adjacent time periods frequently [8, 9]. Thus, using spatio-temporal co-location pattern mining technique [10], we can discover the road spatial transitivity and time transitivity. Meanwhile, using continuous FCD, we can discover congestion propagation patterns including short time congestion transfer patterns.

However, the traffic congestion pattern needs to consider various attributes such as traffic velocity and traffic flow direction. Thus, how to apply the spatio-temporal co-location mining technique on FCD is main challenge in this paper.

Addressed on solving this challenge, this paper focuses on mining STCCPs from FCD. The contributions of our work can be summarized as follows: (1) proposing the new concepts of spatio-temporal co-location congestion pattern(STCCP); (2) proposing a framework of mining STCCPs and an efficient algorithm to discover the STCCPs; (3) evaluating the efficiency and effectiveness of proposed method on real data set.

The remainder of the paper is organized as follows: Sect. 2 introduces STCCP basic concepts and related measures; Sect. 3 gives a STCCP mining framework and the corresponding algorithm; the experimental evaluation is discussed in Sect. 4; Sect. 5 ends this paper with some conclusive remarks.

2 Related Definitions

In this section we will formally define the spatio-temporal congestion co-location pattern (STCCP). Before we introduce the relevant definitions, we first explain Table 1 and Fig. 2. Table 1 is a 4-days records of FCD. In the data set, the direction of the traffic is from the road C to the road B to the road A. Figure 2 shows the spatio-temporal co-location instances of Table 1.

Table 1. 4-days record of FCD

rec_I d	sf_I d	tf_I d	sti_I d	Speed(m/ s)
1	<A,	T1,	1>	7
2	<B,	T1,	1>	3
3	<C,	T1,	1>	6.5
4	<A,	T1,	2>	2
5	<B,	T1,	2>	8
6	<C,	T1,	2>	3
,,,		...		...
22	<A,	T4,	2>	8
23	<B,	T4,	2>	7.5
24	<C,	T4,	2>	7

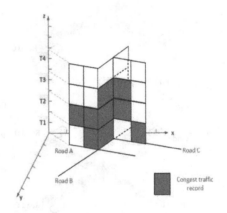

Fig. 2. Spatio-temporal co-location instance of Table 1

According to the discussion in Sect. 1, we firstly introduce the concepts of spatio-temporal co-location pattern as following:

Definition 1 (spatio-temporal features). Given a spatial feature set $SF=\{sf_1,\ldots, sf_n\}$, a temporal feature set $TF=\{tf_1,\ldots, tf_m\}$, the spatio-temporal feature represents a combination of any temporal feature and spatial feature in the set. The spatio-temporal feature $stf_{i,j}$ is represented by a pair of tuples: $stf_{i,j} =<sf_i, tf_j>$ ($tf_i \in TF$, $sf_j \in SF$). The spatio-temporal feature set is a combination of the temporal feature set and the spatial feature set. It is represent as: $STF=\{stf_{1,1},\ldots, stf_{n,m}\}= \{<sf_1,tf_1>,\ldots,<sf_n,tf_m>\}$

Example 1 As shown in Table 1, <A, T1> is a spatio-temporal feature, among them A is a spatial feature and T1 is a temporal feature. We note that given a data set, the set of spatio-temporal features is a Cartesian set of spatial and temporal feature sets, for a given set of spatial features $SF=\{sf_1,...,sf_n\}$ and temporal feature set $TF=\{tf_1,...,tf_m\}$ whose spatio-temporal feature set $STF=\{stf_{1,1},...,stf_{n,m}\}=\{<sf_1,\ tf_1>,...,<sf_n,\ tf_m>\}$, $n*m$ represents the total number of spatio-temporal features.

Definition 2 (Spatio-Temporal Instance). Given a spatial-temporal feature $stf_{i,j}$, the spatio-temporal instance is a certain record in the spatio-temporal feature. A spatio-temporal instance $sti_{i,j,k}$ represented as a triplet$< i$ (the instance belongs to the spatial future id), j (the instance belongs to the temporal feature id), k (the spatio-temporal instance id)$>$, A set of instances of a spatio-temporal feature $stf_{i,j}$ is denoted as $STI_{i,j}=\{$ $sti_{i,j,1},..., sti_{i,j,o}\}$, o represents the total number of spatio-temporal instance.

Example 2 As shown in Table 1, $STI_{A,T1}=\{ sti_{A,T1,1},\ sti_{A,T1,2}\}$ is a set of instances $stf_{A,T1}$, a spatio-temporal instance $sti_{A,T1,2}=<A, T1, 2>$ represents the second instance of Day 1 on the road A.

The STCCP mining framework makes full use of traffic velocity and traffic flow direction to identify traffic congestion correctly. Next we give the definition of congestion instance and spatio-temporal neighborhood relationship.

Definition 3 (Spatio-Temporal Congestion Instance). Given a spatio-temporal feature $stf_{i,j}$, its instance set $STI_{i,j}=\{sti_{i,j,1},..., sti_{i,j,o}\}$, a congestion coefficient p, then the critical velocity of congestion is calculated as follows:

$$v_{jam} = \sum_{w=1}^{n*m*o} \frac{v_w}{n*m*o} * p \qquad (1)$$

In the formula,

v_{jam} —— represents the critical speed of congestion for a given road;

v_w —— represents the speed of the spatio-temporal instance $sti_{i,j,w}$;

p—— given congestion threshold p;

o—— the total number of instances of the spatio-temporal features $stf_{i,j}$;

m—— the total number of temporal features in the data set;

n—— the total number of spatial features in the data set;

If the speed v_w of spatio-temporal instances $sti_{i,j,w}$ satisfies $v_w<v_{jam}$, then the instance $sti_{i,j,w}$ is considered to be an instance where congestion occurs, and is called **spatio-temporal congestion instance**.

Example 3 As shown in Fig. 2, the $sti_{A,1,2}$ is a spatio-temporal congestion instance. Every congest traffic record is a spatio-temporal congestion instance.

Definition 4 (Spatio-Temporal Co-location Congestion Pattern). Spatio-temporal **co-location congestion pattern is a set of spatio-temporal features with the same temporal feature** $sc=\{ stf_{1,j}, stf_{2,j},..., stf_{k,j} \}$, **where** $sc\subseteq STF$, k **is the number of spatio-temporal features included in** sc. **called the size of spatio-temporal pattern** sc.

Example 4 As shown in Fig. 2, $sc=\{ stf_{A,T1}, stf_{B,T1}, stf_{C,T1}\}$ is a 3-size spatio-temporal co-location pattern.

Definition 5 (Spatio-temporal neighborhood relationship). Given two spatio-temporal instances $sti_{i1,j1,l}$ and $sti_{i2,j1,h}$ ($i1{\neq}i2$) and a time interval threshold t_win, if (1) the spatio-temporal features sf_{i1} and sf_{i2} of the two instances are up to ω ($\omega{>}1$) edges on the directed topology diagram And (2) $|h{-}l| \leq |\omega{-}1|{*}t\_win$, called that $sti_{i1,j1,l}$ and $sti_{i2,j1,h}$ satisfy the ω-spatio-temporal neighborhood relationship.

Example 5 As shown in Fig. 2, $sti_{A,T1,2}$ and $sti_{B,T1,1}$ satisfy 2-spatio-temporal neighborhood relationship.

Definition 6 (Row instances and Table instances of spatio-temporal co-location congestion patterns). Given a k-size spatio-temporal co-location congestion pattern $sc{=}\{\ stf_{1,j},\ stf_{2,j},...,\ stf_{k,j}\ \}$, if a spatio-temporal congestion instance set STI' satisfies 1) contains all features of a spatio-temporal co-location pattern sc, and none of the subsets of STI's can contain all the features of sc, 2) Any two spatio-temporal instances in STI' satisfy the ω-spatio-temporal neighborhood relationship, then STI' is called a **row instances**. The set of all row instances of the sc is called a **table instance**, denoted as: $T(sc)$.

For evaluating the interestingness of STCCP, we give the definition of spatio-temporal participation index and participation rate as follows:

Definition 7 (Spatio-Temporal Participation Index and Participation Rate). The **Participation Rate(PR)** of a spatio-temporal feature $stf_{i,j}$ in a spatio-temporal co-location congestion pattern $sc{=}\{\ stf_{1,j},\ stf_{2,j},...,\ stf_{k,j}\ \}$ is represent as PR(sc, $stf_{i,j}$), which is the ratio of the number of non-repeated occurrences of all instances of sc in the spatio-temporal co-location congestion pattern sc to the total number of instances of $stf_{i,j}$. The formula is as follows:

$$PR\left(sc, stf_{i,j}\right) = \frac{\pi_{stf_{i,j}}|T(sc)|}{|T(\{stf_{i,j}\})|} \tag{2}$$

Where

π is a de-duplication relational projection operation.

The **Participation Index(PI)** of the spatio-temporal co-location congestion pattern $sc{=}\{\ stf_{1,j},\ stf_{2,j},...,\ stf_{k,j}\ \}$ is expressed as PI(sc), which is the minimum among PRs of all the features of the co-location patterns sc. The formula is as follows:

$$PI(sc) = min_{i=1}^{k}\left\{PR\left(sc, stf_{i,j}\right)\right\} \tag{3}$$

Given a minimum participation threshold min_prev. When PI(sc) $\geq min\_prev$, we call this pattern is a **prevalent spatio-temporal co-location congestion pattern**.

Example 6 As shown in Fig. 2, the STCCP $sc{=}\{stf_{C,T2},\ stf_{B,T2},\ stf_{A,T2}\}$, PI(sc, $stf_{A,T2}$) = 1, PI(sc, $stf_{B,T2}$) = 0.5, PI(sc, $stf_{C,T2}$) = 0.5,PR(sc) = 0.5, PI(sc) = 0.5. Given a min_prev = 0.4, then the STCCP sc is a prevalent STCCP.

3   The Mining Framework and the Basic Algorithm

In this section we will describe the framework of STCCP mining and then give a basic algorithm of STCCP mining.

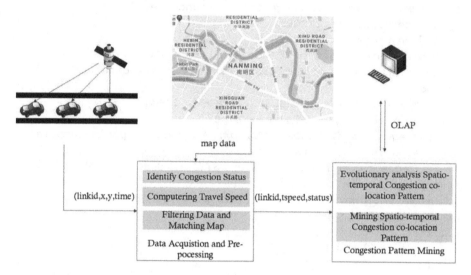

Fig. 3. Framework of mining spatio-temporal co-location congestion patterns

As shown in Fig. 3, we give a framework of mining spatio-temporal co-location congestion patterns. The steps are as follows:

First, we collect the floating car GPS data as input.

Second, we match the input data and map data generate road topology map. Then compute travel time and identify congestion status of each road.

Third, we mine spatial-temporal co-location congestion patterns from processed data. Then analysis mined patterns.

According to Definition 7, we give the proof that the spatio-temporal co-location congestion pattern satisfies the anti-monotone, and design a mining algorithm prevalent spatio-temporal co-location congestion pattern (APSTCCP) based on the anti-monotone.

Theorem 1 (anti-monotone). The participation rate and participation index decrease monotonously with the increase of spatio-temporal co-location pattern size.

Proof. If there is a spatio-temporal co-location pattern $c_1 \cup c_2, c_1 \cap c_2 \neq \emptyset$, $|table\_instance(c_1 \cap c_2)| \leq \min(|table\_instance(c_1)|, |table\_instance(c_2)|)$. According to Definition 7, $\mathrm{PI}(c_1 \cap c_2) \leq \min(\mathrm{PI}(c_1), \mathrm{PI}(\{c_2\}))$. Therefore, the participation Ratio and participation index decrease monotonously with the increase of spatio-temporal co-location congestion pattern size.

The algorithm of prevalent spatio-temporal co-location congestion pattern (APSTCP) mining is designed as follow:

Input:
$STF=\{stf_{1,1},...,stf_{n,m}\}$, $STI=\{sti_{1,1,1},..., sti_{n,m,o}\}$, t_win, p, min_prev
Output:
all prevalent spatio-temporal co-location congestion patterns
Variable:
STI': spatio-temporal congestion instance set
C_k: k-size candidate STCCPs
P_k: k-size prevalent STCCPs
T_k: table instance set of k-size candidate STCCPs
m: the total number of temporal features in the data set
Method:
1. $num\_stf=|STF|$, $c_2\_table\_instance =\emptyset$, $C_2=\emptyset$;
2.   Initialized C_k, P_k, T_k, to be empty
3. $STI'=gen\_jam\_instancs$ (STI, p);// Generate spatio-temporal congestion instance sets
4. $T\_C_2=gen\_table\_ins\_2(E, t\_win, STI')$// Generate table instances of 2-szie candidate STCCPs
5. $P_2=select\_prevalent\_C$ (T_C_2, min_prev) // Generate 2-size prevalent STCCPs
6. $k=2$;
7.   While($P_k\neq\emptyset$ and $k<num\_stf$) do
8.   {
9. $C_{k+1}=gen\_candidate$ (P_k, P_2)// Generate $k+1$-size candidate STCCPs
10. $T\_C_{k+1}=gen\_table\_ins\_k(T\_C_k, T\_C_2)$ // Generate table instances of $k+1$-szie candidate STCCPs
11. $P_{k+1}=select\_prevalent\_C$ (T_C_{k+1}, min_prev) // Generate $k+1$-size prevalent STCCPs
12. $k=k+1$
13. }
14. Return $\cup$ $(P_2,..., Pk)$// return all prevalent STCCPs

4   Experiments

In this section we will show the effectiveness and practicability of the algorithm on the Guiyang traffic data. The Guiyang traffic data set contains a total of 132 roads and 7,662,385 records in 92 days. The data set size is 594MB. In order to study traffic congestion patterns better, we choose the morning peak data from 7 to 9 o'clock for experimentation.

Firstly, we will test the sensitivity of the *min_ prev* and *t_win*. We used control variables method in the experiments. First, we let *min_ prev* = 0.6, and we attempt let *t_win* = 1, *t_win* = 2, *t_win* = 3, and *t_win* = 4. Figure 4 shows the result of the relationship of *t_win* and number of prevalent spatio-temporal co-location congestion patterns. We find that the number of prevalent spatio-temporal co-location congestion patterns increases with the increase of *t_win* and *t_win* is not very sensitive. Similarly, we let *t_win* = 1, and we attempt let *min_ prev* = 0.58, *min_ prev* = 0.6, *min_ prev* = 0.62, and *min_ prev* = 0.64. Figure 5 shows the result of the relationship of *min_ prev* and number of prevalent spatio-temporal co-location congestion patterns. We find that the number of prevalent spatio-temporal co-location congestion patterns reduce with the increase of *min_ prev* and *min_ prev* is more sensitive than *t_win*. The selection of parameters needs expert knowledge or lot of experiments. Finally, we choose *t_win* = 2, min_pre = 0.6 by lots of the experiments.

Secondly, we will illustrate the propagation process of two STCCPs {A, B, D} and {A, B, F} in Fig 6. The Fig 6 shows that road A is jammed in snapshot 1, after 2 snapshots (*t_win* = 2) road B is jammed in snapshot 3. The road D and F are jammed in snapshot 5. The result shows STCCPs are congestion propagation patterns. If the road

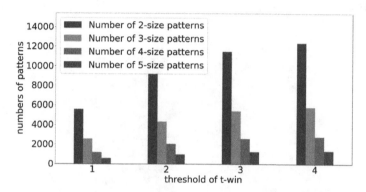

Fig. 4. The relationship of *t_win* and number of prevalent STCCPs.

A become smooth as soon as possible, the traffic will become smooth quickly.

Besides, there is some interesting discovery by comparing the same pattern's average velocity every day. Among Fig. 7, the redder the block, the lower the average velocity. Just as shown in Fig. 7, we can detect that prevalent STCCP {A, B} is a full time pattern, prevalent STCCP {C, E} is a working day pattern.

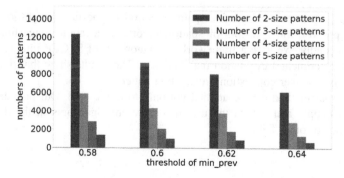

Fig. 5. The relationship of *min_prev* and number of prevalent STCCPs.

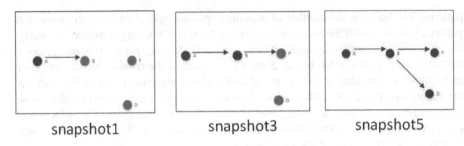

snapshot1   snapshot3   snapshot5

Fig. 6. Propagation process of {A, B, D}, {A, B, F}

Through the above experiments, we verified the effectiveness of the mining algorithm APSTCP that means we can find congestion propagation patterns correctly. In addition, we also found some interesting period patterns by evolutionary analysis mined STCCPs.

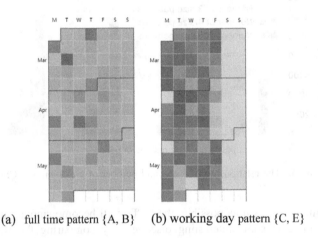

(a) full time pattern {A, B}   (b) working day pattern {C, E}

Fig. 7. Two prevalent STCCPs average velocity

5   Conclusion

We propose a new method to mine spatio-temporal co-location congestion pattern. Firstly, we propose the concepts of spatio-temporal co-location congestion pattern (STCCP). Secondly, we give a framework and algorithm of STCCP mining. Finally, we offer an experimental evaluation real data sets. The results show that our method can effectively discover congestion propagation patterns.

Actually, our method can be applied not only to traffic field. Our potential future research is to apply and extend the use of the present algorithms in the domain of internet traffics or other fields.

Acknowledgements. This work is supported by the National Natural Science Foundation of China (61472346, 61662086, 61762090), the Natural Science Foundation of Yunnan Province (2015FB114, 2016FA026), and the Project of Innovative Research Team of Yunnan Province (2018HC019).

References

1. Nguyen, H., Liu, W., Chen, F.: Discovering congestion propagation patterns in spatio-temporal traffic data. IEEE Trans. Big Data **3**(2), 169–180 (2017)
2. Liu W., Zheng Y., Chawla S., et al.: Discovering spatio-temporal causal interactions in traffic data streams. In: ACM SIGKDD International Conference on Knowledge Discovery and Data Mining, pp. 1010–1018. Elsevier (2011)
3. Pang, L., Chawla, S., Liu, W., et al.: On detection of emerging anomalous traffic patterns using GPS data. Data Knowl. Eng. **87**(9), 357–373 (2013)
4. Chu, V.W., Wong, R.K., Liu, W., Chen, F.: Causal structure discovery for spatio-temporal data. In: Bhowmick, S.S., Dyreson, C.E., Jensen, C.S., Lee, M.L., Muliantara, A., Thalheim, B. (eds.) DASFAA 2014. LNCS, vol. 8421, pp. 236–250. Springer, Cham (2014). https://doi.org/10.1007/978-3-319-05810-8_16
5. Qi, L., Zhou, M., Luan, W.: A Two-level traffic light control strategy for preventing incident-based urban traffic congestion. IEEE Trans. Intell. Transp. Syst. **19**(1), 13–24 (2016)
6. Keler, A., Krisp, J.M., Ding, L.: Detecting traffic congestion propagation in urban environments – a case study with floating taxi data (ftd) in shanghai. J. Locat. Based Serv. **11**(2), 133–151 (2017)
7. Liang Z., Chen H., Song Z., et al.: Traffic congestion incident detection and dissipation algorithm for urban intersection based on FCD. In: IEEE International Conference on Computer and Communications, pp. 2578–2583. IEEE press (2017)
8. Celik, M.: Discovering partial spatio-temporal co-occurrence patterns. In: 1st International Conference on Spatial Data Mining and Geographical Knowledge Services, pp. 116–120. Fuzhou (2011)
9. Yoo, J., Bow, M.: Mining maximal co-located event sets. In: 15th Pacific-Asia Conference on Advances in Knowledge Discovery and Data Mining, pp. 351–362. Elsevier (2011)
10. Wang, L., Bao, X., Chen, H., et al.: Effective lossless condensed representation and discovery of spatial co-location patterns. Inf. Sci. **436**, 197–213 (2018)

Spectroscopy-Based Food Internal Quality Evaluation with XGBoost Algorithm

Lingling Li[1], Caihong Li[1], Yuan Wu[1(✉)], Yi Yang[1,2], Yixuan Zhang[3], Hairui Zhang[1], Bin Wu[4], Li Liu[5,6], and Rong Xie[7]

[1] School of Information Science and Engineering, Lanzhou University, Lanzhou 730000, China
hjf6318567@gmail.com
[2] Silk Road Economic Belt Research Center, Lanzhou University, Lanzhou 730000, China
[3] Division of Biological Science, University of California at San Diego, La Jolla, USA
[4] Communication and Network Center, Lanzhou University, Lanzhou 730000, China
[5] Key Laboratory of Dependable Service Computing in Cyber Physical Society, Ministry of Education, Chongqing 400044, China
[6] School of Software Engineering, Chongqing University, Chongqing 400044, China
[7] School of Computer Science, Wuhan University, Wuhan 430072, China

Abstract. In this paper, the combination of Near-Infrared (NIR) spectroscopy and a novel forecasting algorithm called XGBoost was proposed for food internal quality evaluation. First, the original NIR spectral data was preprocessed by Savitzky-Golay smoothing method to reduce the influence of noises. Secondly, the preprocessed spectra was submitted to PCA to extract essential information. Finally, the model was established by using the XGBoost algorithm. The performance of the proposed model was examined by comparing with different models including back propagation neural network (BPNN) and support vector regression (SVR). The results showed that the new proposed model outperformed other two models and this XGBoost-based tool was suitable for food internal quality control.

Keywords: NIR spectroscopy · Internal quality forecasting · Food XGBoost

This study is supported by the CERNET Innovation Project, China (Grant No. NGII20150603), the Fundamental Research Funds for the Central Universities (Grant No. 2022016zrbr12), Gansu Provincial Science & Technology Department (Grant No. 1506RJZA107), the Natural Science Foundation of PR of China (Grant No. 61300230), the Fundamental Research Funds for the Central Universities (Grant No. lzujbky-2016-br03), the Fundamental Research Funds for the Key Research Program of Chongqing Science & Technology Commission (Grant No. cstc2017rgznzdyf0064), the Chongqing Provincial Human Resource and Social Security Department (Grant No. cx2017092), and the Central Universities in China (Grant No. 2018CDXYRJ0030, CQU0225001104447).

© Springer Nature Switzerland AG 2018
L. H. U and H. Xie (Eds.): APWeb-WAIM 2018, LNCS 11268, pp. 56–64, 2018.
https://doi.org/10.1007/978-3-030-01298-4_6

1   Introduction

With the rapid development of economy and society, people are consuming more and more food. The consumers are paying more attention from the quality of appearance to internal quality and safety. However, traditional internal quality analytical methods need destruct and homogenize the interested samples, these methods are time consuming and they require a considerable amount of manual work and materials [1,2]. Nondestructive food internal quality measurement has been a popular subject of interest for researchers.

NIR spectroscopy has been proposed as an alternative to traditional methods, and it has been widely used for rapid analysis of agricultural products [3]. This technology is a powerful analytical technique because it has advantages of non-destructive, easy-operated, accurate, low-cost and reliable. Moreover, it need no sample preparation and could analyze multiple internal quality attributes simultaneously [4]. NIR spectroscopy is suitable for quality control of food since it is based on the principle that different chemical composition of the analyzed samples have its own specific absorption pattern in NIR region [5,6].

In general, food internal quality evaluation is composed of the following steps: (1) spectral data acquisition; (2) spectral data preprocessing; (3) establishing models with a set of samples; (4) validating established models with another set of independent samples. The spectral data preprocessing step is important and should be implemented initially because of the influences of temperature, light scatter, baseline drift and background noise. These influences could directly affect the forecasting accuracy. In the field of NIR spectroscopy, Savitzky-Golay smoothing method is the most common used method for spectral data preprocessing [2]. Spectral data always has thousands of variables, redundant variables may involve useless or irrelevant information for forecasting tasks, which will result in deterioration for models' forecasting abilities [7]. Furthermore, high dimensional input could make the modeling time become remarkably long. In this study, PCA algorithm was used to extract main features from the spectral data and reduce the input dimensionality. PCA, proposed in 1901 by Pearson [8], is a very effective data mining technique and has been widely applied for dealing with spectral data. PCA has the abilities to reduce dimension and orthogonalize the high dimensional data to obtain a set of values that are linearly uncorrelated, these values are called principal components (PCs). Generally, the first several PCs are qualified to represent most of the whole variances. Therefore, a few PCs could be considered as the new input.

For model establishing step, many reports have been reported for food qualitative and quantitative analysis using different regression algorithms [9,10]. Extreme gradient boosting (XGBoost) algorithm is an outstanding machine learning algorithm which is based on the gradient boosting decision tree (GDBT) [11]. On the basis of GDBT, XGBoost modifies the objective function and the loss function. There are two types of boosted trees in XGBoost: the regression trees and the classification trees. XGBoost is an extension of random forest [12], compared to the random forest algorithm, XGBoost can utilize regularization to further reduce overfitting, improve forecasting accuracy and decrease the time needed to construct trees [13]. XGBoost can be applied to solve multiple machine learning problems in various domains [14].

In this paper, a new method is proposed for food internal quality attributes forecasting based on NIR spectroscopy. This approach is developing through combining Savitzky-Golay smoothing method, PCA, and XGBoost. We will investigate the potential of this new approach for food internal quality forecasting, and compare the performance of the new proposed model with different supervised machine learning algorithms, including BPNN and SVR.

2   Data Processing

In this paper, a new approach with the functions of noise reduction and forecasting modeling capabilities is proposed. The proposed approach consists of Savitzky-Golay smoothing method, PCA and XGBoost. The Savitzky-Golay first-order derivative with a 5-point moving window is used to reduce noises from the original spectral data, the PCA is applied to select main features from the preprocessed spectral data and the XGBoost is applied as the forecasting tool. The specific process of the proposed model was described below and the overall framework of the this model was shown in Fig. 1.

Step1: Noise reduction: The Savitzky-Golay smoothing method is used to reduce the noises from the raw spectral data.

Step2: Main feature extraction: The PCA is used to extract main features of the de-noised spectral data.

Step3: Forecasting: Use the main features and the XGBoost model to get the final forecasting results.

3   Experiments and Results

3.1   Spectral Datasets

The NIR spectral dataset of corn was chosen for this study, which is available on http://www.eigenvector.com/data/Corn/index.html. This dataset has NIR spectra of 80 corn samples measured at Cargill Inc. (Minneapolis, MN, USA) using three different NIR spectrometers (M5, MP5 and MP6).The spectra is collected in 2 nm intervals within the spectral range 1100–2498 nm. Reference values include moisture, oil, protein and starch. Table 1 listed the reference value distributions. This dataset has been widely used for standardization regularization [15]. In this study, the spectral data, collected by using m5 instrument, was used for data analysis. 60 samples were selected randomly as the calibration set while the remaining 20 samples were used as the validation set. The calibration set is used to construct the forecasting models, and the validation set is regarded as an independent test set.

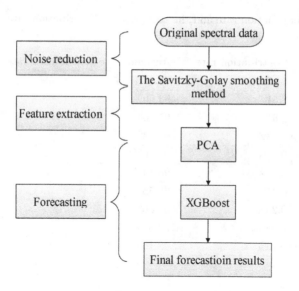

Fig. 1. The overall framework of the proposed model.

Table 1. Summary statistics on reference values of the corn NIR spectral dataset.

Reference value	Min.	Max.
Moisture	9.38%	10.99%
Oil	3.09%	3.83%
Protein	7.65%	9.71%
Starch	62.83%	66.47%

3.2 Spectral Preprocess and Feature Selection

For the dataset, the spectral data was subjected to the Savitzky-Golay first-order derivative with a 5-point moving window (fitted by a polynomial of two degree) to reduce the baseline drifts and enhance the small spectral difference.

Since large number of variables in the input vector may involve irrelevant information and deteriorate the performance of models, PCA was applied to the de-noised spectra for main feature selection. Table 2 listed the contribution rate and the accumulative contribution rate of the first 10 PCs. In this paper, the number of PCs generated by PCA was selected based on the criterion for an increment of explained variance lower than 0.25% [16]. Therefore, the first 8 PCs were used as the input to construct forecasting models.

3.3 Calculation and Performance Evaluation

All computations were performed in Matlab R2014b under Windows 7 with 3.6GHz CPU and 4GB memory. To assess the forecasting capacity of the

Table 2. The contribution rate and the accumulative contribution rate of the first 10 PCs.

PCs	Contribution rate, %	Accumulative contribution rate, %
PC1	84.58%	84.58%
PC2	8.51%	93.09%
PC3	2.61%	95.70%
PC4	0.93%	96.63%
PC5	0.83%	97.46%
PC6	0.68%	98.14%
PC7	0.41%	98.55%
PC8	0.25%	98.80%
PC9	0.20%	99.00%
PC10	0.16%	99.16%

established models, two indices served as the criteria to evaluate the forecasting performance. They were root mean square error of calibration (RMSEC) and root mean square error of prediction (RMSEP). The smaller values of the indices indicated the better forecasting performance. RMSEC and RMSEP are defined as follows:

$$RMSEC, RMSEP = \sqrt{\frac{1}{n}\sum_{i=1}^{n}(\widehat{y_i} - y_i)^2}. \tag{1}$$

Where $\widehat{y_i}$ is the predicted value of the i-th observation, y_i is the measured value of the i-th observation and n is the number of observations in the calibration set or validation set.

3.4   Comparison of Different Models

For BPNN models, the structure of BPNN was three layers, where the number of neurons for the input layer was determined by the number of PCs, that for hidden layer was optimized by using leave-one-out cross validation and that for the output layer was one. Tangent-sigmoid function and linear function were used as the transfer functions in the hidden layer and the output layer respectively. Table 3 showed the optimal number of neurons in the hidden layer.

For SVR models, radial basis function (RBF) was used as the kernel function since it has abilities to reduce the computational complexity of training process. Penalty factor C and RBF width coefficient γ, are two main factors in RBF. Grid search and 4-fold cross validation were used to find the optimal parameter set (C, γ), the best parameter set (C, γ) should guarantee the SVR model has the minimum mean square error. Table 4 listed the optimal parameter set (C, γ) values. For XGBoost models, Table 5 showed the parameter setting for the XGBoost models.

Table 3. The optimal number of neurons in the hidden layer of BPNN models.

Reference value	The number of neurons
Moisture	8
Oil	19
Protein	12
Starch	14

Table 4. The optimal values of parameter set (C, γ) in SVM models.

Reference value	C parameter	γ parameter
Moisture	16	0.031
Oil	8	0.063
Protein	84.45	0.031
Starch	3.03	0.125

Table 5. The optimal number of neurons in the hidden layer of BPNN models.

Parameter	Value
max_depth	100
learning_rate	0.1
n_estimator	100
objective	reg:linear
nthread	−1
gamma	0
min_chile_weight	1
max_delta_step	0
subsample	0.85

Three models (i.e. XGBoost, BPNN and SVR) were established on the dataset. Table 6 presented the forecasting accuracy comparison results. The forecasting accuracies were represented in terms of RMSEC and RMSEP. A good model should have lower RMSEC and RMSEP.

In summary, XGBoost achieved relatively low RMSEC and RMSEP values on all four reference values. BPNN gave the worst performance among the three algorithms. For RMSEP, XGBoost performed better than SVR on oil and protein. For RMSEC, XGBoost gave relatively better forecasting accuracies than the other two on all four reference values. These results revealed that XGBoost could achieve a good forecasting accuracy and have potential for practical applications with a comparable accuracy. By means of different forecasting models, the forecasting accuracy comparisons were presented in Figs. 2, 3, 4 and 5.

Table 6. Forecasting accuracy comparison of three algorithms.

Reference value	BPNN		SVR		XGBoost	
	RMSEC	RMSEP	RMSEC	RMSEP	RMSEC	RMSEP
Moisture	0.12	0.26	0.05	0.12	0.007	0.2
Oil	0.12	0.15	0.06	0.15	0.002	0.12
Protein	0.27	0.31	0.09	0.23	0.006	0.17
Starch	0.25	0.69	0.19	0.41	0.02	0.47

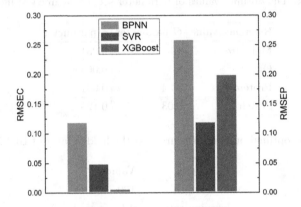

Fig. 2. RMSEC and RMSEP comparisons among different forecasting models on moisture.

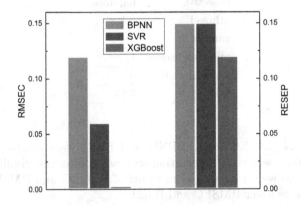

Fig. 3. RMSEC and RMSEP comparisons among different forecasting models on oil.

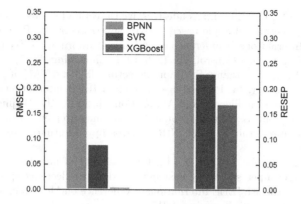

Fig. 4. RMSEC and RMSEP comparisons among different forecasting models on protein.

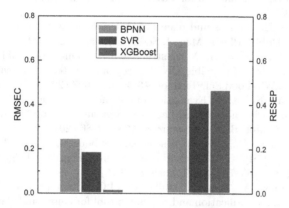

Fig. 5. RMSEC and RMSEP comparisons among different forecasting models on starch.

4   Conclusion

In this paper, a new approach based on XGBoost was proposed for food internal quality evaluation. Also the forecasting performances for BPNN, SVR and XGBoost were compared. The experimental results revealed that the XGBoost model could obtain satisfactory and robust performance than its competitors. Therefore, it may be a promising method for spectroscopy-based food internal quality evaluation with a comparable accuracy.

References

1. Zheng, W., Fu, X., Ying, Y.: Spectroscopy-based food classification with extreme learning machine. Chemom. Intell. Lab. Syst. **139**, 42–47 (2014)
2. Porep, J.U., Kammerer, D.R., Carle, R.: On-line application of near infrared (NIR) spectroscopy in food production. Trends Food Sci. Technol. **46**(2), 211–230 (2015)

3. Shen, Y., Wu, Y., Li, L., Li, L.: Nondestructive Detection for forecasting the level of acidity and sweetness of apple based on NIR spectroscopy. In: Proceedings of the 2nd IEEE International Conference on Advanced Information Technology, Electronic and Automation Control, pp. 1250–1257. Chongqing (2017)

4. Li, L., Wu, Y., Li, L., Huang, B.: Rapid detecting SSC and TAC of peaches based on NIR spectroscopy. In: Proceedings of the 2nd IEEE International Conference on Computational Intelligence and Applications, pp. 312–317. Beijing (2017)

5. Wu, Y., Li, L., Liu, L., Liu, Y.: Nondestructive measurement of internal quality attributes of apple fruit by using NIR spectroscopy. Multimed. Tools Appl., pp. 1–17 (2017)

6. Shen, Y., Tian, J., Li, L., Wu, Y., Li, L.: Feasibility of non-destructive internal quality analysis of pears by using near-infrared diffuse reflectance spectroscopy. In: Proceedings of the 9th IEEE International Conference on Modelling, Identification and Control, pp. 31–36. Kunming (2017)

7. Han, Q.J., Wu, H.L., Cai, C.B., Xu, L., Yu, R.Q.: An ensemble of Monte Carlo uninformative variable elimination for wavelength selection. Anal. Chim. Acta **612**(2), 121–125 (2008)

8. Pearson, K.: LIII. On lines and planes of closest fit to systems of points in space. Lond. Edinb. Dublin Philos. Mag. J. Sci. **2**(11), 559–572 (1901)

9. Liu, Y., Sun, X., Ouyang, A.: Nondestructive measurement of soluble solid content of navel orange fruit by visible CNIR spectrometric technique with PLSR and PCA-BPNN. LWT-Food Sci. Technol. **43**(4), 602–607 (2010)

10. Guo, Y., Ni, Y., Kokot, S.: Evaluation of chemical components and properties of the jujube fruit using near infrared spectroscopy and chemometrics. SpectrochimicaActa Part A: Mol. Biomol. Spectrosc. **153**, 79–86 (2016)

11. Chen, T., Guestrin, C.: Xgboost: A scalable tree boosting system. In: Proceedings of the 22nd ACM SIGKDD International Conference on Knowledge Discovery and Data Mining, pp. 785–794. San Francisco (2016)

12. Svetnik, V., Liaw, A., Tong, C., Culberson, J.C., Sheridan, R.P., Feuston, B.P.: Random forest: a classification and regression tool for compound classification and QSAR modeling. J. Chem. Inf. Comput. Sci. **43**(6), 1947–1958 (2003)

13. Luckner, M., Topolski, B., Mazurek, M.: Application of XGBoost algorithm in fingerprinting localisation task. In: Saeed, K., Homenda, W., Chaki, R. (eds.) CISIM 2017. LNCS, vol. 10244, pp. 661–671. Springer, Cham (2017). https://doi.org/10.1007/978-3-319-59105-6_57

14. Ghosh, R., Purkayastha, P.: Forecasting profitability in equity trades using random forest, support vector machine and Xgboost. In: Proceedings of the 10th International Conference on Recent Trends in Engineering Science and Management, pp. 473–486. Kuala Lumpur (2017)

15. Stout, F., Kalivas, J.H., Héberger, K.: Wavelength selection for multivariate calibration using Tikhonov regularization. Appl. Spectrosc. **61**(1), 85–95 (2007)

16. Urbano-Cuadrado, M., De Castro, M.L., Pérez-Juan, P.M., García-Olmo, J., Gómez-Nieto, M.A.: Near infrared reflectance spectroscopy and multivariate analysis in enology: determination or screening of fifteen parameters in different types of wines. Anal. Chim. Acta **527**(1), 81–88 (2004)

Large Scale UAVs Collaborative Formation Simulation Based on Starlings' Flight Mechanism

Rong Xie[1](✉), Cunfeng Gu[2], Li Liu[3], Lifei Chen[4], and Linyu Zhang[1]

[1] School of Computer Science, Wuhan University, 430072 Wuhan, China
xierong@whu.edu.cn
[2] Shanghai Electro-Mechanical Engineering Institute, 200000 Shanghai, China
13816429275@163.com
[3] School of Big Data & Software Engineering, Chongqing University, 401331 Chongqing, China
dcsliuli@cqu.edu.cn
[4] College of Mathematics and Informatics, Fujian Normal University, 350117 Fujian, China
clfei@fjnu.edu.cn

Abstract. Combining the latest biological results of behavior of intensive flight of starlings, this paper presents topological formation structure and an improved particle swarm algorithm based on the flight mechanism of starlings and its application to the formation of UAV cluster. It also proposes an approach of controling formation behavior of UAVs based on the improved artificial potential fields method. Through simulation experiments, the comparative results are given for verifying the efficiency of formation missions of our methods, as well as simulation of cluster behavior of aggregation and dispersion. The results provide technical assistance for simulation of autonomous collaborative formation of large-scale UAV cluster.

Keywords: UAV cluster · Collaborative formation · Starlings' flight mechanism · Improved particle swarm algorithm (*PSO*) · Improved artificial potential fields (*APF*) · Cluster aggregation behavior · Cluster dispersion behavior

1 Introduction

In order to expand the operational capability of unmanned aerial vehicle (UAV), researchers have proposed the concept of Coordinated Deployment Flight (CFF) of UAV cluster [1]. The UAV cluster is arranged according to a certain way to ensure that UAVs can fly in a certain formation and make adjustments during flight according to environment and missions. The purpose is to improve the coordination capability and operational efficiency of UAVs.

According to density and degree of formation of UAV cluster, formation can be divided into strict formation and non-strict formation. Strict formation, also called fixed formation, makes each UAV stay fixed in the formation strictly. It will not change the

© Springer Nature Switzerland AG 2018
L. H. U and H. Xie (Eds.): APWeb-WAIM 2018, LNCS 11268, pp. 65–78, 2018.
https://doi.org/10.1007/978-3-030-01298-4_7

distance between UAVs even though flight conditions are changed. Such formation is usually used for a high flight requirement; while non-strict formation allows distance between UAVs and structure of the entire formation to change with environment and conditions in flight. This formation is often used in some occasions where flight requirement is not so high. In the paper we study the latter method of formation. The UAV group does not have a fixed formation way, but try to keep distance between UAVs and behavior consistency. In the recent years, research on formation and control of UAV cluster has gradually become a hot direction in the field of UAV applications and many results are achieved. The common UAV formation methods include leader-following method, virtual structure method, behavior-based method, artificial potential fields method and graph-based method [2–6] etc. Jadbabaie et al. [2] proposed a leader-following method and applied it to the formation control of ground robots. This method assumes that there are a long UAV and some wing UAVs. Long UAV tracks the target according to the designated route; while wingman UAV follows long UAV. This method is relatively simple that each wingman UAV gets flight status of long UAV and then calculates its own route. It transfers the formation issue into a relatively simple tracking issue. However, if long UAV fails, formation of the entire UAV will be affected. Kuwata and How [3] improved the leader-following method and proposed a virtual structure method. All UAVs move following a virtual point which is equivalent to the leader in the leader-following method. The method overcomes the limitation of the impact on the whole formation through the virtual point under the condition that long UAV is destroyed. Each UAV receives the same virtual point information. Although there is almost no error in information transfer, calculation and communications volume are relatively large. Kan and Shea [4] then extended a stochastic scenatio of deterministic systems for the solution of the classical leader-following containment control problem. Both leader-following method and virtual structure method are a kind of centralized control method. Da and Wu [5] designed a behavior-based approach. Each UAV in the formation has four basic behaviors, including avoiding obstacles, avoiding collisions, searching for target, and maintaining formation. The overall behavior of each UAV is the weighting of these four behaviors. Bennet et al. [6] proposed a formation method based on *APF* method. But it is easy to fall into a local optimal solution and it is difficult to add constraints in the method. Communication topology plays an important role for UAV formation. Saber and Murray [7] achieved control of UAV formation by designing a coherence agreement related to graph theory.

In the recent years, learning from the behavioral pattern of biological group, the research method through swarm intelligence is gradually becoming a new trend, which can increase robustness and scalability of UAV cluster formation. Qiu et al. [8] proposed a multi-UAVs autonomous formation method based on level mechanism of pigeon behavior, which models topological structure and leadership mechanism in UAV cluster through graph theory and *APF*. This method is similar to leader-following. The difference is that it makes classification on wingman UAV. Low-level wingman is not only effected by long UAV, but also effected by a high-level of wingman UAVs. So this method is more suitable for the application of strict formation, but limited to non-strict formation. In the paper, we study the behavioral mechanism of starlings and establish a kind of topological structure and movement model that simulates the flight mechanism of starlings. Also, we present a method to achieve mutual avoidance among individuals within UAV cluster on the basis of the improvement of traditional *APF* method, which can represent cluster aggregation behavior and cluster dispersion behavior.

The rest of the paper is organized as follows. Section 2 presents starling swarm collaborative mechanism. We present our UAVs formation method and formation behavior control method in Sects. 3 and 4, respectively. Section 5 gives experiment results and simulation. Conclusions are finally presented in Sect. 6.

2 Starlings' Flight Mechanism

Starling's flight characteristics. There are roughly two clustering ways of bird. One is linear formation of one-line or herringbone, such as geese, pigeons etc. The other is dense cluster formation, such as starlings, locusts etc. When starlings fly, thousands of starlings fly together, which covers the sky like a huge cloud. Even if the number of cluster is very large, their actions are highly coordinated with each other. The entire cluster can quickly change their flight direction synchronously. The phenomenon how the whole group maintains coordination and coherence has always attracted the attention of many biologists.

Ballerini et al. [9] recorded three-dimensional positions of each starling individual using computer vision technology and found that there was an anisotropy in individual distribution in starling population. Individual only interacts with its nearest 6–7 neighbors, which is a kind of measurement of topological distance, but not a traditional metric distance. Cavagna et al. [10] obtained high-precision position and velocity of starling population using stereo imaging technology and studied response to external disturbances. Individual can indirectly effected by individuals far away, but such effect is far less than the impact of individuals in the direct scope on it, which further reveals the reason why starlings can fly together highly consistently.

Movement model. According to the flight mechanism of starling, we present our motion model as shown in Fig. 1. Number of starlings is N. Scope of vision of each starling is R_v. The current velocity is denoted by v, and position is denoted by P. The maximum flying velocity is the constant V_m. When there is an invaders, velocity of escape is defined as V_e. Taking the current position of starling as the centre, within a certain range with a radius of R_v, each starling compares the distance between neighbour in the set X and itself, and choose the 7 nearest neighbours as the reference objects. As shown in Fig. 1a, the reference objects are represented in gray. Each neighbor in the set X with 7 objects is represented by X_i. Velocity is represented by v_i and position is P_i. Each starling adjusts its velocity When starlings gather together

v according to the velocity v_i of the neighbors in X. In order to prevent from collisions, safety distance between starlings is set as R_s. When the distance between two starlings is less than R_s, there is a repulsion between two starlings, thus they get away from each other spontaneously.

Cluster behaviors. When starlings gather together, shown in Fig. 1b, a center point C is determined according to the position P_i of each object in the set X, and starling adjusts its own velocity to get closer to point C. On the other hand, when starlings are scattered, shown in Fig. 1c, starling determines the position E of the center point of a cluster according to the current state of the cluster, and then immediately turns and flees away from E. And its velocity gradually accelerates during the escape until the velocity reaches the escape velocity V_e.

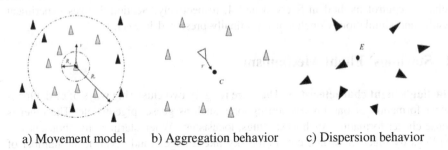

a) Movement model   b) Aggregation behavior   c) Dispersion behavior

Fig. 1. Starling's movement model and cluster behavior.

3 UAV Cluster Formation Method

3.1 Topological Formation Structure

Based on biologists' observation on starling, Montes et al. [11] introduced topological structure to particle swarm algorithm (PSO) and presented that the topology had a significant impact on the PSO. Young et al. [12] theoretically demonstrated the correctness of 6–7 neighbours, which provides theoretical support for the formation modelling via simulating the flying mechanism of starlings.

The flight characteristic of starlings is that it always maintains specific topology in which each individual is effected by its neighbors and adjusts its flight in time. In the traditional PSO, there is a lack of information exchange among individuals. Each individual only communicates with the best individual. As a result, cluster easily loses its omnipotence and falls into local extreme. Therefore, we consider adding topology to formation on the basis of the PSO, which enables each individual to interact with its 7 nearest neighbors. Through the improvement, it can increase the capability of global search and also ensure accuracy of local search. Figure 2 shows our method of topological structure and its example we made for simulating 300 starlings.

Fig. 2. Topological structure of simulated starling flight mechanism.

As shown in Fig. 2, a kind of topology is created with 7 neighbor UAVs for cluster formation. Each UAV can obtain information of neighbor's velocity and acceleration.

3.2   Velocity and Location Update

Velocity update. The flight mechanism of starling is introduced to UAVs formation to make UAV cluster have the flight characteristics of starlings. The topological mechanism is added to the item of velocity in the velocity update equation of the traditional PSO. Velocity update v_i in the $k + 1$ iteration is defined in Eq. (1).

$$\begin{cases} v_i^{(k+1)} = \omega \times v_i^{(k)} + c \times rand \times v\_tp_i^{(k)} \\ v\_tp_i^{(k)} = \frac{1}{N_X} \sum_{n \in X} v_n^{(k)} \end{cases} \tag{1}$$

where ω is inertia weight coefficient. c is topological learning factor. *rand* represents a random number between [0, 1]. N_X is number of neighbors, defaults to 7. X is a set of neighbors that has a topology effect with the current UAV. 7 nearest neighbors to UAV are selected. Due to the introduction of topological factor to UAV cluster, each UAV maintains a certain degree synchronization in velocity with its neighbors. Although this is not a strict formation, the entire group will fly like a cluster.

Location update. Location update is based on the updated velocity, position update x_i in the $k + 1$ iteration is defined in Eq. (2).

$$x_i^{(k+1)} = x_i^{(k)} + v_i^{(k+1)} \times \tau \tag{2}$$

where τ is unit of time, normally set to 1.

Taking into account the actual flying situation of starling, effects by neighbors are not static. In order to ensure to maintain a certain safety distance during the flight, topological action factor c is set in the paper as shown in Eq. (3).

$$c = \begin{cases} 1 - \omega, \ k \le \frac{2}{3} iter_{max} \\ 0, \ others \end{cases} \tag{3}$$

where $iter_{max}$ is the largest evolutionary algebra. When ω decreases, population gradually loses its diversity. The topological function is needed to be strengthened so that the information interaction between the individual and other individual is enhanced to avoid the population from converging to the local point. As finding the local optimal solution after the default two-thirds algebra, it will no longer maintain the diversity of the population. At this time, set $c = 0$ to ensure accuracy.

3.3   Inertia Weight Coefficient

The kinetic energy of UAV at the k-th iteration is defined as $E_i = \frac{1}{2} m v_i^2$, where m is quality of UAV, assuming as 1. v is velocity of movement. Therefore, the total kinetic energy of the entire cluster is calculated as $E = \sum_{i=1}^{N} E_i$. The total energy reflects the evolution of the algorithm. If the kinetic energy is large, which means that evolution is fast, it is required to reduce ω; otherwise, meaning that evolution is slow, it is required to add ω. Here, ω is adjusted by Eq. (4).

$$\omega = \omega_{max} - (\omega_{max} - \omega_{min}) \times \frac{k}{iter_{max}} + \alpha \times rand \times e^{-E} \tag{4}$$

where k is evolutionary algebra. α is control factor. ω_{min} and ω_{max} is dynamic range of weight, representing the minimum weight and maximum weight, respectively. Under the influence of kinetic energy, ω plays a buffer role, which allows inertia weight to be adjusted adaptively according to the evolutionary velocity of the algorithm.

4   UAVs Formation Behavior Control Method

4.1   *APF* Improvement

APF [13] finds a collisionless path by searching for descending direction of potential function under the help of combining the gravitation field at the destination and repulsion field around obstacle. Because of its performances of fast calculation and good real-time, it is often used in the applications of track planning of UAV. However, there are the issues of unreachable destination and local extremum. Fax and Murray [14] solved the problem of unreachability by designing a new potential field function. Kelasidi et al. [15] solved the problem of local extremum. When path planning falls into local extremum, a virtual obstacle was added in the environment of potential field to break the current balance of potential field to jump out of local extreme. Based on these research results, we further improve the *APF* method and propose the following method for controlling the formation behavior of UAVs.

Shown in Fig. 3, O and G represent obstacle and target, respectively. Obstacle repel UAV and target appeals to long-distance UAV. UAV is effected by the resultant force due to repulsive force and attractive force. According to Eq. (5), we can obtain attractive function F_a, repulsive force function F_r and potential field angle A_f, respectively.

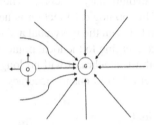

Fig. 3. *APF* diagram.

$$\begin{cases} F_a = \frac{1}{2}\xi D_{rg}^2 \\ F_r = \frac{1}{2}\mu(\frac{1}{D_{ro}} - \frac{1}{D^*})^2 \\ A_f = \angle(\sum F_a + \sum F_r) \end{cases} \tag{5}$$

where ξ and μ are gravity coefficient and repulsion coefficient, respectively. D_{rg} is the distance between UAV and the target. D_{ro} is the distance between UAV and obstacle. D^* is influence radius of obstacle.

Collision analysis. After setting the parameters of D_{rg} and D_{ro} through appropriate experiments, it can implement the function for UAV to avoid static obstacles. However, if there are some moving objects, it is easy for UAV to collide with these moving objects and the UAV's adaptability will become very poor. Figure 4 gives a schematic of three collision scenarios. Figure 4a shows a kind of rear-end collision, i.e. the collision when velocity or acceleration of a UAV is greater than that of another UAV in front of it. Avoidance is generally based on *APF* method. Figure 4b shows a kind of intercept collision. Although UAV adopts *APF* method for collision avoidance, it will crash for no time to avoid because another UAV's velocity or acceleration is too large. Figure 4c shows a kind of head-on collision. The situation is more complicated than the situation in Fig. 4b. Although collisions, the two cases shown in Fig. 4b and Fig. 4c, can be avoided by increasing the repulsive force or increasing repulsive force coefficient, it will cause unnecessary judgements for many collision avoidance behavior from long distance. The traditional *APF* method is designed from the perspective of static obstacle avoidance, not considering velocity and acceleration of moving objects, which has some certain limitations.

a) Rear-end collision   b) Intercept collision   c) Head-on collision

Fig. 4. Three collision scenarios.

Improved *APF*. The *APF* method is improved and is introduced to the UAV formation control to prevent from collisions between UAVs. Known velocity and acceleration of UAV at the current time t, it is possible to predict the position P and the position P_i of other UAVs in the formation after the time interval t_0. For two UAVs, if the distance between P_i and P is lower than the safety distance threshold R_s after t_0, avoidance will be taken in advance. The function of repulsive force F_r is modified as shown in Eq. (6).

$$F_r = \begin{cases} \frac{1}{2}\mu(\frac{1}{D_{ro}} - \frac{1}{D^*})^2, & d(P_i, P) < R_s \\ 0, & \text{others} \end{cases} \tag{6}$$

where $d(P_i, P)$ represents the distance between the current predicted position and the predicted position of other UAV after the time interval t_0.

As two UAVs predict each other's location and take evasive measures in advance, to a certain extent, the collisions shown in Fig. 4a and b can be avoided. If it is predicted that the distance between UAVs is greater than the safe distance, no evasive measure needs to be taken. To some extent, the collision behavior shown in Fig. 4c can be also avoided. In order to avoid collisions, such prediction method is introduced to the traditional *APF*, i.e. transforming the original static *APF* to a dynamic *APF*. Part of the Java implementation code is as follows.

```
public void avoidCrash(Vector<UAV> swarm) {
  Vector<UAV> swarmInVision=getSwarmInVision(swarm);
  if(calDistance(this.loc,target)>arrivalDistance) {
    if(companion.id!=this.id) {
      double[] companionNextLoc=new double[]
{companion.next[0],companion.next[1]};
      double[] myNextLoc=new double[] {this.next[0],
this.next[1]};
      double predictedDistance=
calDistance(companionNextLoc,myNextLoc);
      // When distance between UAVs is less than safe
distance, add repulsion
      if(predictedDistance<(companion.SAFE_RADIUS+this.S
AFE_RADIUS)) {
        vdirection[0]=vdirection[0]-(50/predictedDistance);
        vdirection[1]=vdirection[1]-(50/predictedDistance);
        getNextLoc(loc,vdirection);}}}}
```

4.2   Cluster Aggregation Behavior

When UAVs converge, the centre point between UAV and its neighbour is calculated according to the current location and the state of its neighbour, and then move closer to the centre. The centre point gives a gravitational force to UAV. But this central point is constantly changed. From the perspective of the entire cluster, the distance between all centre points will also gradually decrease.

The UAV cluster takes an evasive mechanism internally. If the distance between UAV and its neighbor is less than the safe distance, they will be mutually exclusive. When aggregation occurs, the distance between UAV continues to shrink. If resultant force of gravity between center point and single UAV and repulsion between UAVs is zero, the entire UAV formation maintains a relatively constant state in a collection area. Part of the Java implementation code is as follows.

```java
public void aggregation(Vector<UAV> swarm) {
  Vector<UAV> swarmInVision=getSwarmInVision(swarm);
  double[] centerLoc=new double[]{this.loc[0],
this.loc[1]};
  if(swarmInVision.size()!=0) {
    for(int i=0; i<swarmInVision.size(); i++) {
    UAV companion=swarmInVision.get(i);
    centerLoc[0]+=companion.loc[0];
    centerLoc[1]+=companion.loc[1];
    }
    // Get coordinates of center point
    centerLoc[0]=centerLoc[0]/((double)(swarmInVision.si
ze()+1));
    centerLoc[1]=centerLoc[1]/((double)(swarmInVision.si
ze()+1));
    seek(centerLoc);
  }
  avoidCrash(swarm);
  getNextLoc(loc, vdirection);
  loc[0]=next[0], loc[1]=next[1];
  vdirection[0]=next[2], vdirection[1]=next[3];
}
```

4.3   Cluster Dispersion Behavior

When dispersion is happened, the coordinates of the center point of the entire UAV cluster are obtained. UAV gets a unit vector pointing from the center point to itself according to the position of the center point and itself, and then UAV turns according to the direction of the vector. Thus, the entire UAV cluster spontaneously spread around, which can achieve the purpose of formation dispersion. Part of the Java implementation code is as follows.

```
public void dispersion(Vector<UAV> swarm) {
    double
vLength=Math.sqrt(vdirection[1]*vdirection[1]+vdirectio
n[0]*vdirection[0]);
    double[] centerLoc=new double[] {this.loc[0],this.loc[1]};
    double vx, vy;
    double x=loc[0], double y=loc[1];
    double xlast, ylast;
    // Get coordinates of center point
    centerLoc=getCenterLoc();
    for(int i=0; i<swarm.size(); i++) {
        double[] eDirection=new double[2];
        eDirection=getUnitVector(centerLoc, this.loc);
        // Do not change the velocity, change the
direction of velocity away from the center point
        eDirection[0]=(vLength*eDirection[0]*1.1);
        eDirection[1]=(vLength*eDirection[1]*1.1);
        vdirection=eDirection;
        getNextLoc(loc, vdirection);
    }
    avoidCrash(swarm);
    loc[0]=next[0], loc[1]=next[1];
    vdirection[0]=next[2], vdirection[1]=next[3];
}
```

5 Simulation

5.1 Experiment

In order to verify the effectiveness of our method, we simulate formation flight of UAV cluster. For easy observation, two-dimensional plane is used for simulation, not considering height. A simple search task for UAV cluster is arranged in simulation.

We test our method from the aspects of formation performance and its impact on execution task of UAV cluster. The efficiency of the formation method is proved by comparing the effectiveness between traditional PSO and improved PSO with starling flight mechanism (starling-PSO). Also, the effectiveness of aggregation behavior and dispersion behavior is proved by observing formation changes of UAV cluster.

When the program starts, UAV cluster randomly appear in a specific area. The initial position, velocity, and direction are randomly generated. The initial parameters are set as follows. Control factor α is set to 1. The weight dynamic ranges ω_{max} and ω_{min} are chosen as 1 and 0.7, respectively. ω is set to 1 initially. Topological effect factor c is defined as 1. The maximum number of iterations is defined as 5000. The threshold of unit safety distance R_s is defined as 50. When the entire UAV cluster is located within a unit distance range of 500 from the target location, the task is regarded to be completed.

5.2   Test Results and Simulation

We test the efficiency of UAV cluster using different numbers of UAVs. The number of UAVs is 100, 300, 500 and 1000, respectively. We also test the behavior of aggregation and dispersion of UAV cluster under the number of UAVs is 300.

Formation efficiency comparison. The comparison result of the efficiency of formation missions is shown in Fig. 5. Compared with the formation method based on the traditional PSO, using our method which joins the flight mechanism of starling, the number of iterations of performing tasks of UAVs is significantly reduced, even only the original half. It is proved that our method has obvious effect on the formation of UAV cluster, which can greatly reduce the number of iterations and improve the efficiency of the entire cluster.

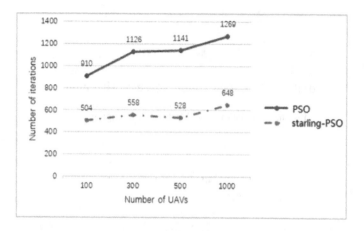

Fig. 5. Comparison of the efficiency of UAV formation.

Formation aggregation simulation. The simulation results of UAV formation aggregation are shown in Fig. 6. Figure 6a shows the formation before aggregation. After the 10, 20, 30 and 40 iterations, the changes of UAV formation are shown in Fig. 6b–e, respectively, representing the obvious aggregation of UAV formation. Each UAV constantly adjusts the position of the virtual center according to the status of itself and its neighbors. It is also shown that the formation of the entire cluster is relatively unchanged after the completion of the aggregation.

Formation dispersion simulation. The simulation results of UAV formation dispersion are shown in Fig. 7. Figure 7a shows the formation before dispersion. After the 10, 20, 30 and 40 iterations, the changes of UAV formation are shown in Fig. 7b–e, respectively, representing the individual UAV spreads out and the entire group maintains the status of formation no longer. In the dispersion behavior, since the only one center point is calculated, the center point will not change. Therefore, there is no phenomenon that spreads outward from the center.

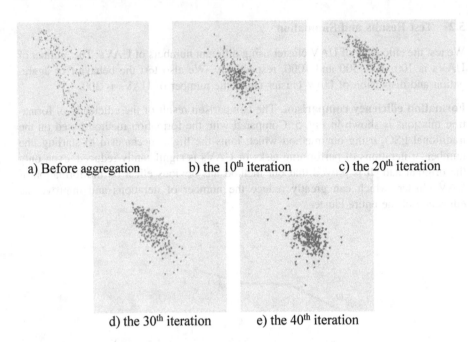

a) Before aggregation   b) the 10th iteration   c) the 20th iteration

d) the 30th iteration   e) the 40th iteration

Fig. 6. Simulation of UAV formation aggregation.

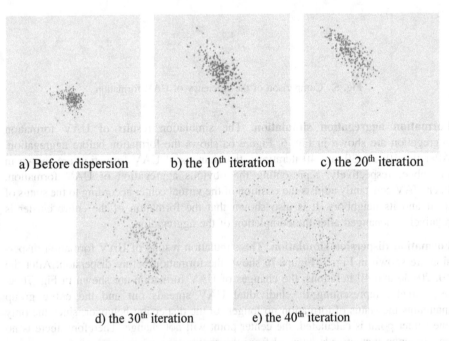

a) Before dispersion   b) the 10th iteration   c) the 20th iteration

d) the 30th iteration   e) the 40th iteration

Fig. 7. Simulation of UAV formation dispersion.

In the simulation process, there may be some situations happened that some individuals may be far from the cluster, leaving behind the group, but the individual will soon return to the cluster. To some extent, it is proved that our method is robust to UAV cluster.

6 Conclusion

Our main work is summarized as follows.

1. With the increasing number of UAVs in UAV applications, many traditional methods are not suitable for UAV concentrated formation. According to the latest biological research results of intensive flight of starlings, combining bionic mechanism as well as swarm intelligence algorithm, we propose modeling method for autonomous formation of UAV cluster which can simulate flight mechanism of starlings and have self-organizing and adaptive features.
2. We present an improved PSO method for autonomous formation of UAVs. Our experimental results show that the number of iterations of performing flight mission has been significantly reduced or even reduced by half after joining the starling flight mechanism to the formation algorithm, which can greatly improve the efficiency of performing tasks. Also, it will not affect task execution of the entire cluster when one or some individuals within cluster is/are left behind or fail, which can increase robustness of the cluster.
3. We improve the traditional *APF* and present an implementation method for the aggregation behavior and dispersion behavior, which can solve the issue of avoidance in UAV cluster formation.

   In the simulation, the parameters debugging is actually very complicated. If the setting of parameters is not appropriate, it will directly affect the execution task or even the task cannot be completed. For the further research, it is required for us to adjust the parameters adaptively, such as inertia weight etc. Also, the proposed topology is needed to be optimized to have better efficiency and robustness of information dissemination, which can be proved to be superior to other classic complex network structures.

Acknowledgements. This work was partially supported by Shanghai Academy of Spaceflight Technology under grant No. sast2017-03 and the Fundamental Research Funds for the Central Universities under grant No. 2042017gf0070.

References

1. Zhang, X., Miao, L., Shang, T.: Utility analysis of network coding for coordinated formation flight in unmanned aerial vehicles. In: International Conference on Networking and Network Applications (NaNA), Hakodate, Japan, 23–25 July (2016)
2. Jadbabaie, A., Lin, J., Morse, A.S.: Coordination of groups of mobile autonomous agents using nearest neighbor rules. IEEE Trans. Autom. Control **48**(6), 988–1001 (2003)

3. Kuwata, Y., How, J.: Robust cooperative decentralized trajectory optimization using receding horizon MILP. In: Proceedings of the American Control Conference, pp. 522–527. IEEE, New York, 9–13 July 2007
4. Kan, Z., Shea, J.M., Dixon, W.E.: Leader–follower containment control over directed random graphs. Automatica. **66**, 56–62 (2016)
5. Cai, D., Sun, J., Wu, S.: UAVs formation flight control based on behavior and virtual structure. In: Xiao, T., Zhang, L., Fei, M. (eds.) AsiaSim 2012. CCIS, pp. 429–438. Springer, Heidelberg (2012). https://doi.org/10.1007/978-3-642-34387-2_49
6. Bennet, D.J., McInnes, C., Suzuki, M., et al.: Autonomous three-dimensional formation flight for a swarm of unmanned aerial vehicles. J. Guid. Control Dyn. **34**(6), 1899–1908 (2011)
7. Saber, R.O., Murray, R.M.: Consensus problems in networks of agents with switching topology and time-delays. IEEE Trans. Autom. Control **49**(9), 1520–1533 (2004)
8. Qiu, H., Duan, H., Fan, Y.: Multiple unmanned aerial vehicle autonomous formation based on the behavior mechanism in pigeon flocks. J. Control Theory Appl. **10**, 1298–1304 (2015) (in Chinese)
9. Ballerini, M., Cabibbo, N., Candelier, R., et al.: Interaction ruling animal collective behavior depends on topological rather than metric distance: evidence from a field study. PNAS **105**(4), 1232–1237 (2008)
10. Cavagna, A., Cimarelli, A., Giardina, I., et al.: Scale-free correlations in starling flocks. PNAS **107**(26), 11865–11870 (2010)
11. Montes, M.A., Stützle, T., Birattari, M., et al.: Frankenstein's PSO: a composite particle swarm optimization algorithm. IEEE Trans. Evol. Comput. **13**(5), 1120–1132 (2009)
12. Young, G.F., Scardovi, L., Cavagna, A.: Starling flock networks manage uncertainty in consensus at low cost. PLoS Comput. Biol. **9**(1), 1–7 (2013)
13. Khatib, O.: Real-time obstacle avoidance for manipulators and mobile robots. Int. J. Robot. Res. **5**(1), 90–98 (1986)
14. Fax, J.A., Murray, R.M.: Information flow and cooperative control of vehicle formations. IEEE Trans. Autom. Control **49**, 1465–1476 (2004)
15. Kelasidi, E., Pettersen, K.Y., Gravdahl, J.T.: A waypoint guidance strategy for underwater snake robots. In: 22nd Mediterranean Conference of Control and Automation (MED), Palermo, Italy, pp. 1512–1519, 16–19 June 2014

Measuring the Spatio-Temporal Similarity Between Users

Hongmei Chen, Peizhong Yang, Lizhen Wang, and Qing Xiao[✉]

Department of Computer Science and Engineering,
School of Information Science and Engineering,
Yunnan University, Kunming 650091, China
xiaoqing@ynu.edu.cn

Abstract. A large volume of user check-in data (check-ins) generated from location-based social networks enable a number of important location-aware services such as grouping users and recommending point-of-interests (POIs). Measuring the similarity between users according to check-ins is a key issue in many technologies for location-aware services such as clustering and collaborative filtering. Some works convert check-ins into vectors and compute the similarity between vectors, such as Cosine similarity and Pearson similarity, as the similarity between users. However, these similarity measurements do not exploit well the spatio-temporal gather and decay of check-ins. It can be easily observed that users tend to visit nearby places at nearby times. In this paper, we define co-occurrence patterns based on the time similarity and the location similarity. Then, we propose the spatio-temporal similarity by utilizing the most similar co-occurrence patterns. Finally, we verify the spatio-temporal similarity is effective by applying it to time-aware POI recommendation.

Keywords: Check-ins · Location-aware service · Spatio-temporal similarity

1 Introduction

With the rapid development of information technology such as mobile terminal and mobile internet, location-based social networks (LBSNs) such as Foursquare and Gowalla have become increasingly popular in recent years. Users can share their trajectories and activities in LBSNs by check-in which represents a user visited a point-of interest (POI) at a specific time. Due to rich and useful information in a large volume of user check-in data (check-ins) such as 'who', 'when' and 'where', these check-ins enable a number of important location-aware services such as grouping users and recommending POIs, which are the efficient ways of helping service providers and users to acquire the needed information from massive check-ins [1–3].

Measuring the similarity between users according to check-ins is a key issue in many technologies for location-aware services such as clustering and collaborative filtering (CF). Take user-based CF for traditional POI recommendation as an example. User-based CF methods think that users who have more similar check-in behaviors have more similar POI preferences. The methods firstly find similar users by measuring the similarity between users according to check-ins. Then, the methods estimate scores

L. H. U and H. Xie (Eds.): APWeb-WAIM 2018, LNCS 11268, pp. 79–89, 2018.
https://doi.org/10.1007/978-3-030-01298-4_8

for POIs by a weighted combination of check-ins of similar users. At last, the methods recommend top-N POIs which were not visited before [1–4].

When measuring the similarity between users according to check-ins, some works convert check-ins into vectors and compute the similarity between vectors, such as Cosine similarity and Pearson similarity, as the similarity between users. [1–4] construct a User-POI matrix to represent check-ins, and the similarity between users is measured by the similarity between vectors. A User-POI matrix represents check-ins regardless of time. However, it can be observed that users tend to visit different places at different times, and users tend to periodically visit the same places at the same times. Based on the above observation, [5–7] focus on time-aware POI recommendation. In [5], a day is split into multiple equal time slots based on hour, the time dimension is introduced into the User-POI matrix, a User-Time-POI cube is used to represent check-ins, and the similarity between users is measured by Cosine similarity between matrixes.

It is noted that the above similarity measurements do not exploit well the spatio-temporal gather and decay of check-ins. We argue that the spatio-temporal gather and decay of check-ins are important because it can be easily observed that users tend to visit nearby places at nearby times. For example, a user usually has lunch in restaurants near to working place during nearby lunch hours. Motivated by the above, we propose the spatio-temporal similarity between users.

Generally, the main contributions of the paper can be summarized as follows.

- We define co-occurrence patterns based on the time similarity of time slots and the location similarity of POIs.
- We propose the spatio-temporal similarity between users by utilizing the most similar co-occurrence patterns.
- We verify the spatio-temporal similarity is effective by applying it to user-based CF for time-aware POI recommendation.

The rest of the paper is organized as follows. Section 2 introduces related work. Section 3 details the spatio-temporal similarity between users. Section 4 shows the results of experiments. Section 5 concludes the paper.

2 Related Works

Some works for measuring the similarity between users construct a User-POI (UP) matrix to represent check-ins in which an element UP(u, p) is non-zero if user u visited POI p; otherwise, UP(u, p) is zero. The similarity between users u and v is measured by the similarity between vectors UP(u,.) and UP(v,.) [1–4]. The work in [5] splits a day into multiple equal time slots based on hour, introduces the time dimension into the User-POI matrix, and uses a User-Time-POI (UTP) cube to represent check-ins in which an element UTP(u, ts, p) is one if user u visited POI p at time slot ts; otherwise, UTP(u, ts, p) is zero. The similarity between users u and v is measured by the similarity between matrixes UTP(u,.,.) and UTP(v,.,.). Then some similarity functions, such as Cosine similarity and Pearson similarity, are adopted to compute the similarity between vectors or matrixes [1–5].

Another works design directly the similarity measurement for check-ins by considering the hierarchy or the sequence of locations. [8] presents the similarity between semantic locations by using the hierarchical location category. [9] presents the maximal semantic trajectory pattern similarity to measure the similarity between semantic trajectories. [10] formalizes the basic principles of similarity measurements and presents the mobility similarity. [11, 12] take into account both the hierarchy and the sequence of locations. [13] extracts routine activities and calculates hierarchically the similarity based on routine activities.

However, the above similarity measurements do not exploit well the spatio-temporal gather and decay of check-ins. We follow User-Time-POI cube presented in [5] to represent check-ins. But distinct from Cosine similarity used in [5], we propose the spatio-temporal similarity between users.

3 The Spatio-temporal Similarity between Users

Let $u \in U$ denote a user in the user set U, t denote a time in a day where $0 \leq t < 24$, $p \in P$ denote a POI in the POI set P where p is located at a geo-location (x, y). A check-in $c(u, t, p(x, y))$ records user u visited POI p located at geo-location (x, y) at time t.

In [5], a day is split into 24 equal time slots denoted as $T = \{0, 1, ..., 23\}$ based on hour, a User-Time-POI (UTP) cube is used to represent check-ins in which an element $UTP(u, ts, p) = 1$ if there is a check-in $c(u, t, p(x, y))$ where time t lies in time slot $ts \in T$; otherwise, $UTP(u, ts, p) = 0$. The similarity between users u and v is measured by Cosine similarity between two slice matrixes of the UTP cube, which are the matrixes $TP_u(ts, p)$ for user u and $TP_v(ts, p)$ for user v. That is to say, the similarity between users u and v is computed as

$$\text{UserSimilarity}_{\text{cosine}}(u, v) = \frac{\sum_{ts \in T} \sum_{p \in P} TP_u(ts, p)}{\sqrt{\sum_{ts \in T} \sum_{p \in P} TP_u^2(ts, p)} \sqrt{\sum_{ts \in T} \sum_{p \in P} TP_v^2(ts, p)}} \tag{1}$$

In this paper, we propose the spatio-temporal similarity between users by exploiting the spatio-temporal gather and decay of check-ins.

3.1 Co-occurrence Pattern

Definition 1. Given two time slots ts and rs in T, a time threshold tt $(0 \leq tt \leq 12)$. The time similarity of ts and rs is defined as

$$\text{TimeSimilarity}(ts, rs) = \begin{cases} f|ts - rs| & |ts - rs| \leq 12 \text{ and } |ts - rs| \leq tt \\ f(24 - |ts - rs|) & |ts - rs| > 12 \text{ and } (24 - |ts - rs|) \leq tt \\ 0 & \text{others} \end{cases} \tag{2}$$

where $\forall x \geq 0, f(x): x \rightarrow (0, 1]; \forall x_1, x_2 \geq 0, f(x_1) > f(x_2)$ if $x_1 < x_2; f(x) = 1$ if $x = 0$. In this paper, $f(x) = e^{-x}$ is adopted [6].

Definition 2. Given two POIs p and q in P, a distance threshold dt $(0 \leq dt)$. The location similarity of p and q is defined as

$$\text{LocationSimilarity}(p, q) = \begin{cases} g(Distance(p,q)) & Distance(p,q) \leq dt \\ 0 & others \end{cases} \quad (3)$$

where Distance(p, q) is the distance between p and q which can be computed according to the coordinates of p and q; $\forall x \geq 0, g(x): x \rightarrow (0, 1]; \forall x_1, x_2 \geq 0, g(x_1) > g(x_2)$ if $x_1 < x_2$; $g(x) = 1$ if $x = 0$. In this paper, $g(x) = e^{-x}$ is adopted [6].

According to Definitions 1 and 2, the function $f(x)$ $(g(x))$ represents that the time similarity (the location similarity) of time slots (POIs) decreases with the increase of their distance when their distance is not larger than the threshold, and the maximum is 1. The time similarity (the location similarity) is 0 when the distance of time slots (POIs) is larger than the threshold. Based on the time similarity (the location similarity) of time slots (POIs), we define co-occurrence patterns and the pattern similarity.

Let TP_u $(\forall u \in U)$ be the matrix generated from the UTP cube by slicing on the user dimension. It is noted that $\text{TP}_u(ts, p) = 1$ if $UTP(u, ts, p) = 1$; $\text{TP}_u(ts, p) = 0$ if $UTP(u, ts, p) = 0$.

Definition 3. Given two users u and v in U. Two non-zero $\text{TP}_u(ts, p)$ and $\text{TP}_v(rs, q)$ are called a pair of co-occurrence patterns between u and v if and only if they satisfy TimeSimilarity$(ts, rs) > 0$ and LocationSimilarity$(p, q) > 0$.

Definition 4. Given a pair of co-occurrence patterns $\text{TP}_u(ts, p)$ and $\text{TP}_v(rs, q)$ between u and v. The pattern similarity of $\text{TP}_u(ts, p)$ and $\text{TP}_v(rs, q)$ is defined as

$$\begin{aligned} \text{PatternSimilarity}(\text{TP}_u(ts,p), \text{TP}_v(rs,q)) = \\ \alpha \times \text{TimeSimilarity}(ts, rs) + (1 - \alpha) \times \text{LocationSimilarity}(p, q) \end{aligned} \quad (4)$$

where $0 \leq \alpha \leq 1$ is a turning parameter.

3.2   The Spatio-Temporal Similarity

Intuitively, if two users have more check-ins at more nearby times and more nearby places, they should have higher similarity. By utilizing the most similar co-occurrence patterns and the pattern similarity, we propose the spatio-temporal similarity between users.

For a non-zero $\text{TP}_u(ts, p)$ in TP_u, there may be no non-zero element in TP_v which is a co-occurrence pattern with $\text{TP}_u(ts, p)$, or there may be several non-zero elements in TP_v each of which is a co-occurrence pattern with $\text{TP}_u(ts, p)$. We focus on the most similar co-occurrence patterns with $\text{TP}_u(ts, p)$ which may be also more than one.

Definition 5. Given two users u and v in U. For a non-zero $TP_u(ts, p)$ in TP_u, the set of the most similar co-occurrence patterns with $TP_u(ts, p)$ w.r.t. non-zero elements in TP_v is defined as

$$MSPatternSet(TP_u(ts,p), TP_v) =$$
$$\begin{cases} \left\{ TP_v(rs,q) | \text{argmax}_{TP_v(rs,q)} (PatternSimilarity(TP_u(ts,p), TP_v(rs,q))) \right\} \\ \quad \text{if there are co} - \text{occurrence patterns in } TP_v \text{with } TP_u(ts,p) \\ \emptyset \quad \text{if there is no co} - \text{occurrence pattern in } TP_v \text{with } TP_u(ts,p) \end{cases} \quad (5)$$

Definition 6. Given two users u and v in U. For a non-zero $TP_u(ts, p)$ in TP_u, the similarity of $TP_u(ts, p)$ and its most similar co-occurrence pattern in TP_v is defined as

$$MSPatternSimilarity(TP_u(ts,p), TP_v) =$$
$$\begin{cases} PatternSimilarity(TP_u(ts,p), TP_v(rs,q)) \forall TP_v(rs,q) \in MSPatternSet(TP_u(ts,p), TP_v) \\ \quad \text{if} \quad MSPatternSet(TP_u(ts,p), TP_v) \neq \emptyset \\ 0 \quad \text{if} \quad MSPatternSet(TP_u(ts,p), TP_v) = \emptyset \end{cases} \quad (6)$$

It is noted that the most similar co-occurrence patterns may be not symmetrical. That is to say, the most similar co-occurrence pattern with $TP_u(ts, p)$ is $TP_v(rs, q)$, but the most similar co-occurrence pattern with $TP_v(rs, q)$ may be not $TP_u(ts, p)$. Furthermore, the proportion of the most similar co-occurrence patterns in check-ins should be considered when measuring the spatio-temporal similarity between users.

Definition 7. Given two user u and v in U. The spatio-temporal similarity between u and v is defined as

$$UserSimilarity_{spatio-temporal}(u, v) =$$
$$\frac{\sum_{TP_u(ts,p)} MSPatternSimilarity(TP_u(ts,p), TP_v) + \sum_{TP_v(rs,q)} MSPatternSimilarity(TP_v(rs,q), TP_u)}{2\sqrt{m}\sqrt{n}}$$
$$(7)$$

where m and n are the number of non-zero elements in TP_u and TP_v respectively.

In Definition 7, the weight of each non-zero $TP_u(ts, p)$ in TP_u is set as $\frac{1}{m}$ which supposes that the probability of each non-zero element is same. Similarly, the weight of each non-zero $TP_v(rs, q)$ in TP_v is $\frac{1}{n}$. For each non-zero $TP_u(ts, p)$ in TP_u, the weight of MSPatternSimilarity($TP_u(ts, p)$, TP_v) is set as $\frac{1}{\sqrt{m}\sqrt{n}}$ which is the geometric mean. Similarly, the weight of MSPatternSimilarity($TP_v(rs, q)$, TP_u) is $\frac{1}{\sqrt{n}\sqrt{m}}$. Finally, the arithmetic average of the sum of all weighted MSPatternSimilarity($TP_u(ts, p)$, TP_v)s and the sum of all weighted MSPatternSimilarity($TP_v(rs, q)$, TP_u)s is defined as the spatio-temporal similarity UserSimilarity$_{spatio-temporal}(u, v)$ between u and v.

It is noted that UserSimilarity$_{spatio-temporal}(u, u) = 1$ and UserSimilarity$_{spatio-temporal}(u, v) = $ UserSimilarity$_{spatio-temporal}(v, u)$. And the spatio-temporal similarity is equal to Cosine similarity on condition that (1) time slots satisfy: $|ts-rs| > 0$ if $ts \neq rs$; (2) POIs satisfy: Distance(p, q) > 0 if $p \neq q$; (3) time threshold $tt = 0$ and distance threshold $dt = 0$. In fact, compare with Cosine similarity, the spatio-temporal similarity considers the similar check-ins besides the same check-ins when $tt > 0$ or $dt > 0$. To some extent, it not only exploits the spatio-temporal gather and decay of check-ins but also alleviates the sparsity issue of check-ins.

4 Experiments

The goal of experiments is to evaluate the effect of the spatio-temporal similarity between users by applying it to user-based CF for time-aware POI recommendation which aims at recommending top-N POIs for a user u at a time slot ts [5].

4.1 Experiment Setup

In experiments, we use the check-ins from Foursquare and Gowalla which are used and provided by [5]. Foursquare dataset contains 194108 check-ins made by 2321 users at 5596 POIs within Singapore between Aug. 2010 and Jul. 2011. Gowalla dataset contains 456988 check-ins made by 10162 users at 24250 POIs within California and Nevada between Feb. 2009 and Oct. 2010. The datasets are divided into train, test and tune datasets. We use train and test datasets.

We follow the metrics in [5] to evaluate the effect of the methods, i.e. precision@N and recall@N where N is the number of recommended top-N POIs. Given a user u and a time slot ts, the set of POIs in test dataset is denoted as $T_{u,\ ts}$ and the set of recommended top-N POIs is denoted as $R_{u,\ ts}$. Then the precision and recall for time slot ts are calculated as follows.

$$\text{precision}(ts) = \frac{\sum_{u \in U} \left| R_{u,ts} \cap T_{u,ts} \right|}{\sum_{u \in U} \left| R_{u,ts} \right|} \text{ and } \text{recall}(ts) = \frac{\sum_{u \in U} \left| R_{u,ts} \cap T_{u,ts} \right|}{\sum_{u \in U} \left| R_{u,ts} \right|} \qquad (8)$$

The overall precision and recall are calculated as follows.

$$\text{precision} = \frac{1}{|T|} \sum_{ts \in T} \text{precision}(ts) \text{ and } \text{recall} = \frac{1}{|T|} \sum_{ts \in T} \text{recall}(ts) \qquad (9)$$

It is noted that, following [5], we focus on the relative improvement the method has achieved instead of the absolute value due to the low density of the datasets which results in low precision and recall.

We compare the spatio-temporal similarity with Cosine similarity and Pearson similarity. The formula of Cosine similarity is formula (1) [5]. The formula of Pearson similarity is as follows [11].

$$\text{UserSimilarity}_{\text{Pearson}}(u, v) = \left| \frac{\sum_{ts \in T} \sum_{p \in P} (\text{TP}_u(ts, p) - \bar{u})(\text{TP}_v(ts, p) - \bar{v})}{\sqrt{\sum_{ts \in T} \sum_{p \in P} (\text{TP}_u(ts, p) - \bar{u})^2} \sqrt{\sum_{ts \in T} \sum_{p \in P} (\text{TP}_v(ts, p) - \bar{v})^2}} \right| \qquad (10)$$

where $\bar{u} = \frac{\sum_{ts \in T} \sum_{p \in P} \text{TP}_u(ts, p)}{|T||P|}$.

4.2   Performance of Methods

Effect of the methods. In the spatio-temporal similarity, $tt = 1$ (time slot), $dt = 0.5$ (km), $f(x) = g(x) = e^{-x}$, $\alpha = 0.5$. All methods recommend top-N POIs to a user at a time slot based on top-k similar users. The results are shown in Figs. 1 and 2. From Figs. 1 and 2, we can see that precision@N decreases but recall@N increases when N increases. Precision@N and recall@N do not obviously change when k increases. This implicates that not all similar users need be used when recommending, which will reduce the computing time. It is noted that the precision@N and the recall@N of the spatio-temporal similarity outperform that of Cosine similarity and Pearson similarity. On Foursquare dataset, the improvement of precision@N is about 6%–20%, and the improvement of recall@N is about 7%–21%. On Gowalla dataset, the improvement of precision@N is about 6%–11%, and the improvement of recall@N is about 8%–16%.

Further, we analyze the precision(ts) and recall(ts) of all methods for different time slots. We take the results on Foursquare dataset shown in Fig. 3 as a case, where $tt = 1$ (time slot), $dt = 0.5$ (km), $f(x) = g(x) = e^{-x}$, $\alpha = 0.5$, $k = 500$, $N = 5$. From Fig. 3, we can see that the changes of precision(ts)@5 and recall(ts)@5 of all methods are similar. When $ts = 2$ or 3, precision(ts)@5 and recall(ts)@5 reach the minimums. When $ts = 5$ or 6, precision(ts)@5 and recall(ts)@5 reach the maximums. It is noted that the precision(ts)@5 and the recall(ts)@5 of the spatio-temporal similarity outperform that of Cosine similarity and Pearson similarity on 22 time slots except $ts = 0, 2$. The improvement of precision(ts)@5 is more than 10% on 20 time slots, and the improvement of recall(ts)@5 is more than 10% on 19 time slots.

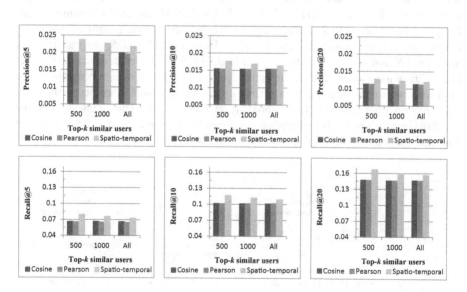

Fig. 1. Effect of the methods for Precision@N and Recall@N on Foursquare dataset

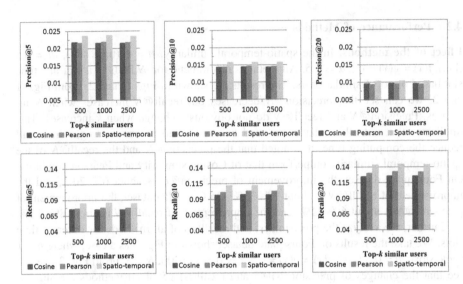

Fig. 2. Effect of the methods for Precision@N and Recall@N on Gowalla dataset

Effect of the thresholds and the parameter. To evaluate the effect of the time threshold and the distance threshold of the spatio-temporal similarity, we design two groups of experiments. In the first group, $tt = 1$ (time slot), $f(x) = g(x) = e^{-x}$, $\alpha = 0.5$, $k = 500$. The method recommends top-N POIs when $dt = 0, 0.5, 1, 1.5$ (km). In the second group, $dt=0.5$ (km), $f(x) = g(x) = e^{-x}$, $\alpha = 0.5$, $k = 500$. The method recommends top-N POIs when $tt = 0, 1, 2, 3$ (time slot). Here, we report two groups of results on Foursquare dataset shown in Figs. 4 and 5. From Figs. 4 and 5, we can see that precision@N and recall@N with $tt = 1$ and $dt = 0.5$ are better than that with other thresholds in most cases. The reason is that the most similar co-occurrence patterns are used to compute the spatio-temporal similarity. The most similar co-occurrence patterns may not change when the thresholds reach some values. This implicates that the thresholds needn't be too big in application, for example, $tt = 1$ and $dt = 0.5$ correspond with that users tend to visit places away 0.5 km within an hour.

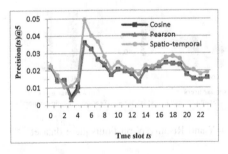
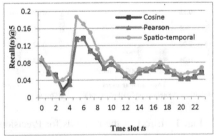

Fig. 3. Effect of the methods for Precision(ts)@5 and Recall(ts)@5 on Foursquare dataset

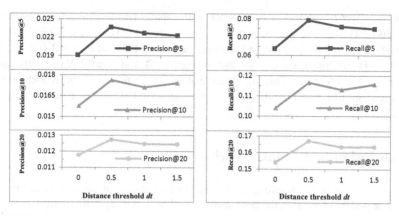

Fig. 4. Effect of the distance threshold for Precision@N and Recall@N on Foursquare dataset

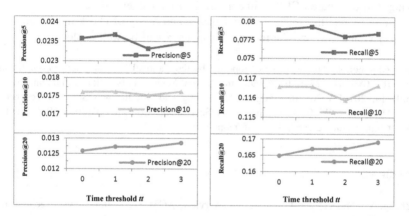

Fig. 5. Effect of the time threshold for Precision@N and Recall@N on Foursquare dataset

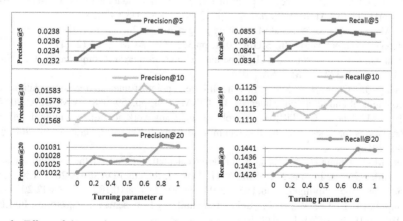

Fig. 6. Effect of the turning parameter for Precision@N and Recall@N on Gowalla dataset

For the effect of the turning parameter of the spatio-temporal similarity, we report the results on Gowalla dataset shown in Fig. 6, where $tt = 1$ (time slot), $dt = 0.5$ (km), $f(x) = g(x) = e^{-x}$, $k = 500$. The method recommends top-N POIs when $\alpha = 0, 0.2, 0.4, 0.5, 0.6, 0.8, 1$. From Fig. 6, we can see that precision@N and recall@N reach the maximums when $\alpha = 0.6$ or 0.8. This implicates that the spatio-temporal gather and decay of check-ins should be considered, and the spatio-temporal similarity between users are reasonable.

5 Conclusion and Future Works

The paper considers the spatio-temporal gather and decay of check-ins, defines co-occurrence patterns based on the time similarity of time slots and the location similarity of POIs, proposes the spatio-temporal similarity between users by utilizing the most similar co-occurrence patterns, and verifies the spatio-temporal similarity is effective by applying it to user-based CF for time-aware POI recommendation.

In future works, the spatio-temporal gather and decay of check-ins including the functions $f(x)$ and $g(x)$ and the turning parameter α will be well explored.

Acknowledgements. This work was supported by the National Natural Science Foundation of China (61662086, 61472346, 61762090), the Natural Science Foundation of Yunnan Province (2015FB114, 2016FA026), the Program for Young and Middle-aged Skeleton Teachers of Yunnan University (WX069051), the China Scholarship Council (201708535025), and the Project of Innovation Research Team of Yunnan Province.

References

1. Zheng, Y.: Trajectory data mining: an overview. ACM Trans. Intell. Syst. Technol. **6**(3), Article 29 (2015)
2. Bao, J., Zheng, Y., Wilkie, D., et al.: Recommendations in location-based social networks: a survey. Geoinformatica **19**, 525–565 (2015)
3. Yu, Y., Chen, X.: A survey of point-of-interest recommendation in location-based social networks. In: AAAI 2015 Workshop, pp. 53–60. AAAI, USA (2015)
4. Ye, M., Yin, P., Lee, W., et al.: Exploiting geographical influence for collaborative point-of-interest recommendation. In: SIGIR 2011, pp.325–334. ACM, USA (2011)
5. Yuan, Q., Cong, G., Ma, Z., et al.: Time-aware point-of-interest recommendation. In: SIGIR 2013, pp.363–372. ACM, USA (2013)
6. Yuan, Q., Cong, G., Sun, A.: Graph-based point-of-interest recommendation with geographical and temporal influences. In: CIKM 2014, pp.659–668. ACM, USA (2014)
7. Zhang, J., Chow, C.: TICRec: a probabilistic framework to utilize temporal influence correlations for time-aware location recommendations. IEEE Trans. Ser. Comput. **9**(4), 633–646 (2016)
8. Lee, M.-J., Chung, C.-W.: A user similarity calculation based on the location for social network services. In: Yu, J.X., Kim, M.H., Unland, R. (eds.) DASFAA 2011. LNCS, vol. 6587, pp. 38–52. Springer, Heidelberg (2011). https://doi.org/10.1007/978-3-642-20149-3_5
9. Ying, J., Lu, E., Lee, W., et al.: Mining user similarity from semantic trajectories. In: LBSN 2010, pp.19-26. ACM, USA (2010)

10. Chen, X., Lu, R., Ma, X., Pang, J.: Measuring user similarity with trajectory patterns: principles and new metrics. In: Chen, L., Jia, Y., Sellis, T., Liu, G. (eds.) APWeb 2014. LNCS, vol. 8709, pp. 437–448. Springer, Cham (2014). https://doi.org/10.1007/978-3-319-11116-2_38
11. Li, Q., Zheng, Y., Xie, X., et al.: Mining user similarity based on location history. In: GIS 2008, Article 34. ACM, USA (2008)
12. Zheng, Y., Zhang, L., Ma, Z., et al.: Recommending friends and locations based on individual location history. ACM Trans. Web **5**(1), Article 5 (2011)
13. Lv, M., Chen, L., Chen, G.: Mining user similarity based on routine activities. Inf. Sci. **236**, 17–32 (2013)

Gesture Recognition Based on Accelerometer and Gyroscope and Its Application in Medical and Smart Homes

Huaizhou Su[1] , Yande Li[2] , and Li Liu[1]([⊠])

[1] School of Big Data and Software Engineering,
Chongqing University, Chongqing, China
860746485@qq.com, dcsliuli@cqu.edu.cn
[2] College of Information Science and Engineering,
Lanzhou University, Lanzhou, China
liyd2016@lzu.edu.cn

Abstract. In recent years, with the rapid development of science and technology, artificial intelligence has gradually entered various fields, and medicine is no exception. For the patients with hand or leg disability after hand injury surgery, the rapid development of artificial intelligence undoubtedly also contributes to improving their life a lot.

Lately, gesture-based human-computer interaction has further accelerated its research due to its natural and intuitive interaction, but building a powerful gesture recognition system is still based on traditional visual methods such as the one proposed in [1] based on multiple cameras which uses color matching for motion detection and tracking and applies it to vehicle control; and another gesture recognition algorithm using stereo imaging and color vision proposed in [2].

The popularity of accelerometers and gyroscopes opens up a new path for gesture recognition. This paper presents a gesture recognition algorithm based on neural network using accelerometer and gyroscope, and applies it to a simple mobile game and smart socket. The mobile game will be used in the hand rehabilitation training of patients after hand injury surgery, leg disability patients inconvenience, smart socket will help them more easily and quickly manage household electricity.

Keywords: Gesture recognition · Accelerometer · Gyroscope
Smart bands · Neural network · Recovery exercise · Smart home

1 Introduction

In the context of pervasive computing, gestures have increasingly become attractive in the human-computer interaction of consumer electronics and mobile devices. With gestures, we can not only deploy a wide range of applications

© Springer Nature Switzerland AG 2018
L. H. U and H. Xie (Eds.): APWeb-WAIM 2018, LNCS 11268, pp. 90–100, 2018.
https://doi.org/10.1007/978-3-030-01298-4_9

in complex computing environments such as virtual reality systems and interactive gaming platforms, but also can help facilitate our daily lives. Potter L E describes the use of gesture recognition can be for hearing or visually impaired people to provide assistance in [3], Khnel C also mentioned the gesture recognition related technologies to achieve smart home.

The relationship between medicine and artificial intelligence has gradually become inseparable. Hand functional exercise is particularly important for hand injuries. Hand injury patients begin hand functional training 4 weeks after the operation, which not only can increase the patient's blood supply, prevent tendon adhesion and joint stiffness, but also for the prevention of muscle fibrosis or disuse atrophy has a positive meaning, but the traditional Hand function exercise has the following disadvantages:

(1) relatively monotonous movements, the frequently same operation will cause patients feel irritable.

(2) the stage of exercise cannot be standardized; the patient cannot control their movements well.

The mobile phone games proposed in this paper have obvious advantages for the recovery of hand trauma surgery:

(1) The mobile phone games are simple in operation, simplifying the traditional hand-function exercises into game-based exercises not only exercises the hand joints, but also the patient is not bored by frequent and monotonous operations.

(2) In the stage of exercise, we can set different exercises in different games. Patients can play different games in different stages of recovery, thus ensuring the standardization of the stage.

For patients with leg disabilities, mobility is the biggest problem. Therefore, the household safety electricity has become particularly important, gesture-controlled smart socket proposed in this paper will be an effective solution to this problem. Patients do not have to move the body next to the socket to control the power, just make the corresponding gestures of the appropriate switch in the position. This simplified operation not only helps patients with leg disability to manage their electricity more conveniently, but also enables their families to relocate to work comfortably. Currently the most effective tool for capturing gestures is the electromechanical or magnetic sensing device (data glove) [5], which uses a sensor connected to a glove to convert a finger's motion into an electrical signal to determine the gesture, which method provides the most full gesture real-time measurement, but it also has the following disadvantages: (1) the cost is too high; (2) gloves must be wore, not particularly convenient; (3) complicated calibration and setting are required to obtain accurate measurement.

Vision-based gesture recognition is an effective alternative based on data glove because it provides a more natural and convenient interaction, however, due to the limitations of the optical sensor, the captured image requires high illumination conditions so it is difficult to detect the gesture information stably, which largely limits the performance of gesture recognition.

In recent years, there have also been more accelerometer-based gesture recognition technologies. Accelerometers have achieved a wide range of achievements in the research of gesture recognition because of the cheapness and portability. Accelerometer-based personalization Gesture recognition and its related applications mentioned in [6], however, its shortcomings are also gradually exposed: The accelerometer recorded a relatively simple information, only a simple three-dimensional acceleration information, which directly increases the recognition phase of the algorithm design difficulty, so come to a correct gesture is relatively difficult.

In this article we will discuss a gesture recognition technique based on accelerometers and gyros, both of which are widely installed in most mobile devices such as smartphones and smartbands. The advantages of using accelerometers and gyroscopes over the above techniques are that accelerometers and gyroscopes can collect information synchronously. By accurately analyzing these information, correct gestures can be obtained without being affected by external environmental factors, such as light conditions [1].

The rest of the paper is organized as follows. We discuss the gesture recognition in the second part, and then in the third part we explain the experimental results . Next we will show in detail in Part 4 two applications implemented with this gesture recognition system: the mobile game Flappy bird and the smart socket, and finally we will conclude in the fifth section.

2   Gesture Recognition

2.1   Data Collection

In our experiment, we used the Microsoft band to obtain relevant information (see Fig. 1). Microsoft Band is a health tracking wearable device made by Microsoft to improve our daily lives. The device is equipped with accelerometers and gyroscopes that collect data in real time, And luckily Microsoft gave developers various APIs for the band. and we noticed these advantages so chose the device for data acquisition.

To make the data we get more accurate, we used the Xbox Kinect (see Fig. 1) to record the video. Xbox Kinect is a component of the Xbox that records video and time stamp information.

Four volunteers from Chongqing University participated in this data connection, they are all boys. First, the volunteer wore a Microsoft band and sat in front of the Xbox Kinect camera, awaiting the start of the acquisition signal that the band and the Xbox Kinect began collecting data at the same time. When the signal was received, volunteers began to make a series of the wave of action: up and down, the action lasted 47 s. After the action was recorded, all the volunteers were asked to make some gestures that they would use in their daily life, which also lasted 47 s (Fig. 2).

(a) 1   (b) 2

Fig. 1. Microsoft band and Xbox kinect.

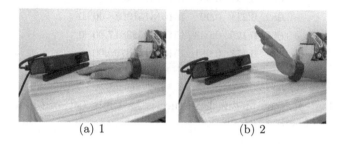

(a) 1   (b) 2

Fig. 2. All the participants were told to make a series of up and down gesture for 47 s.

2.2 Data Preprocessing

We need to mark the data recorded by the Xbox Kinect. Because the time to complete each gesture is not the same. Some gestures may require eight frames of data, some may require four frames, and the data for each gesture is flagged as a corresponding number: flagged as 1 when the gesture is raised and flagged as 0 when the gesture is dropped, Table 1 shows a period of data.

The fourth axis means the time of recoding this accelerometer data. One gesture was done in this data piece and was labeled as one.

2.3 Gesture Segmentation and Feature Extraction

The purpose of gesture segmentation is to identify the start and end of each movement from the information acquired by the accelerometer and gyroscope. Gesture recognition must have a reliable and accurate segmentation capability. There are many ways to achieve gesture segmentation. In [7,8], the gesture is divided directly by pressing and releasing the button. However, this technique has obvious disadvantages. It does not suffice for our mobile games because we do not have time to press and release a button to split the gesture during the game. In [9], Euclidean distance is used to measure the distance between time frame data. If the distance is above the threshold, it is considered the beginning of the gesture. In order to solve this problem effectively and simply, we propose a new segmentation scheme. A fixed-length sliding frame is designed to contain $DATA = a[1], a[2], ..., a[L]$, where $a[n] = (ax[n], ay[n], az[n], gx[n], gy[n], gz[n])$.

Table 1. In this piece of accelerometer data the first three vertical axes mean the magnitude of acceleration of XYZ axis.

X	Y	Z	Time	Label
0.425	0.152	1.116	1510131514837.00	0
0.055	0.182	0.957	1510131514720.00	0
0.09	0.131	0.985	1510131514582.00	1
0.092	0.154	0.985	1510131514468.00	1
0.26	0.138	0.964	1510131514368.00	1
0.057	0.24	1.015	1510131514234.00	1
−0.069	0.147	0.98	1510131514057.00	1
−0.082	0.211	0.99	1510131513918.00	0
−0.067	0.125	0.983	1510131513817.00	0
−0.1	0.177	0.981	1510131513700.00	0

For most volunteers, completing a gesture is basically between four and eight frames, so we define L = 7.

2.4   Classcifier Construction

Fully connected neural network has been widely used in e-mail spam detection, financial fraud detection, emotion recognition and other fields. Fully connected neural networks can make predictions very fast because all the knowledge it learns is encoded in weights. In this paper, we set up a three-layer fully connected neural network. The network consists of an input layer, two hidden layers and an output layer, as shown (Fig. 3):

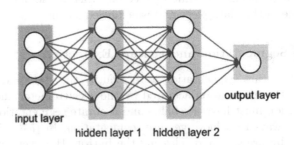

Fig. 3. Topology of a three-layer Fully-connected network classifier.

Assumed that $n[i]$ are the numbers of nodes in the ith layer and m is the number of sample. The output layers is expressed as:

$$y = f(w[3]f(w[2]f(w[1]X + b[1]) + b[2] + b[3]) \tag{1}$$

where X is $a(n[0], m)$ matrix in the input layer, $w[i]$ is the connective weight between nodes in the $(i-1)$th layer and the ith layer, $b[i]$ are bias terms [12]. For hidden layer1 and hidden layer2 each node is a neuron with a nonlinear activation function call ReLu which is described by $f(z) = max(0, z)$. And for output layer we use activation function call sigmoid which is described by $f(z) = 1/(1 + exp(-z))$ (Figs. 4 and 5).

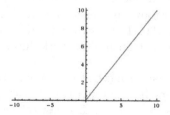

Fig. 4. Rectified Linear Unit (ReLu) activation function, which is zero when x < 0 and then linear with slope 1 when x > 0.

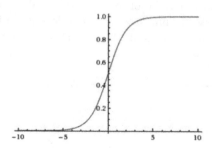

Fig. 5. Sigmoid non-linearity squashes real numbers to range between [0, 1] therefore, in output value f (Z) of each node is between [0, 1]. By using forward propagation algorithm and backpropagation algorithm [10], we can train neural networks effectively. After fully-connected network is trained, it can be used to recognize hand gesture.

3   Experimental Results

In order to reduce the differences between participants, it is necessary to standardize the features. By mapping to the range of each participant's feature [0, 1], the characteristics of each participant are normalized separately. The result is then divided by the standard deviation.

Tensorflow 1.4 was employed to classify the using features we have extracted. For fully connected networks, the hyper parameter iteration is set to 1000. The parameter means that the learning rate per iteration is set to 0.0005. To speed up the gradient descent, we used the mini batch gradient descent method and set each batch to 32. We also used Adam, an algorithm for first-order gradient-based optimization of stochastic objective functions, based on adaptive estimates of lower-order moments to speed up gradient descent. Adam is computationally efficient and requires little memory. When we do not use regularization, the fully connected network has obvious over-fitting problems. In order to solve this problem, dropout was used [11]. The key to exit is to randomly drop units (along with their connections) from the neural network during training. The drop out parameter is set to 0.85.

We use a 4x cross-validation method to evaluate the performance of fully connected network classifiers. In the evaluation, 300 samples from all participants were randomly selected as the test set. The rest of the gesture samples are set as training sets.

Figure 6 shows the cost value per five iterations. Figure 7 shows the recognition rate every 5 iterations. In Fig. 6, the cost value decreases rapidly at the beginning of the iteration. Then the cost curve is minimized. In other words, we do not iterate over all the data at one iteration. We train 32 data at a time, some of the training data may be of low quality, resulting in a minimized cost value. This is a weakness of mini batch gradual decline. In Fig. 7, the accuracy increases rapidly in a short period of time and remains high all the time.

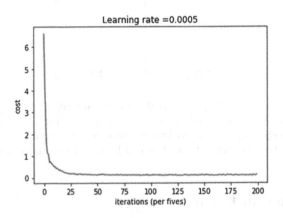

Fig. 6. This fig shows the cost value per five iterations.

The recognition accuracy rates and F1score for each data set are given in the Table 2. And the confusion matrix for gesture recognition is shown in Table 3.

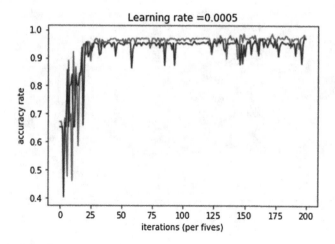

Fig. 7. This fig shows the cost value per five iterations.The red curve represents the accuracy rate of training set and the blue curve represents the accuracy rate of cross-validation set.

Table 2. Gesture Recognition Accuracies (%).

	Training set	Cross-validation set	F1score
Accuracy rate	96.7	96.5	0.74

Table 3. Confusion matrix for gesture recognition.The horizontal axis means actual class and the vertical axis means predicted class.

	True	False
True	124	2
False	9	222

4 Application

4.1 Mobile Games-Flappy Bird

Flappy bird is a simple and difficult mobile game in which players have to control a chubby bird that spans obstacles made up of pipes of various lengths. Easy to get started but want to customs clearance is not easy. Flappy bird was launched at the Apple App store in May 2013 and topped the list in over 100 countries in February 2014. Despite there are no elaborate animations, no fun rules of the game and no numerous levels, it has suddenly become very popular, downloads exceeded 50 million times (Fig. 8).

Our discussion shows that the combination of this game and gesture recognition will provide a new way for the recovery exercise of patients with hand injury surgery. After wearing a Microsoft wristband, the patient can use the gestures to

Fig. 8. Flappy bird.

control the bird's flight: When the arm is lifted, the bird will go up and when the arm falls, the bird will fall rapidly. Throughout the game, patients waving their arms rhythmically do not let the bird fall, which will effectively help patients exercise rehabilitation exercises. There are some advantages to help patients with hand injury surgery recovery exercise by playing this game:

(1) Flappy bird swept the world, the patient in the hands rehabilitation exercise during the game will not feel boring, which can speed up hand rehabilitation.

(2) different game can be set in different stages of the rehabilitation, which can effectively improve the speed of late recovery.

Similar to Flappy bird, there are some simple mobile games, such as Fruit Slice, Angry Birds, these games can be made into a gesture operation of the game for hand injury surgery, the patient can choose any games they want to play for recovery exercise, which not only can avoid boring caused by playing a single game too long, but also can exercise different hand joints through different games.

4.2   Smart Home-Smart Socket

Smart socket is currently the most popular smart home products, it is different from the traditional socket, in addition to the power interface in addition to the internal USB interface and a WIFI connection device, you can control a variety of smart home appliances. However, the current smart socket has the following disadvantages:

(1) control operation is too dependent on APP, operation and experience is not good, you need a few steps to control the appliance switch.

(2) smart socket control Home appliances need WIFI, if there is no good WIFI signal, the outlet can not be well controlled (Fig. 9).

The smart socket discussed in this article is based on a Bluetooth connection. After the user wears the Microsoft wristband and performs Bluetooth interfacing

Fig. 9. Smart Socket.

with the smart socket, the user can control the socket switch by gesturing: the user swings the wrist socket on and swings the wrist again The socket is closed. The advantages of using the gestures to control the smart socket are as follows:

(1) There is no limitation of APP, just connect the Bluetooth to control the socket, convenient and quick.

(2) For people with disabilities on the leg, just gently swipe wrist to control the outlet, which greatly facilitates the lives of such people.

Smart sockets are just a small part of the smart home where we can extend the gestures recognition to more smart homes, such as smart switches, that make it easy to manage lighting, curtains, or other household items using gestures, This will undoubtedly bring a whole new world to people with a great degree of convenience, especially for those with inconvenient mobility.

5   Conclusion

In this paper, we propose a gesture recognition algorithm based on accelerometer and gyroscope. Accelerometer and gyroscopeoli are installed on many smart wearable devices. The core algorithm of the gesture recognition system discussed in this paper is a three-layer neural network model, The input value is the accelerometer value, the gyroscope value and the given label when the data is preprocessed. The classifier obtained after learning can directly convert the gesture information into the output label 0 or 1 in the application to make different the response to. This simple and efficient method can be implemented on a variety of devices.

We use this gesture recognition system in two environments, one in a simple mini-game for rehabilitation exercise in patients with hand injury surgery and one in a smart socket for the convenience of people's lives. Both applications have different hardware features and system resources, with high recognition accuracy and recognition speed.Both applications using gestural recognition systems help the appropriate population. We should also believe that the significance of gesture recognition is not only the case, the rehabilitation of patients with hand injury surgery is a good example, we are constantly exploring the use of human-computer interaction can also bring any unexpected surprises, which is endless. We just hope our research can help people to further understand the role that artificial intelligence plays in this era. That is our ultimate goal.

Acknowledgement. This work was supported by grants from the Fundamental Research Funds for the Key Research Programm of Chongqing Science & Technology Commission (grant no. cstc2017rgzn-zdyf0064), the Chongqing Provincial Human Resource and Social Security Department (grant no. cx2017092), the Central Universities in China (grant nos. 2018CDXYRJ0030, CQU0225001104447).

References

1. Paul, G.V., Beach, G.J., Cohen, C.J., et al.: Tracking and gesture recognition system particularly suited to vehicular control applications: US, US7050606[P] (2006)
2. Grzeszczuk, R., Bradski, G.R., Chu, M.H., et al.: System and method for gesture recognition in three dimensions using stereo imaging and color vision: US, US6788809[P] (2004)
3. Potter, L.E., Araullo, J., Carter, L.: The Leap Motion controller: a view on sign language[J] **17**(4), 175–178 (2013)
4. Khnel, C., Westermann, T., Hemmert, F., et al.: I'm home: defining and evaluating a gesture set for smart-home control[J]. Int. J. Hum.-Comput. Stud. **69**(11), 693–704 (2011)
5. Foxlin, E.: Motion tracking requirements and technologies. Handbook of Virtual Environment Technology, p. 163C210 (2002)
6. Liu, J., Zhong, L., Wickramasuriya, J., et al.: uWave: accelerometer-based personalized gesture recognition and its applications[C]. In: IEEE International Conference on Pervasive Computing and Communications, pp. 1–9. IEEE (2009)
7. Akl, A., Feng, C., Valaee, S.: A novel accelerometer-based gesture recognition system. Ieee Trans. Signal Process. **59**(12), 6197–6205 (2011)
8. Wu, X.: Trajectory-based view-invariant hand gesture recognition by fusing shape and orientation. Iet Comput. Vis **9**(6), 797–805 (2015)
9. Zhou, S.: 2D human gesture tracking and recognition by the fusion of MEMS inertial and vision sensors. Ieee Sens. J. **14**(4), 1160–1170 (2014)
10. Kingma, D.P., Ba, J.: Adam: A Method for Stochastic Optimization. Computer Science (2014)
11. Srivastava, N.: Dropout: a simple way to prevent neural networks from overfitting. J. Mach. Learn. Res. **15**, 1929–1958 (2014)

Smartphone-Based Human Activity Recognition Using CNN in Frequency Domain

Xiangyu Jiang[1], Yonggang Lu[1(✉)], Zhenyu Lu[1,2], and Huiyu Zhou[3]

[1] School of Information Science and Engineering, Lanzhou University,
730000 Lanzhou, Gansu, China
ylu@lzu.edu.cn
[2] College of Computer Science and Engineering, Northwest Normal University,
730070 Lanzhou, Gansu, China
[3] Institute of Electronics, Communications and Information Technology,
Queens University of Belfast, BT3 9DT Belfast, UK

Abstract. Human activity recognition (HAR) based on smartphone sensors provides an efficient way for studying the connection between human physical activities and health issues. In this paper, three feature sets are involved, including tri-axial angular velocity data collected from gyroscope sensor, tri-axial total acceleration data collected from accelerometer sensor, and the estimated tri-axial body acceleration data. The FFT components of the three feature sets are used to divide activities into six types like walking, walking upstairs, walking downstairs, sitting, standing and lying. Two kinds of CNN architectures are designed for HAR. The one is Architecture A in which only one set of features is combined at the first convolution layer; and the other one is Architecture B in which two sets of the features are combined at the first convolution layer. The validation data set is used to automatically determine the iteration number during the training process. It is shown that the performance of Architecture B is better compared to Architecture A. And the Architecture B is further improved by varying the number of the features maps at each convolution layer and the one producing the best result is selected. Compared with five other HAR methods using CNN, the proposed method could achieve a better recognition accuracy of 97.5% for a UCI HAR dataset.

Keywords: Human activity recognition · Convolutional neural network
Smartphones · Accelerometer · Gyroscope

1 Introduction

With the rapid advancement of technology, smart phones become an integral part of human's daily life. Smartphones are usually embedded with various sensors gathering data for smart security, user authentication, intelligent health monitoring, human activity recognition (HAR) and so on. HAR has been widely used in intelligent wearable device, human computer interaction, athletic training and competition, military, healthcare domains such as health assisted diagnosis and treatment, cognitive disorder recognition systems, elder care support, rehabilitation assistance and so on [1].

© Springer Nature Switzerland AG 2018
L. H. U and H. Xie (Eds.): APWeb-WAIM 2018, LNCS 11268, pp. 101–110, 2018.
https://doi.org/10.1007/978-3-030-01298-4_10

HAR system is a typical pattern recognition system, which could be divided into several parts, including sensing, segmentation, feature extraction and classification. [2] In the process of the HAR based on smartphones, data collection from built-in sensors is the first step. So, it is very important to choose the proper kinds of sensors in research. Akram Bayat et al. [3] only use acceleration data generated by users' smartphones to recognize six kinds of activities. In the works of Wanmin Wu et al. and Yongjin Kwon et al. [4, 5], accelerometer and gyroscope embedded in a smartphone are both used. They found that combining these two complementary sensors can improve the recognition accuracy. Shinya Matsui et al. [6] use three sensors including accelerometers, magnetometers, and gyroscopes for HAR. Their experiments have proved that different kinds of data from different sensors do improve the recognition accuracy of activities by offering extra information.

The feature extraction in traditional method takes a lot of effort because it has to be done manually. Features extracted in traditional methods can be divided into two types: time-domain features and frequency-domain features. Time-domain features include mean value, standard deviation, kurtosis, skewness, Inter-quartile-Range (IR), correlation between axes, zero crossing rate, etc. Frequency-domain features includes frequency-domain entropy, Fast Fourier Transform (FFT) coefficients and Discrete Cosine Transform (DCT) coefficients, etc. Some others approaches such as Principal Component Analysis (PCA), Autoregressive Model and Haar filters are also used in HAR researches [6, 7].

HAR is a classic multi-classification problem that uses one-dimensional sensor signals and extracts discriminative features to recognize human activities by a classification method [8]. The quality of the classifier has significant influence on the HAR system. Traditional classification methods used in HAR include Support Vector Machines (SVM), k-Nearest Neighbor (k-NN), Decision Tree (DT), Naive Bayesian (NB), Hidden Markov Model (HMM), etc. So, for the traditional methods, it is important to combine an effective feature extraction method with a good classification method in HAR.

Deep learning could transform raw data into more abstract expressions of higher level through some simple but nonlinear models. Complex functions could be also learned by means of enough combinations of transformations [9]. Deep learning carries automatic features extraction instead of manual heuristic operation. That is to say, the core aspect of deep learning is that the features of each layer are not designed for artificial engineering, but rather a universal learning process from the data [9]. For the past several years, deep learning has demonstrated a strong and excellent learning ability in many fields such as computer vision, speech recognition, and machine translation. In the field of HAR, deep learning can enhance HAR efficiency by automatic feature extraction. Convolutional neural network (CNN) is a kind of deep feedforward artificial neural network; it has remarkable performance for large image processing [10–12]. Recently, CNN is also applied to the HAR, and the results are dramatic and encouraging [1, 2, 13–17].

2 Related Works

The previous works related to HAR using deep learning techniques have shown promising results. Yuqing Chen et al. use a CNN contains 3 convolution layers and 3 pooling layers for HAR [2]. They set the width of convolution kernel to 2 and set up a validation set to locate the best epochs. The average classification accuracy of their method is about 93.8%. A wider time span of temporal local correlation (1×9–1×14) with a low pooling size (1×2–1×3) is exploited and is shown to be beneficial to HAR by Charissa Ann Ronao et al. [16]. This work uses the time-series sensor data and additional information of FFT of the HAR data separately as the input to CNN. They show that the accuracy rate reaches 94.79% for the time-series data, and reaches 95.75% for the FFT of the HAR data. It can be seen that using the data in frequency domain produces better results than using the data in time domain. Ming Zeng et al. [13] design a simple network architecture, including an input layer, a convolution layer, a max-pooling layer, a 1024 neuron fully-connected layer and a soft-max layer to for HAR. Daniele Rav et al. [17] propose a deep learning architecture: a filter is applied to the input, and the weighted sums are computed in the temporal convolution layer which is followed by a fully-connected layer and a soft-max layer for classification. In their works, they decrease the computation cost by limiting the connections from the input nodes in order to extract features efficiently through fewer nodes and levels. The works of Tahmina Zebin et al. show that the performance is noticeably enhanced when the third convolution layer is added [1].

3 Design of the CNN Architectures

From the previous works, it can be seen that the CNN architecture is vital for the final recognition results. So the influence of the network architectures on the recognition accuracy is studied. In the experiments, two different CNN architectures are designed, which are called Architecture A and Architecture B, and they are shown in Figs. 1 and 2 respectively. The differences between the two architectures are the size of input maps and the size of filter maps in the first convolution layer. So the only difference of the two architectures is in the first convolution layer. As shown in Figs. 1 and 2, 1-channel and 2D convolution are performed and three convolution layers are applied in our work because the three convolution layers are shown to be a proper setup for HAR [2, 16]. It is also found in our experiments that using three convolution layers in CNN can produce better results than using one or two convolution layers, while using four convolution layers produces similar results.

The details of the Architecture B are shown in Fig. 1, where the input maps contain tri-axial angular velocity data from the gyroscope sensor, tri-axial total acceleration data from the accelerometer sensor and the tri-axial estimated body acceleration data. The filter size is 13×3 and the step size is 1×3.

The details of the Architecture A are shown in Fig. 2. The tri-axial angular velocity data from the gyroscope sensor is duplicated at the end of the input in architecture B for including all pairs of feature sets in the convolution. The filter size is 13×6 and the step size is also 1×3. The advantage of doing this is that the information among three different sets of features is combined at the first convolution layer: the tri-axial angular velocity information from the gyroscope sensor and the tri-axial total acceleration information from the accelerometer sensor are combined, the tri-axial total acceleration information from the accelerometer sensor and the tri-axial estimated body acceleration information are combined, the tri-axial angular velocity information from the gyroscope sensor and the tri-axial estimated body acceleration information are combined respectively. In this case, each filter map covers two different sensor information in architecture B.

After the input layer, 3 CNN layers are included in the two architectures. Each CNN layer includes convolution, batch normalization, clipped-Relu [18] and max-pooling. After that, dropout, fully connected and soft-max are applied in the two architectures. Then a dropout layer is added to randomly set input elements to zero with a given probability, which is set to 50% in our experiments. This operation corresponds to temporarily dropping a randomly chosen unit and all its connections from the network during the training. The dropout could prevents overfitting [19, 20].

The validation data set is also used in the experiments to control the number of epochs during the training process. The addition of validation set could prevent the over-fitting of the training.

For different optimizers, many different hyper-parameters are also analyzed in these experiments. In the experiments, the order of the training data is randomized before each training epoch. One common solution is to use a back propagation algorithm to train the network with SGDM [21] and Adam [22].

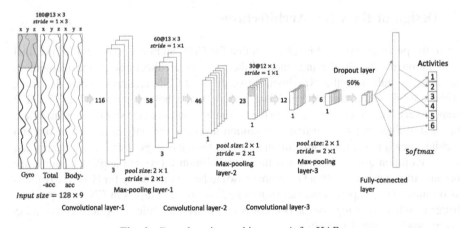

Fig. 1. Deep learning architecture A for HAR.

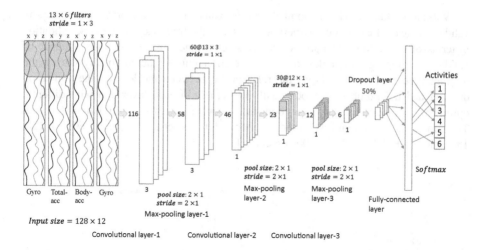

Fig. 2. Deep learning architecture B for HAR.

4 Experiments

4.1 Data Set

The UCI HAR dataset [23] are downloaded. The UCI dataset includes tri-axial total acceleration data from the accelerometer sensor, the estimated body acceleration and tri-axial angular velocity data from the gyroscope sensor for six kinds of activities including walking, walking upstairs, walking downstairs, sitting, standing and lying. The UCI dataset is collected from a group of 30 volunteers in the 19 to 48 age range. The sensor signal is preprocessed by the noise filter, and then sampled in the fixed width sliding window of 2.56 s with 50% overlap. The acceleration sensor signal has gravity and body parts, which are separated by a Butterworth low pass filter to body acceleration and gravity acceleration in the UCI dataset. There are total 7352 examples for the training data and 2947 examples for the test data in the raw date set. For the convenience of calculation, the training set samples is extended to 8000 with random repetition, then 1000 examples are randomly picked from the training set to serve as validation data set, and the test set samples is also extended to 3000 with random repetition. As shown in the previous studies, data in the frequency domain works better for HAR than the data in time domain [24–26]. So only the frequency-domain data is used in our experiments. More specifically, the FFT of the input data are used as the input to the network.

4.2 Experimental Results

It is obvious that the choice of hyper-parameters is significant to the performance of the network when CNN is used to solve practical problems. Therefore, this experiment shows different accuracies with different values of Hyper-parameter settings.

Validation data set also is applied and the values of two important parameters in validation are evaluated in our experiments. One is the frequency of network validation in terms of the number of epochs, which is set to one. The other is the patience of validation stopping which determine the iteration number automatically during the training process. Figure 3 shows the relationship between the loss on the validation data set and the number of epochs with 30 kernels in each layer and with different optimizers. It is found that 15 is a good setting for the patience of validation stopping. The similar result is found when the number of filter is set to 10, 20, 30, 60, 90, 120, 150, 180, 210, 240, 270 and 300 respectively.

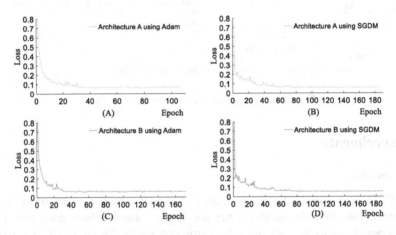

Fig. 3. (A) and (B) are the results for architecture A. (C) and (D) are the results for architecture B. Adam optimizer is used in (A) and (C). SGDM optimizer is used in (B) and (D).

For comparing the two different CNN architectures, the experiments have been done ten times. For the fairness and objectivity, an accuracy rate is obtained by averaging the results of 10 random experiment results. When the Adam optimizer is used in this CNN architecture, the average accuracy for architecture A is 96.30% and the average accuracy for architecture B is 96.72%. When the SGDM optimizer is used in this CNN architecture, the average accuracy for architecture A is 95.91% and the average accuracy for architecture B is 96.49%.

The results of the 10 experiments are also plotted in Fig. 4. It can be seen that for most cases the final recognition accuracy of architecture B is higher than that of architecture A with both optimizers.

The number of filter maps is very important for the CNN network architectures. For seeking the most appropriate number of filter maps per layer for Architecture B, ten experiments are done with different number of filter maps in different layers. The curves about the average accuracy of the Architecture B under different filter map numbers in the different layers are plotted, where the number of filter maps is set to 10, 20, 30, 60,

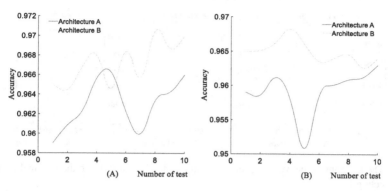

Fig. 4. The result of 10 random tests in the frequency-domain. The Adam optimizer is used in (A) and the SGDM optimizer is used in (B).

90, 120, 150, 180, 210, 240, 270 and 300 respectively. First of all, we only change the number of the filter maps in the first layer when all other parameters are fixed. As shown in Fig. 5(A), 180 is the best number for the first layer. Second, we only change the number of the filter maps in the second layer when the number of filter maps in the first layer is set to 180 and other parameters remain the same. As shown in Fig. 5(B), 60 is the best number in the second layer. Finally, we set the number of filter maps in the first and second layer to 180 and 60 respectively, and only change the number of filter maps in the third layer. As shown in Fig. 5(C), 30 is the optimal number of filter maps in third layer. So, a good setting for the numbers of filter maps in the three layers is (180, 60, 30).

Table 1 shows the confusion matrix produced using the Adam optimizer with the good setting found above for Architecture B. It is found that the recognition accuracy of all six activities is over 93%. And the accuracy of laying is up to 100%, the accuracy of walking and walking upstairs is over 99%. The accuracy of walking downstairs is up to 98.1%. The accuracy of sitting and standing is lower than other activities, but still reaches 95.5% and 93.3% respectively. A probable cause is that sitting and standing both are static movements, so the data collected for the two activities are similar, thus the local features extracted by CNN are similar as well.

The two proposed CNN architectures including Architecture A and Architecture B are also compared with CNN architectures for HAR proposed in other papers for the same UCI HAR dataset. Both Architecture A and Architecture B uses the good setting for the numbers of filter maps which is (180, 60, 30). The recognition accuracies produced with different CNN architectures are compared in Table 2. From Table 2, it can be seen that using the proposed Architecture B produces the highest recognition accuracy, while using the proposed Architecture A produces the second highest recognition accuracy among all the 7 CNN methods.

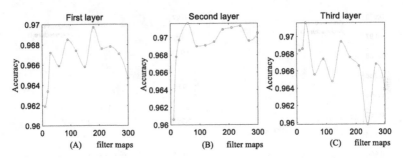

Fig. 5. The relationship between the number of filter maps in every layer and the final accuracy of architecture B.

5 Conclusion

In this paper, Architecture A and Architecture B are designed for HAR. From the experimental results, it is easily found that the accuracy of the Architecture B is higher than that of Architecture A under the same conditions. It is found that the CNN network architecture which uses the combinations of different kinds of signal sources at the first convolution layer, as in Architecture B, can produce better HAR results. The reason may be that the kind of architecture can extract more discriminative features for classification. Our future work is to apply the Architecture B on more datasets and to improve the recognition rate by adding more features to input data.

Table 1. Confusion matrix of the architecture B

Target Class	Output Class						Precision
	Walking	Upstairs	Downstairs	Sitting	Standing	Laying	
Walking	500	0	4	0	0	0	99.2%
Upstairs	1	474	4	0	0	0	99%
Downstairs	5	3	423	0	0	0	98.1%
Sitting	0	0	0	462	22	0	95.5%
Standing	0	0	2	35	517	0	93.3%
Laying	0	0	0	0	0	548	100%
Recall	98.8%	99.4%	97.7%	93%	95.9%	100%	97.5%

Table 2. Comparison with other methods

Method	Accuracy on test data set
CNN [2]	93.8%
CNN [25]	95.31%
Deep CNN [1]	97.01%
tFFT + Convnet [16]	95.75%
CNN + Stat.features [25]	96.06%
Proposed Architecture A	97.08%
Proposed Architecture B	97.50%

Acknowledgements. This work is supported by the National Key R&D Program of China (Grants No. 2017YFE0111900).

References

1. Zebin, T., Scully, P.J., Ozanyan, K.B.: Human activity recognition with inertial sensors using a deep learning approach. In: SENSORS, 2016 IEEE, pp. 1–3. IEEE (2016)
2. Chen, Y., Xue, Y.: A deep learning approach to human activity recognition based on single accelerometer. In: 2015 IEEE International Conference on Systems, Man, and Cybernetics (SMC), pp. 1488–1492. IEEE (2015)
3. Bayat, A., Pomplun, M., Tran, D.A.: A study on human activity recognition using accelerometer data from smartphones. Proc. Comput. Sci. **34**, 450–457 (2014)
4. Wu, W., Dasgupta, S., Ramirez, E.E., Peterson, C., Norman, G.J.: Classification accuracies of physical activities using smartphone motion sensors. J. Med. Internet Res. **14** (2012)
5. Kwon, Y., Kang, K., Bae, C.: Unsupervised learning for human activity recognition using smartphone sensors. Expert Syst. Appl. **41**, 6067–6074 (2014)
6. Matsui, S., Inoue, N., Akagi, Y., Nagino, G., Shinoda, K.: User adaptation of convolutional neural network for human activity recognition. In: Signal Processing Conference (EUSIPCO), 2017 25th European, pp. 753–757. IEEE (2017)
7. Hanai, Y., Nishimura, J., Kuroda, T.: Haar-like filtering for human activity recognition using 3d accelerometer. In: Digital Signal Processing Workshop and 5th IEEE Signal Processing Education Workshop, 2009. DSP/SPE 2009. IEEE 13th, pp. 675–678. IEEE (2009)
8. Plötz, T., Hammerla, N.Y., Olivier, P.: Feature learning for activity recognition in ubiquitous computing. In: IJCAI Proceedings-International Joint Conference on Artificial Intelligence, pp. 1729 (2011)
9. LeCun, Y., Bengio, Y., Hinton, G.: Deep learning. Nature **521**, 436 (2015)
10. Simonyan, K., Zisserman, A.: Very deep convolutional networks for large-scale image recognition. arXiv preprint arXiv:1409.1556 (2014)
11. Sharif Razavian, A., Azizpour, H., Sullivan, J., Carlsson, S.: CNN features off-the-shelf: an astounding baseline for recognition. In: Proceedings of the IEEE Conference on Computer Vision and Pattern Recognition Workshops, pp. 806–813 (2014)
12. An, D.C., Meier, U., Masci, J., Gambardella, L.M., Schmidhuber, J., rgen: Flexible, high performance convolutional neural networks for image classification. In: IJCAI 2011, Proceedings of the International Joint Conference on Artificial Intelligence, Barcelona, Catalonia, Spain, pp. 1237–1242. (2011)
13. Zeng, M., et al.: Convolutional neural networks for human activity recognition using mobile sensors. In: 2014 6th International Conference on Mobile Computing, Applications and Services (MobiCASE), pp. 197–205. IEEE (2014)
14. Wang, J., Chen, Y., Hao, S., Peng, X., Hu, L.: Deep learning for sensor-based activity recognition: a survey. arXiv preprint arXiv:1707.03502 (2017)
15. Hammerla, N.Y., Halloran, S., Ploetz, T.: Deep, convolutional, and recurrent models for human activity recognition using wearables. arXiv preprint arXiv:1604.08880 (2016)
16. Ronao, C.A., Cho, S.-B.: Human activity recognition with smartphone sensors using deep learning neural networks. Expert Syst. Appl. **59**, 235–244 (2016)
17. Ravi, D., Wong, C., Lo, B., Yang, G.-Z.: Deep learning for human activity recognition: A resource efficient implementation on low-power devices. In:2016 IEEE 13th International Conference on Wearable and Implantable Body Sensor Networks (BSN), pp. 71–76. IEEE (2016)

18. Hannun, A., et al.: Deep Speech: scaling up end-to-end speech recognition. Computer Science (2014)
19. Srivastava, N., Hinton, G., Krizhevsky, A., Sutskever, I., Salakhutdinov, R.: Dropout: A simple way to prevent neural networks from overfitting. J. Mach. Learn. Res. **15**, 1929–1958 (2014)
20. Krizhevsky, A., Sutskever, I., Hinton, G.E.: Imagenet classification with deep convolutional neural networks. In: Advances in Neural Information Processing Systems, pp. 1097–1105 (2012)
21. Robert, C.: Machine learning, a probabilistic perspective. Taylor & Francis (2014)
22. Kingma, D.P., Ba, J.: Adam: a method for stochastic optimization. arXiv preprint arXiv: 1412.6980 (2014)
23. http://archive.ics.uci.edu/ml/datasets/Human+Activity+Recognition+Using+Smartphones
24. Huynh, T., Schiele, B.: Analyzing features for activity recognition. In: Proceedings of the 2005 Joint Conference on Smart Objects and Ambient Intelligence: Innovative Context-Aware Services: Usages and Technologies, pp. 159–163. ACM (2005)
25. Ignatov, A.: Real-time human activity recognition from accelerometer data using convolutional neural networks. Appl. Soft Comput. **62**, 915–922 (2018)
26. Wang, X., Lu, Y., Wang, D., Liu, L., Zhou, H.: Using jaccard distance measure for unsupervised activity recognition with smartphone accelerometers. In: Asia-Pacific Web (APWeb) and Web-Age Information Management (WAIM) Joint Conference on Web and Big Data, pp. 74–83. Springer, Berlin (2017)

Text Classification Methods Based on SVD and FCM

Ning Yang[1], Shuaibing Li[1], Rong Sun[3], and Yi Yang[1,2](✉)

[1] School of Information Science and Engineering, Lanzhou University,
Gansu 730000, China
yy@lzu.edu.cn

[2] Silk Road Economic Belt Research Center of Lanzhou University,
Gansu 730000, China

[3] School of Mathematics, Jilin University, Jilin 130000, China

Abstract. In order to find key and useful messages among massive online resources, this paper propose a method to classify documents about soybean metabolism based on Singular Value Decomposition (SVD) and Fuzzy c-Means(FCM). Singular Value Decomposition (SVD) is an important way of matrix decomposition, which can represent a complex matrix by dividing it into smaller and simpler submatrices that describe important properties of matrices. After the dimension reduction, the Fuzzy c-Means (FCM) is used for clustering, which makes the objects divided into the same cluster have the highest similarity, while the object between different clusters have the lowest similarity. Besides, term frequency (TF) and entropy weight method (EWM) can also be used to construct matrix.

Keywords: Soybean metabolism · Text classification · SVD · FCM
TF · EWM

1 Introduction

A rapid growth of the online users inspires plenty of internet resources and data about soybean metabolism, which is a topic worth paying attention to. To quickly identify the article about soybean metabolism, it becomes essential to find out the useful and key information among a massive of internet resources and data.

Nowadays, a number of approaches have been developed to find out the articles about soybean metabolism, and one of the most classic methods is Vector Space Model(VSM) [2], which is founded in terms of rigorous machine learning theory to understand and analysis, besides it also represents the contents of documents with a set of index terms. For this reason, the Vector Space Model(VSM) has been widely used in text categorization. However, this approach requires a high-dimensional space to represent a document without taking into account semantic relationship between the terms, which could lead to poor classification

© Springer Nature Switzerland AG 2018
L. H. U and H. Xie (Eds.): APWeb-WAIM 2018, LNCS 11268, pp. 111–120, 2018.
https://doi.org/10.1007/978-3-030-01298-4_11

performance [5]. Besides, a small number of features can be considered as irrelevant [1]. In articles about soybean metabolism, there will be a large number of words, so the Vector Space Model(VSM) maybe is not a good way to classify.

To represent the relationship between a lot of words and a mass of texts about soybean metabolism, the Singular Value Decomposition(SVD) is used, which relies on the fact that Singular Value Decomposition(SVD) reduces noise to improve the accuracy of text classification [9]. Meanwhile, it could not only reduce the dimensional space that Vector Space Model(VSM) cannot resolve, but also improve the classification accuracy.

In this paper, we implement a large number of words and a mass of texts by using Singular Value Decomposition(SVD). The unsupervised methods TF and Fuzzy c-Means(FCM) are also used. The unsupervised method TF approach can capture the relevancy among words and text documents [13], and the Fuzzy c-Means(FCM) can group data into some small categories. After that, we use confusion matrix to judge the accuracy of classification.

The rest of this paper is organized as following: the related work about text mining are described in Sect. 2. Section 3 explains the background about the detail of some algorithm we used. The detail of some algorithms, such as Singular Value Decomposition and Fuzzy c-Means, is described in Sect. 3.1. Section 4 presents the experiment results and analysis. Finally, we conclude this paper in Sect. 5.

2   Related work

At present, with the increasing demand for improving soybean production, soybean metabolism has been paid more and more attention to. Therefore, it is increasingly important to quickly find articles about soybean metabolism. Today, there are many mature technologies for text mining. However, the labeled data provides less information than unlabeled data most of the time. Traditional K-means cluster technique can divide a class into several sub-classes to bring more similar documents into the same group, which can largely strengthen the relationship between the words and sub-classes [8]. However, K-means cannot get a posteriori probability that a sample belongs to this cluster. To solve this problem, Expectation-Maximization(EM) is used. It is a class of iterative algorithms for maximum likelihood or maximum a posteriori estimation in problems with incomplete data [3], which can be combined with a naive Bayes classier to learn from labeled and unlabeled documents [7]. But Expectation-Maximization(EM) can only get local extremum rather than global extremum.

In cluster algorithms, due to its simplicity and ability to detect clusters of data, Density-Based Spatial Clustering of Application with Noise(DBSCAN) has been widely used [10]. The drawback is that when detecting boundary objects of neighboring clusters, it will become unstable. BIRCH has poor anti-interference ability to abnormal data. A new clustering algorithm called CURE identifies clusters having non-spherical shapes and wide variances in size, which execution times lower than BIRCHs [4].

In addition, clustering and region oriented queries are common problems in spatial data mining [12]. A new clustering method called CLAHANS is developed and designed for large data sets, which is based on randomized search and is more efficient than PAM [6]. Another active spatial data mining algorithm that can effectively support user-defined triggers for dynamically evolving spatial data appears, called STING+, which utilizes the advantage of using active database systems and effective tracking in STING [11].

In this paper, we cluster articles on soybean metabolism by combining Singular Value Decomposition and Fuzzy c-Means, which can distinguish articles about soybean, metabolism, soybean metabolism and others.

3   Backgroud

3.1   Singular Value Decomposition

The Singular Value Decomposition(SVD) of term-by-document A(m*n) can be calculate in formula

$$A = USV^T \tag{1}$$

In formula, the U and V are matrixes respectively named as left singular matrix and right singular matrix.

$$S_{m \times n} = diag\left(\delta_1, \delta_2, \cdots, \delta_m\right) \tag{2}$$

It is a diagonal matrix, which is arranged from big to small.

3.2   Dimension Reduction

In all these documents, there will be a mass of words, meaning that n must be a very large number. For this reason, we need to reduce dimension. Take the first k, which is far less than both m and n, as a nonzero singular value, then we can get

$$A_k = U_k S_k V_k^T \tag{3}$$

In the formula, U_k is formed by the first k columns of U, V_k^T is constituted by the first k rows of V_T, comprised by the first k factors, which is largely represents original matrix. The product of the three matrices on the right results in a matrix that is close to A. The closer k is to n, the closer the product is to A. Then we get inner product, named $N_{m \times k}$ by taking the inner product of A and V_k^T.

3.3   Fuzzy c-Means

Fuzzy c-Means(FCM) algorithm is a clustering method based on partition, whose purpose is to maximize the similarity among objects divided into the same cluster and minimize the similarity among different clusters. It is a kind of flexible fuzzy partition and diffusely used in classification.

The core concept of FCM is to divide a data set X to c classes. Then there will be C centers in c cluster. The membership matrix u_{ij} means the value each sample belongs to each class. The object function of FCM is

$$J == \sum_{i=1}^{c} \sum_{j=1}^{n} u_{ij}^{m} \|x_j - c_i\|^2 \tag{4}$$

The constraint condition is

$$\sum_{i=1}^{c} u_{ij} = 1, j = 1, 2, \cdots, n \tag{5}$$

Then we will get

$$minJ(U,V) = min\left\{ \sum_{k=1}^{n} \sum_{i=1}^{c} (u_{ik}^{m})(d_{ik})^2 \right\} \tag{6}$$

And

$$minJ(U,V) = \sum_{k=1}^{n} \left\{ min\left[\sum_{i=1}^{c} (u_{ik})^{m}(d_{ik})^2 \right] \right\} \tag{7}$$

In order to find the extreme value of the objective function under constraint conditions, we construct a new function by using Lagrange multiplier method

$$F = \sum_{i=1}^{c} (u_{ik})^{m}(d_{ik})^2 + \lambda(\sum_{i=1}^{c} u_{ik} - 1) \tag{8}$$

$$d_{ik}^2 = \|X_k - V_i\|^2 \tag{9}$$

The optimum conditions for finding the extreme value of F function are as follows

$$\frac{\partial F}{\partial \lambda} = (\sum_{i=1}^{c} u_{ik} - 1) = 0 \tag{10}$$

$$\frac{\partial F}{\partial u_{ik}} = \left[m(u_{ik})^{m-1}(d_{ik})^2 - \lambda \right] = 0 \tag{11}$$

$$\frac{\partial F}{\partial v_i} = \sum_{k=1}^{n} (u_{ik})^{m} x_k - v_j \sum_{i=1}^{n} (u_{ik})^{m} = 0 \tag{12}$$

The necessary conditions to solving extreme conditions 10 are as follows

$$u_{ik} = \frac{1}{\sum_{j=1}^{c} (\frac{d_{ik}}{d_{jk}})^{\frac{2}{m-1}}} \tag{13}$$

$$v_j = \frac{\sum_{k=1}^{n} (u_{ik})^{m} x_k}{\sum_{k=1}^{n} (u_{ik})^{m}} \tag{14}$$

The first step to use FCM is setting number of types c and parameter m. Secondly, define initial membership matrix U(0). Thirdly, calculate the new cluster centers V_j through 12 and new membership matrix using 11. Finally, a matrix norm is used to compare the membership matrix between two iterations, and stop iterating when $\|U(k+1) - U(k)\| \le e$. Then we will get a membership matrix about classes and samples.

4 Experiments

4.1 Experiment Design

The flow chart is shown in Fig. 1

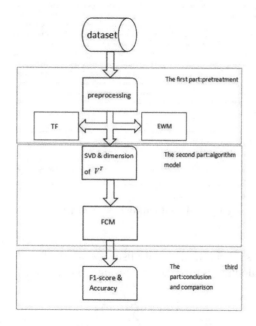

Fig. 1. Flow chart of the experiment.

In this paper, we collect 600 articles whose subjects are mainly about soybeans metabolism and extract the titles, key words and abstracts as data sets, which is divided into four categories as follow Table 1.

And a corpus D of soybeans metabolism is defined by $D = \{D_1, D_2, \cdots, D_m\}$, where each article D_i contains with amount of words $W_i = \{w_1, w_2, \cdots, w_n\}$. Obviously, in an article there will be lots of words, usually named as stop-words, which appear frequently but with no real meaning, such as a, an, one and so on. Besides, there will be some punctuation mark and special character. For this reason, we delete stop-words and useless words, then construct a document-by-word text. Take the soybean metabolism for example, which is shown in Fig. 2

Table 1. Data set and the number of each class.

Main content	Number of articles
Metabolism	150
Soybean	150
Soybean metabolism	150
Other	150

Hidden Nickel Deficiency? Nickel Fertilization via Soil Improves Nitrogen Metabolism and Grain Yield in Soybean Genotypes

Nickel (Ni)-a component of urease and hydrogenase-was the latest nutrient to be recognized as an essential element for plants. However, to date there are no records of Ni deficiency for annual species cultivated under field conditions, possibly because of the non-appearance of obvious and distinctive symptoms, i.e., a hidden (or latent) deficiency. Soybean, a crop cultivated on soils poor in extractable Ni, has a high dependence on biological nitrogen fixation (BNF), in which Ni plays a key role. Thus, we hypothesized that Ni fertilization in soybean genotypes results in a better nitrogen physiological function and in higher grain production due to the hidden deficiency of this micronutrient. To verify this hypothesis, two simultaneous experiments were carried out, under greenhouse and field conditions, with Ni supply of 0.0 or 0.5 mg of Ni kg(-1) of soil. For this, we used 15 soybean genotypes and two soybean isogenic lines (urease positive, Eu3; urease activity-null, eu3-a, formerly eu3-e1). Plants were evaluated for yield, Ni and N concentration, photosynthesis, and N metabolism. Nickel fertilization resulted in greater grain yield in some genotypes, indicating the hidden deficiency of Ni in both conditions. Yield gains of up to 2.9 g per plant in greenhouse and up to 1,502 kg ha(-1) in field conditions were associated with a promoted N metabolism, namely, leaf N concentration, ammonia, ureides, urea, and urease activity, which separated the genotypes into groups of Ni responsiveness. Nickel supply also positively affected photosynthesis in the genotypes, never causing detrimental effects, except for the eu3-a mutant, which due to the absence of ureolytic activity accumulated excess urea in leaves and had reduced yield. In summary, the effect of Ni on the plants was positive and the extent of this effect was controlled by genotype-environment interaction. The application of 0.5 mg kg(-1) of Ni resulted in safe levels of this element in grains for human health consumption. Including Ni applications in fertilization programs may provide significant yield benefits in soybean production on low Ni soil. This might also be the case for other annual crops, especially legumes.

Fig. 2. The document-by-word text.

In Fig. 2, the word soybean and metabolism are marked in different colors. For one word W_i, we calculate its frequency through the method TF which used the formula

$$tf_{i,j} = \frac{n_{i,j}}{\sum_k n_{k,j}} \tag{15}$$

In formula the numerator $N_{i,j}$ means the number of occurrences of the word W_i in text D_j, while the denominator is the sum of occurrences of all words in text document D_j. Then we can construct a document-by-word matrix named A with m rows and n columns,every row of which means a document, and its columns are the frequency of words appears in document. By this way, we could get a matrix named data with 600 rows and 8891 columns.

To emphasize the contribution or importance of four characteristics, we put a weight on each classes though the entropy weight method(EWM), which means the smaller the entropy value, the more discrete the index is, the greater influence of the index on the comprehensive evaluation.

The three matrices are obtained after using singular value decomposition to A. The matrix U based on row vector with m rows and m columns. The diagonal matrix S is a singular value matrix of A with m rows and n columns. The matrix V^T based on column vector with n rows and n columns.

In this paper, for further dimension reduction, We extract k rows from matrix V and get a matrix N, which based on the dimension of article vector, with k rows and n columns via the formula

$$N_{m \times k} = A \cdot V_k^T \tag{16}$$

Which k is the most proper to represent the matrix N? We can get solution by counting how many rows in matrix S can greatly represent matrix S, the formula is

$$1 - \frac{\sum_1^k S_i}{\sum_1^m S_i} \leq 0.01 \tag{17}$$

It means k rows of matrix can represent matrix S, so that we can reduce dimension to K rows.

4.2   Experiment Result

Figure 3 shows 200-dimensional data in two data. In our experiment, the FCM is used to cluster the matrix N. And we compress the dimension of matrix N to 200 rows. The data is divided to four classes, four different colors describe four categories. It can be seen that our approach can divide data into four categories.

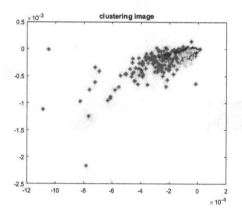

Fig. 3. Four types data are depicted in different colors.

Figure 4 describe the value of the target function varies with the increase of the number of iteration. When the number of iteration increase bigger than 2, the curve gradually flattens out. So, the parameter of iteration should be 2, which will get a better results.

In Fig. 5, it can be see four classes membership matrix value changes with the number of samples. The value of membership is bigger, the higher the degree

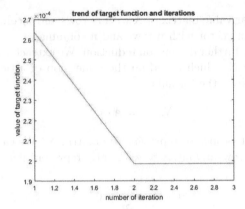

Fig. 4. The change curve of the function with the number of iterations.

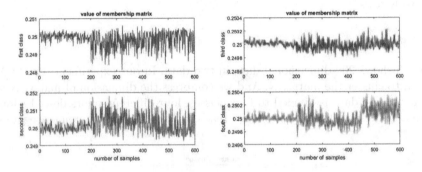

Fig. 5. The change trends in four classes membership matrix value.

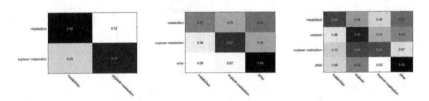

Fig. 6. The matrix N is divided into three kinds of categories.

that the sample point belongs to the category, and the closer the membership value is to 0, the lower the degree that the sample point belongs to the category. According to the maximum membership principle in fuzzy sets, we can be determine which class each sample point belongs to. For better observation, the confusion matrix can be shown in Fig. 6.

In confusion matrix, the greater the element on the diagonal, the smaller the value on the non-diagonal, the better the performance of the algorithm. We can see that our approach have good performance in two categories. While in three categories, the kind of metabolism texts cannot be distinguished well between the kind of soybean texts and other texts. In four categories, the kind of other texts and soybean texts can be divided well.

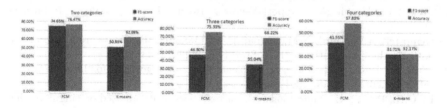

Fig. 7. F1-score and accuracy of FCM and K-Means of the three classification results.

In Fig. 7, we compared F1-score between FCM and K-means in 2 clusters, 3 clusters and 4 clusters, the accuracy of clustering and F1-score decrease with the increase of classification number. Besides, the number of accuracy and F1-score in FCM have better performance than in K-means, which proves our approach has better performance than K-means.

5   Conclusion and Future Work

In this paper, we use TF and EWM to construct a terms-by-documents matrix about soybean metabolism. Then, SVD is used to reduce the dimension of original matrix. Finally, FCM is used to cluster matrix. By using above methods, we can quickly find the relevant texts about soybean metabolism in the massive text data. In experimental results, four categories cannot be divided well, which may have two reasons. First, the selection of data sets is not good enough so that the data is too smooth and the data categories arenot distincted. Second, the algorithm is not accurate enough. In future works, we will try some other data in the model, and improve the accuracy of the model for both two reasons.

Acknowledge. This work was supported by grants from the Fundamental Research Funds for the Key Research Programm of Chongqing Science & Technology Commission (grant no. cstc2017rgzn-zdyf0064), the Chongqing Provincial Human Resource and Social Security Department (grant no. cx2017092), the Central Universities in China (grant nos. 2018CDXYRJ0030, CQU0225001104447). The authors would like to express their gratitude to all the subjects that participated in the experiments. This study is supported by Science and Technology Innovation Project of Foshan City, China (Grant No. 2015IT100095), the Fundamental Research Funds for the Central Universities (Grant No. lzujbky-2016-br03), CERNET Innovation Project (Grant No. NGII20150603) and Science and Technology Planning Project of Guangdong Province, China (Grant No. 2016B010108002).

References

1. Carrera-Trejo, V., Sidorov, G., Miranda-Jimnez, S., Ibarra, M.M., Martnez, R.C.: Latent dirichlet allocation complement in the vector space model for multi-label text classification. Cancer Biol. Ther. **7**(7), 1095–1097 (2015)
2. Cortes, C., Vapnik, V.: Support-Vector Networks. Kluwer Academic Publishers (1995)
3. Dempster, A.P., Laird, N.M., Rubin, D.B.: Maximum likelihood from incomplete data via the EM algorithm. J. R. Stat. Soc. **39**(1), 1–38 (1977)
4. Guha, S., Rastogi, R.: Cure: an efficient clustering algorithm for large database. Inf. Syst. **26**(1), 35–58 (2001)
5. Li, C.H., Park, S.C.: Neural network for text classification based on singular value decomposition. In: IEEE International Conference on Computer and Information Technology, pp. 47–52 (2007)
6. Ng, R.T., Han, J.: Efficient and Effective Clustering Methods for Spatial Data Mining. University of British Columbia (1994)
7. Nigam, K., Mccallum, A.K., Thrun, S., Mitchell, T.: Text classification from labeled and unlabeled documents using EM. Mach. Learn. **39**(2–3), 103–134 (2000)
8. Roul, R.K., Sahay, S.K.: K-means and wordnet based feature selection combined with extreme learning machines for text classification. In: International Conference on Distributed Computing and Internet Technology, pp. 103–112 (2016)
9. Symeonidis, P., Kehayov, I., Manolopoulos, Y.: Text classification by aggregation of SVD eigenvectors. In: East European Conference on Advances in Databases and Information Systems, pp. 385–398 (2012)
10. Tran, T.N., Drab, K., Daszykowski, M.: Revised DBSCAN algorithm to cluster data with dense adjacent clusters. Chemom. Intell. Lab. Syst. **120**(2), 92–96 (2013)
11. Wang, W., Yang, J., Muntz, R.: Sting+: an approach to active spatial data mining. In: ICDE, p. 116 (1999)
12. Wang, W., Yang, J., Muntz, R.R.: Sting: a statistical information grid approach to spatial data mining. In: Proceedings of the 23rd Very Large Database Conference, pp. 186–195 (1997)
13. Zhang, Y.T., Gong, L., Wang, Y.C.: An improved TF-IDF approach for text classification. J. Zhejiang Univ. Sci. A **6A**(1), 49–55 (2005)

Bigdata Analytics for Healthcare

An Image Processing Method via OpenCL for Identification of Pulmonary Nodules

Genlang Chen[1(✉)], Jiajian Zhang[2], Yuning Pan[3], and Chaoyi Pang[1]

[1] Ningbo Institute of Technology, Zhejiang University, Ningbo 315100, China
cgl@zju.edu.cn, chaoyi.pang@qq.com
[2] College of Computer Science, Zhejiang University, Hangzhou 310058, China
403522093@qq.com
[3] Department of Radiology, Ningbo First Hospital, Ningbo Hospital, Zhejiang
University, Ningbo 315100, China
panyuning@163.com

Abstract. Lung cancer is one of the most diagnosable form of cancer worldwide. Recent researches have showed that the diagnoses of pulmonary nodules in Computed Tomography (CT) chest scans based on deep learning have made a significant progress for the medical diagnoses. However, the existence of many false positives or the high costs of processing time make it impossible to apply to clinical practice. Toward this purpose, this paper proposed a new image processing method to improve the performance by exploiting the power of acceleration technologies via OpenCL. We use parallel programming and pipeline models to parallelize the CT image preprocessing, and classify them by 3D CNNs according to the significant differences between nodules and non-nodules in 3D shapes. Extensive experimental results have shown that image processing can be accelerated significantly on GPU. In addition, the experiments on 500 patients indicate that our proposed method improved the performance by 12.5% and achieved 97.78% sensitivity rate for segmentation.

Keywords: CT images · Pulmonary nodules · OpenCL

1 Introduction

Lung cancer is one of the most diagnosable form of cancer worldwide. It has been estimated that there will be approximately $234,030$ new diagnosed cases of lung and bronchus cancer and $154,050$ deaths in the USA in 2018 [1]. Recent years, the use of Computed Tomography (CT) scans for automatically identifying pulmonary nodules has attracted great attentions in research and applications at computer-aided diagnoses. Early diagnosis of pulmonary nodules in Computed Tomography chest scans have a huge impact on the quality and length of life of lung cancer patients. However, it is a hard work for a radiologist to evaluate hundreds of scanned results with poor sensitivity and specificity daily [2].

© Springer Nature Switzerland AG 2018
L. H. U and H. Xie (Eds.): APWeb-WAIM 2018, LNCS 11268, pp. 123–133, 2018.
https://doi.org/10.1007/978-3-030-01298-4_12

Hence, implementing computer-aided diagnosis (CAD) of pulmonary nodules can effectively diminish the work-load of clinicians.

To identify pulmonary nodules, most of the early researches use machine learning methods to extract the nodules features and then classify these features for the detection. SVM, Boosting, Decision trees, k-nearest neighbor, LASSO, neural networks are used as classify algorithms [3]. However, in conjunction with additional information such as demographic data and morphological features, the methods also required the high time costs on processing. Recently, several research groups have developed techniques in order to reduce the high computational cost of image processing in other areas, such as seafloor classification, face recognition, etc [4]. Chen et al. proposed a parallel sparse coding algorithm for seafloor image analysis [5]. Rajat et al. developed general principles for massively parallel unsupervised learning tasks using GPUs [6]. However, these improved algorithms mentioned above need to apply additional restrictions, especially not for CT images.

In this work we intend to develop an image processing method based on OpenCL to meet the following requirements. First, the method should to be suitable for identification of pulmonary nodules in CT images. Second, the learning models should have a good applicability with as few restrictions as possible. Third, the computational cost should be reasonable so that it can be adopted in clinical practice. In recent years more and more programs with parallel data processing use the GPU computing. As an open standard, OpenCL can execute programs on the CPU and GPU computing platforms. Therefore, in this paper OpenCL is selected to speed the algorithm. The rest of the paper is organized as follows. A redesigned parallel learning algorithm and our proposed methods based on OpenCL are introduced in Sect. 2. The experimental results and evaluation using the CT images are presented in Sect. 3. We conclude this work in Sect. 4.

2   Method

To process CT images, our identification approach consists of three stages in functioning: Data pre-processing, Data segmentation, and Removal of false positives. First, the raw CT images in the format of Communications in Medicine (DICOM) standard is denoised and transformed into Digital JPG format and the associated tag data of XML is changed into PNG format. Second, the clean normalized data are then segmented to extract the suspected nodular lesions. An optimized 2D FCNs is used to segment and extract the suspected nodular lesions from CT images. However, the use of 2D FCNs can bring in many false positives, which will be removed in the next stage. Finally, nodule candidates detected by the FCNs are fed into 3D CNNs to remove the false positives and the final results are output.

2.1   Image Preprocessing

At first, a data set needs to be denoised and transformed to fit the learning network architecture. Pydicom package is used to read a series of DICOM images into arrays about Hounsfield Units (HU) [7]. HU scale is a linear transformation of the original linear attenuation coefficient measurement. The Hounsfield scale, is a quantitative scale for describing radiodensity. We can identify the nodules in lung tissue based on the different values of HU. The value in HU is from 1024 to about 3200 on the Hounsfield scale. The specific attenuation of various tissues in the chest are shown in Table 1.

Table 1. The HU values corresponding to substance

Substance	HU
Air	+1000
Bone	Over +200
Blood	From +13 [8] to +50 [9]
Lung	From −700 to +600 [10]
Nodule	From 0 to +150

The entire preprocessing process includes four steps, as follows:

Step 1. Load the DICOM file, which is used as a reference and named as "pixelArray", to extract metadata. We then use the "SliceThickness" attributing to calculate the spacing between pixels in Z-axis to link the nodules in the adjacent slices. The CT DICOM images are not directly in HU. The raw metadata is transformed to HU format according to Eq. 1.

$$dateHU = metadataValue * slope + intercept. \tag{1}$$

Step 2. By locating the apex pulmonis and basis pulmonis in the CT slices, we can reserve those slices which contains the lung tissue. The slices that locate above the apex pulmonis or below the basis pulmonis can be removed. We then extract the regions of interest.

This can be done by either segmenting the contour or using HU threshold only with sensitivities of 98.9% and 97.78% accordingly. Since the contour method is much more expensive in time than that of HU threshold method, we would focus on HU threshold method in this article and set the HU threshold within [−800, 300] to distinguish lung tissue and bone and reduce the sample space significantly. These extracted regions are then normalized by the MinMax Scaling into an image format.

Step 3. The tagged data is generated and stored in a XML document. For each item in the tagged data, four experienced chest radiologists perform two-round diagnoses. In the first round, each physician independently diagnoses, marks the location of nodules, and labels a nodule in three categories:

C1:   unblindedReadNodule   if diameter of the nodule ≥ 3 mm;
C2:   unblindedReadNodule   if diameter of the nodule < 3 mm;
C3:   non-nodules   if the diameter of the non-nodule ≥ 3 mm.

In the subsequent unblinded-read round, each radiologist independently review their own marks along with the anonymized markers from the three other radiologists to render a final decision. Such a two-round labeling can label all results completely, avoiding forced consensus. We select all of the nodules marked with "unblindedReadNodule" for training and evaluation. As such, all the nodules in XML form are transformed into the images expressed in pixels.

Step 4. In order to make the adjacent slices have an equal distance, new slices need to be created and added from the existed slices and some existed slices may need to be removed. The insertion of a new slice can be done by using Cubic Spline Interpolation, a special case of Spline interpolation that is often used to avoid Runge's phenomenon. In this step, the normalized slices have the uniform distances (1 mm) between adjacent slices.

2.2 Parallelization via OpenCL

However, when faced with large amounts of data, the image preprocessing will be time-consuming. In order to speed the preprocessing, parallel implementations are considered. There are two levels of parallelism we can take advantages of. In the first level (level-1) of parallelism, it can be found that all the CT images can be optimized independently. The whole data set is divided into several subsets. Each CT image is calculated by a set of threads. In the second level (level-2) of parallelism, the preprocessing can be parallelized as well. In our implementation, the entire process is a combination of sequential stages and parallel stages.

In order to take advantage of GPUs parallel architecture, the learning task is implemented to fit the two levels of parallelism: blocks and threads. Blocks are used to achieve the data parallelism by working on separate subsets. Inside a block, each thread computes just 2 kernels so as to exploit more fine-grained parallelism. In this method, all threads within a block can synchronize with each other by using a pipeline model to share memory very quickly. As Fig. 1 shows, sequential stages and parallel stages are organized as a pipeline model to work. The algorithm is described as Algorithm 1. The mainly part is to distribute the tasks and transfer the data between CPU and GPU.

Since the HU threshold method is expensive in time, as Eq. 1, the *HU value* is calculated as Algorithm 2 inside a block. In this work, we did not use multiple GPUs in our implementation.

2.3 Segmentation and Filtering

After preprocessing, the normalized data should be segmented and filtered to identify the pulmonary nodules. In papers [11,12], in order to locate the pulmonary nodules, it first uses three-sized sliding windows ($5 * 5, 7 * 7, 9 * 9$) to frame a part in the map as a candidate area. It then extracts visual features

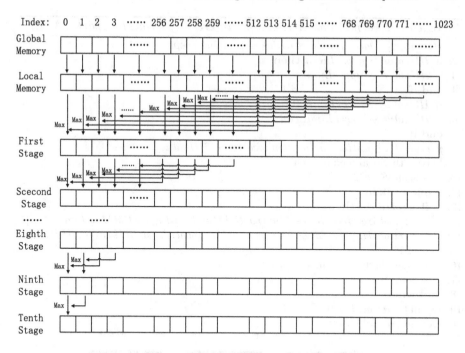

Fig. 1. The pipeline model to parallel computing.

Algorithm 1 Pipeline $(dataBuffer_1, dataBuffer_2, intercept, partialMax,$ $partialMin)$

1: $pointBuffer[2] \leftarrow dataBuffer_1, dataBuffer_2$
2: $pointBuffer[cursor] \leftarrow readFromDisk$
3: **while** $True$ **do**
4:    $GPUMemory \leftarrow CPUMemory$
5:    Launch the kernel
6:    Enqueue the command Read memory from GPU (No Block)
7:    $pointBuffer[!cursor] \leftarrow readFromDisk$
8:    Wait GPU Calculating and Transforming
9:    Writing to disk
10:   $cursor \leftarrow !cursor$
11:   **if** ReadTextOver **then**
12:       break
13:   **end if**
14: **end while**

---

associated with candidate regions. This approach can be time consuming and produce considerable miss detections for the nodules adhering to lung walls or blood vessels as showed in Fig. 2. In this paper, the proposed method by segmenting suspected nodules in FCNs helps to obtain high sensitivity. We modified

Algorithm 2 $kernel_1(dataArray, slope, intercept, partialMax, partialMin)$

1: $HUvalue \leftarrow dataArray[globalId] * slope + intercept$
2: **if** $HUvalue > upperLimit$ **then**
3:     $HUvalue \leftarrow upperLimit$
4: **end if**
5: **if** $HUvalue < lowerLimit$ **then**
6:     $HUvalue \leftarrow lowerLimit$
7: **end if**
8: $partialMax[localId], partialMin[localId] \leftarrow HUvalue$
9: barrier the local memory fence
10: $i \leftarrow groupSize/2$
11: **while** i ¿ 0 **do**
12:     **if** $localId < i$ **then**
13:         $partialMax[localId] \leftarrow Max(partialMax[localId], partialMax[localId + i])$
14:         $partialMin[localId] \leftarrow Min(partialMin[localId], partialMin[localId + i])$
15:     **end if**
16:     barrier the local memory fence
17:     $i \leftarrow i >> 1$
18: **end while**
19: Output $partialMax[0]$
20: Output $partialMin[0]$

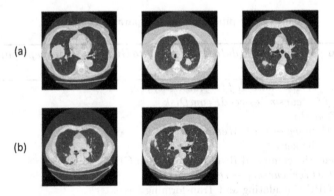

Fig. 2. (a) The images are recognizable in traditional methods. (b) The images where nodules adhere to lung tissue are not recognizable traditionally.

the network architecture as depicted in Table 2 to fit less complicated classifiers than that of Çiçek et al. [13] methods.

Since the cross sections of pulmonary nodules are very similar to the blood vessels in shapes, it is difficult to clearly identify the nodules in 2D networks. The method of 3D CNNs is adopted to construct three-dimensional shapes and distinguish the nodules from result of FCNs. We first integrate the 2D suspected nodules into a three-dimensional shape. According to the standard of medical requirements on pulmonary nodules, the small-sized nodules where the areas less than 10 need to be removed. We then dilate the results of FCNs into a

Table 2. The best network architecture

Layer	Kernel Size	Activation	Layer	Kernel Size	Activation
Conv1_1	3*3	Relu	Conv5_1	3*3	Relu
Conv1_2	3*3	Relu	Conv5_2	3*3	Relu
MaxPol	2*2		Conv5_3	1*1	Relu
Conv2_1	3*3	Relu	MaxPol	2*2	
Conv2_2	3*3	Relu	Conv6	3*3	Relu
MaxPol	2*2 (denoted as out1)		dropout	0.85	
Conv3_1	3*3	Relu	Conv7	1*1	Relu
Conv3_2	3*3	Relu	dropout	0.85	
Conv3_3	1*1	Relu	Conv8	1*1	
MaxPol	2*2 (denoted as out2)		deconv	4*4	
Conv4_1	3*3	Relu	+=out1		
Conv4_2	3*3	Relu	deconv	4*4	
Conv4_3	1*1	Relu	+=out2		
MaxPol	2*2		deconv	16*16 stride=8	

$3 * 3$ rectangular structured element since there can exist many scattered results referring to the same nodule. Finally, these three-dimensional shapes are fed into the Convolutional Neural Networks for classifications.

3   Experiment

In order to evaluate the performance of the preprocessing and segmentation under different parameters and network architectures, a series of experiments were carried out on a large lung Image data set. In this section, we present a brief discussion of out experimental data set, experiment platform and the results.

3.1   Data Set

Our experiments dataset is the Lung Image Database Consortium image collection (LIDC-IDRI) [14]. It consists of lung cancer screening thoracic computed tomography scans with marked-up annotated lesions and contains 1018 cases. The original DICOM images (anonymized and uncompressed) are associated with XML files for all 1018 CT scans.

3.2   Experiment Platform

All the experiments are conducted on a GPU framework. We implemented the GPU implementation on a workstation that contains an Nvidia Tesla K40 with 8 GB system memory. The host CPU is Intel Core i7 − 3820 CPU at 3.60 GHz with 16 GB memory. The compiler version is CUDA 5.5.

3.3   Results and Discussion

We conduct the experiments and present the detailed experimental results on CT images preprocessing and the sensitivity rate of segmentation. In our experiments, we mainly evaluate the following two aspects:the performance of image processing, and the segmentation performances under different parameters and network architectures. In training the models, each experiment requires $10,000$ iterations.

3.3.1   Image Processing Performance

We implemented the parallel execution via OpenCL. We applied the optimization on the random data set using Algorithm 1 and Algorithm 2. The dataset contains 500 patients and $105,606$ CT images. In this experiment, we conduct a comparison between three technologies (CPU only, GPU without pipeline and GPU with pipeline) at the full capacity of a single device. The base case implementation is a single thread implementation on Intel Xeon E5606 2.13 GHz CPU with 8 GB memory. The GPU implementation on a single Nvidia K20 GPU device. In the GPU implementation, we scheduled a 1-dimensional grid consisting of 1,024 blocks. Each thread block is 1-dimensional and contains 512 threads.

Table 3. Computation time of three parts

	Without pipeline	With pipeline
Load data	651(81.07%)	643(91.60%)
Compute	137(17.06%)	42(5.98%)
Others	15(1.87%)	17(2.42%)
Total	803	702

From Fig. 3 we can find that GPU has tremendous performance advantage for the image processing. A single K40 device can achieve more than 8 times speedup than the CPU implementation. On the other hand, the overwhelmingly dominant computational effort is spent on loading data, as shown in Table 3. By contrast, the pipeline model is efficient and it can improve the performance of 12.5%.

3.3.2   Sensitivity Rate

To find the suspected nodules precisely, we proposed a segmentation method through adjusting the architectures of FCN. Taking this into consideration, lager sized networks need to learn much more parameters, require much more iterations, and increase overfitting and the time costs potentially. Furthermore, deep structured networks might increase the generalization error. To segment nodules and non-nodules, our first intention was to use a minimal model that was capable

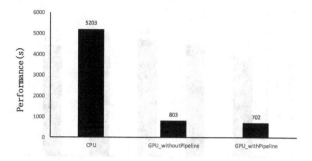

Fig. 3. Performance on single device.

of learning the exact architectures of the network. In the experiment, the network structures were different on convolutional layers. The part of convolutional was divided into four blocks which included several convolutional layers and a max-pooling layer.

In order to further optimize the network structure, different network structures are used to compare with the others. As shown in Table 4, the number of every lays in Category2, Category4 and Category5 is the same. However, in Category4, the one by one $(1 * 1)$ convolution is used in the last layer of Block3 and Block4. In Category5, the dilated convolution with the dilation rate $k = 3$ is used in the last layer of Block1 and Block2 and in last two layers of Block3 and Block4.

In the experiment, the *sensitivityRate* is used as an evaluation. It is computed as follows:

$$sensitivityRate = TP/(TP + FN) \tag{2}$$

where TP means true positive, FN means false negative.

As illustrated in Fig. 4, the results showed that appropriately increased the number of network layers can improve the performance. Given the same number of network layers, it showed that $(1 * 1)$ convolution is useful and the dilated convolution is still not need.

Table 4. Optimize the network architectures

	Block1	Block2	Block3	Block4
Category1	1	1	1	1
Category2	2	2	3	3
Category3	2	2	2	2
Category4	2	2	3	3
Category5	2	2	3	3

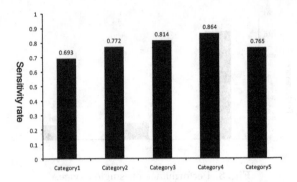

Fig. 4. The sensitivity rates on optimized network architectures.

Finally, based on Category4 of Table 4, the loss weight was gradually tuned to fit the general cases in lung CT images. The results demonstrated that when the loss weight was set to 0.004790, the *sensitivityRate* would achieve the best result of 0.9778.

4   Conclusion

In order to automatically identify of the pulmonary nodules, a novel method for CT images processing is proposed. In this work, we conduct a detailed study regarding the performance and sensitivity rate via OpenCL. The results on CT images show that it is very important to use GPU to achieve the best performance. In addition, pipeline model demonstrates the better performance. In future, the work will focus on the scalability of multicore GPU to adapt to the application of large amount of CT images.

Acknowledgments. The authors would like to thank the data providers of [14] for the testing data sets. This work was partially supported by the Natural Science Foundation of Zhejiang, China (Grant $No.Y15F020113$) and the Ningbo eHealth Project ($No.2016C11024$).

References

1. Siegel, R.L., Miller, K.D., Jemal, A.: Cancer statistics 2018. CA: Cancer J. Clin. **68**(1), 7–30 (2018)
2. Dey, R., Lu, Z., Hong, Y.: Diagnostic classification of lung nodules using 3D neural networks, 2018. cite arxiv:1803.07192Comment: Accepted for publication in IEEE International Symposium on Biomedical Imaging (ISBI) 2018, IEEE
3. Choi, W.-J., Choi, T.-S.: Automated pulmonary nodule detection based on three-dimensional shape-based feature descriptor. Comput. Methods Programs Biomed. **113**(1), 37–54 (2014)
4. Lee, H., Battle, A., Raina, R., Ng, A.Y.: Efficient sparse coding algorithms. In: Schölkopf, B., Platt, J.C., Hoffman, T., (eds.) Advances in Neural Information Processing Systems, vol. 19, pp. 801–808. MIT Press (2007)

5. Chen, G., Lai, C., Huang, M.: Parallel sparse coding for seafloor image analysis (2015)
6. Raina, R., Madhavan, A., Ng, A.Y.: Large-scale deep unsupervised learning using graphics processors. In: Proceedings of the 26th Annual International Conference on Machine Learning, pp. 873–880. ACM (2009)
7. Urata, M., et al.: Computed tomography hounsfield units can predict breast cancer metastasis to axillary lymph nodes. BMC Cancer **14**(1), 730 (2014)
8. Orth, D., Fosbinder BA RT(R), R.A.: Essentials of Radiologic Science. LWW (2011)
9. Wright, F.W.: Radiology of the Chest and Related Conditions. CRC Press, Boca Raton (2001)
10. Kazerooni, E.A., Gross, B.H.: The Core Curriculum: Cardiopulmonary Imaging (The Core Curriculum Series). LWW (2003)
11. Chan, T.F., Vese, L.A.: Active contours without edges. IEEE Trans. Image Process. **10**(2), 266–277 (2001)
12. Comaniciu, D., Ramesh, V., Meer, P.: Kernel-based object tracking. IEEE Trans. Pattern Anal. Mach. Intell. **25**(5), 564–577 (2003)
13. Çiçek, Ö., Abdulkadir, A., Lienkamp, S.S., Brox, T., Ronneberger, O.: 3D U-Net: learning dense volumetric segmentation from sparse annotation. In: Ourselin, S., Joskowicz, L., Sabuncu, M.R., Unal, G., Wells, W. (eds.) MICCAI 2016. LNCS, vol. 9901, pp. 424–432. Springer, Cham (2016). https://doi.org/10.1007/978-3-319-46723-8_49
14. Tracyn Bvendt01. LIDC-IDRI, 2012. Accessed 21 March 2012

Pulmonary Nodule Segmentation Method of CT Images Based on 3D-FCN

Yan Nie[1], Deyun Zhuo[2], Guanghui Song[3], and Shiting Wen[3(✉)]

[1] College of Science and Technology, Ningbo University, Ningbo 315212, China
`nieyan@nbu.edu.cn`
[2] College of Computer Science, Zhejiang University, Hangzhou 310058, China
`3514569758@qq.com`
[3] Ningbo Institute of Technology, Zhejiang University, Ningbo 315100, China
`499700647@qq.com`, `wensht@nit.zju.edu.cn`

Abstract. In this work, we use a 3D Fully Convolutional Network (FCN) architecture for pulmonary nodule segmentation. Our method integrates FCN and Conditional Random Field(CRF) into an end-to-end network. Using this approach, the spatial features of CT image series can be better utilized to obtain the three-dimensional global features of pulmonary nodules according to the context. The model includes pulmonary nodule segmentation and classification recognition and the noise is reduced by effective image preprocessing. We achieved competitive results during the testing phase of the LIDC/IDRI dataset for segmentation and detection with sensitivity of 0.918 using 3D-FCN and VGG19.

Keywords: Pulmonary nodule · Semantic segmentation · Fully convolutional network

1 Introduction

Lung cancer is the leading cause of global cancer death. The main reason for the close proximity between morbidity and mortality is the low recognition rate in the early stage of lung cancer. Therefore, early diagnosis and early treatment of lung cancer are very important for improving lung cancer survival and reducing lung cancer mortality.

Before the appearance of deep learning, the work of image semantic segmentation includes the simplest pixel-level 'threshold method', pixel-based clustering-based segmentation method, and segmentation method based on 'map partitioning', etc. Most of these methods perform image segmentation based on the low-order visual information of the image pixels [1]. The calculation complexity of these methods is not high without training, however, it is not effective on difficult complex segmentation.

After computer vision has entered the era of deep learning, Semantic segmentation also entered a new stage of development. A series of semantic segmentation methods based on convolutional neural networks have been proposed represented by FCN.

L. H. U and H. Xie (Eds.): APWeb-WAIM 2018, LNCS 11268, pp. 134–141, 2018.
https://doi.org/10.1007/978-3-030-01298-4_13

2   Related Work

FCN is mainly transformed from Convolutional Neural Network (CNN) model, replacing the traditional fully connected layers fc6 and fc7 with deconvolution layers [2]. The SegNet and DeconvNet models belonging to Encoder-Decoder architecture proposed later are similar with the FCN. The biggest differences between them are the upsampling methods. In FCN, the upsampling method is to add the feature vector after deconvoluting to the high resolution layer.In FCN, the upsampling method is to add the feature vector after deconvoluting to the high resolution layer. The latter two map the values of the feature to the new one based on the location of the pooled pool, and then connect to result in deconvolution layer [3, 4]. One of the defects of FCN is that upsampling cannot restore all the lost information losslessly. In this regard, Dilated Convolution is a good solution , that can greatly improve the recognition of the semantic categories and the fineness of the segmentation [5, 6].

The method of Deep Learning + Probabilistic Graph Model (GPM) is proposed considering the space context of each pixel for settling the defects of image semantic segmentation not taking into account the spatial context of each pixel as a whole [7, 8]. The GPM is integrated into the deep learning framework to form a fully automatic end-to-end system, which is used to explain connection between the nature of image pixel [9, 10].

The above FCN model based on natural image segmentation has also achieved important breakthroughs in the segmentation of medical images. U-Net network architecture is a Semantic Separation Network based on FCN, which has been proved to be suitable for the segmentation of medical images [11]. As research continues to deepen, U-Net has made some progress that PSP-Net splits the entire image into four equal parts and pools it to get global information [12]. Mask R-CNN integrates the advantages and methods of object recognition and has also achieved good results [13].The above methods are all based on two-dimensional medical image segmentation, inspired by the successful application of 3D-CNN model in human behavior recognition. In recent years, researchers have also used the two-dimensional sequence characteristics of medical images to apply them to the field of medical image recognition and segmentation [14].For example, a network model called DeepMedic combines three-dimensional multi-scale CNN and full-connected CRF on the basis of the above studies, and has been used for the segmentation of brain tumors [15].The 3D U-Net model directly expands the U-Net model and achieves stereoscopic segmentation of the kidney under sparse sample conditions [16]. The 3D FractalNet model enables the stereo-scopic segmentation of cardiovascular MRI medical images based on FCN [17]. The above-mentioned methods all extend the two-dimensional FCN network to three-dimension, and do not fully consider the contextual relationship of the pixels in the three-dimensional space, nor do they study the three-dimensional stereoscopic segmentation of the CT image.

In general, the semantic segmentation method based on FCN is a very popular research topic. Scholars are trying to apply the FCN model more deeply to the field of medical image segmentation in multi-scale feature integration, structure prediction and network structure aspects.

The segmentation and labeling of pulmonary nodules in CT images has made great strides. After segmented for lung parenchymal structure and preprocessed, the data task is to search for pulmonary nodule candidate regions, and then classify and identify the ROIs, in order to distinguish between benign and malignant tumors in further. Due to the lack of the Ground Truth, its a very difficult problem to find pulmonary nodules because some of the nodules were attached to the blood vessels, trachea, and pleura. In previous studies, the segmentation methods such as active contour model, fuzzy clustering method, threshold method, and template matching were successively proposed [18]. However, due to complex lung structure, the types of nodules which is difficult to accurately segmented include: micronodules, adhering vascular nodules, adherent pleural nodules, and ground-glass nodules. At present, no model established by the algorithm can be applied to all types of nodules [19,20].

With the application of deep learning method in the field of medical image processing, pulmonary nodule segmentation mainly adopts two modes: image segmentation-based method and target-based detection method. For example, it gets practical in using U-Net for cutting, Faster R-CNN for detection, and then using 3D CNN for target classification and identification.

The Faster R-CNN model is based on the object detection model R-CNN. Compared with the traditional CNN model, R-CNN proposes candidate regions for detection, and then uses CNN for detection. This method can not only identify objects, but also provide target location information [21]. Fast R-CNN speeds up and simplifies the R-CNN model, and optimizes the way of generating regional proposals in the model and further improves system performance [22,23].

The Mask R-CNN model inherits the framework of Faster R-CNN. It also introduces the fully-connected segmentation subnet, performs object detection and pixel segmentation at the same time. This method is currently a research hotspot in the field of medical image processing.

In summary, the image semantic segmentation method based on deep convolutional neural network is widely used in the segmentation of suspected lung nodules and has excellent performance.

3   Method

3.1   3D-FCN

The FCN architecture proposed for semantic segmentation improves CNN architectures for dense predictions without any fully connected layers. Enjoying great popularity on the low time cost and the segmentation maps to be generated for image of any size, the FCN architecture can be spread to almost any models.

However, due to the incapability of grasping the connections from the context of adjacent image slices. So the appropriate way is to feed a combination of image slices into networks. The methods proposed is take a handful of into a combination, which differentiates from the general 3D-FCN. Nevertheless, due to the tiny scale, the cost of time and memory considerably lower than the general one.

For segmentation, the network architecture are designed four blocks for convolution, three blocks for deconvolution and four concatenation of two result with the convolution and deconvolution in Fig. 1. In the first convolution block, the stride is 2 in first dimension for gasp the features in Z-axis. Due to the tiny scale for learning, the vector will not be convoluted in the first dimension. The convolution kernels uniformed $3 \times 3 \times 3$ requires the value of "Extend" from pixels outside of the image boundaries, which guarantee the size of out result with original same. The concatenation layers which connect with the convolution and deconvolution are designed for taking multiple inputs that have the same height, width and depth and concatenates them along the fourth dimension.

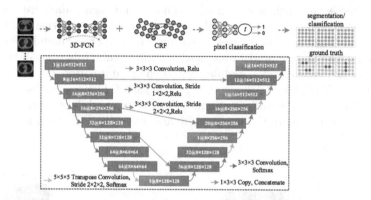

Fig. 1. Network structure of the 3D-FCN + CRF model.

3.2 CRF

For each pixel i, there is category label x_i and corresponding observation value y_i. Each pixel point is taken as a node, and the relationship between pixels and pixels is taken as an edge, which constitutes a condition random field. By observing the variable y_i, we can deduce the category label x_i corresponding to pixel i. The CRF conforms to gibbs distribution, and x in the formula is the observed value above

$$P(X = x|I) = \frac{1}{Z(I)} exp(-E(X|I)) \qquad (1)$$

Where $E(X|I)$ is the energy function, the unary potential function comes from the output of 3D-FCN, binary potential function is used to describe the relationship between the pixels and the pixels, encourage similar pixels assigned the same label, the definition of the distance between the pixel is associated with color value and the actual relative distance, the CRF can then segment the image as far as possible at the boundary. The binary potential function of fully connection CRF describes the relationship between each pixel and all other pixels. For 3D-FCN, the relationship between each pixel and the pixels in adjacent slice images needs to be further processed.

4   Experiments and Results

4.1   Dataset and Pre-processing

This study used the LIDC/IDRI dataset, consisting of 1,018 helical thoracic CT scans collected retrospectively from seven academic centres. The LIDC/IDRI employed a two-phase image annotation process. In the first phase, four radiologists independently reviewed all cases. In the second phase all annotations of the other three radiologists were made available and each radiologist independently reviewed their marks along with the anonymized marks of their colleagues. Findings were annotated and categorized into nodule≥3 mm, nodule<3 mm, or non-nodule. Non-nodule marks were used to indicate abnormalities in the scan, which were not considered a nodule. All labeling information for each case is recorded in an XML file. As shown in Fig. 2, in each CT image, we integrate the labeling of four radiologists and combine them into a nodule. By preprocessing the dataset, we eliminate the image noise and unify the CT series images into $16 \times 512 \times 512$.

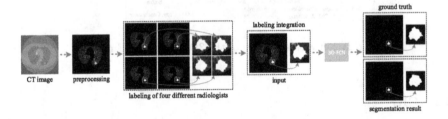

Fig. 2. CT image series preprocessing process.

4.2   Evaluation Measures

To determine whether the nodule is correctly detected according to the coordinate information. If the nodule falls into a sphere with a center radius of R according to the reference standard, the detection is considered correct. FROC curve was calculated based on the final nodule detection probability. Sensitivity is used as the final criterion. All experiments are performed over five train/test folds.

4.3   Setup and Parameters

The Caffe framework is deployed to train on a single NVIDIA Tesla K40c card. We adopt the method of segmentation + classification, and use 3D-FCN for segmentation to find out the suspected nodes, and then use the classification network to judge whether each suspected node is true positive, and give the probability. The suspected nodes detected by the segmentation network contain a large number of false positives, the ratio is about 1:20. Therefore, a classification network is required for false positive attenuation. We compared CaffeNet, VGG16 and VGG19 network structures. The initial learning rate of the segmentation network is set to 0.0002, and it is decreased by a factor of 10 every 15 K iterations, for a total of 50 K iterations. The initial learning rate of the classification network is set to 0.0001, and it is decreased by a factor of 10 every 15 K iterations, for a total of 40 K iterations.

4.4   Experimental Results

Figure 3 shows learning curves of the sensitivity for the CaffeNet, VGG16 and VGG19 for one of the 5-fold cross validation experiments. We see that the experiment results in stable training and that the model does not overfit. As shown in the comparison in Table 1, the performance based on 3D-FCN and VGG19 is the best. And the performance based on 3D-FCN is superior to that based on the conventional methods using different algorithms. The experimental results show that the model we proposed is strong and stable for segmentation and detection of pulmonary nodule on the LIDC/IDRI dataset.

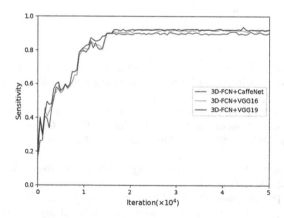

Fig. 3. Tracking sensitivity during training.In this plots we show the training curves for one of the five cross-validation experiments.

Table 1. Sensitivity of different methods

	3D-FCN	U-Net
CaffeNet	90.27%	88.45%
VGG16	91.32%	89.90%
VGG19	91.80%	90.08%

5   Conclusion

In this study, we evaluate the performance of the 3D-FCN on pulmonary nodule detection problem. In view of the similarity and continuity characteristics of adjacent slices of 2D CT medical image series, a segmented network consisting of 3D-FCN and CRF is used to obtain the suspected nodule area, then the deep convolution neural network is used to classify them. The experiment results show that our proposed method is very effective and outperforms the conventional pulmonary nodule detection method on the LIDC/IDRI dataset. Next, we will further obtain the three-dimensional structural features of pulmonary nodules on the basis of the above work to improve the performance of the system. Furthermore, we will also extend the experiment to other medical image series dataset to test the generalization performance of this method.

Acknowledgements. This work has been supported by the Ningbo eHealth Project (No.2016C11024) and the Humanities and Social Sciences Foundation of the Ministry of Education with Grant No.16YJCZH112. This work also supported a Project Supported by Scientific Research Fund of Zhejiang Provincial Education Department Y201553788.

References

1. Badrinarayanan, V., Kendall, A., Cipolla, R.: SegNet: a deep convolutional encoder-decoder architecture for image segmentation. IEEE Trans. Pattern Anal. Mach. Intell. **39**(12), 2481–2495 (2017)
2. Noh, H., Hong, S., Han, B.: Learning deconvolution network for semantic segmentation. In: Proceedings of the 2015 IEEE International Conference on Computer Vision (ICCV), ICCV 2015, pp. 1520–1528. IEEE Computer Society, Washington, DC, USA (2015)
3. Yu, F., Koltun, V.: Multi-scale context aggregation by dilated convolutions. CoRR, arXiv:abs/1511.07122 (2015)
4. Chen, L.C., Papandreou, G., Kokkinos, I., Murphy, K., Yuille, A.L.: Deeplab: semantic image segmentation with deep convolutional nets, atrous convolution, and fully connected CRFS. CoRR, arXiv:abs/1606.00915 (2016)
5. Chen, L.C., Papandreou, G., Kokkinos, I., Murphy, K., Yuille, A.L.: Semantic image segmentation with deep convolutional nets and fully connected CRFS. CoRR, arXiv:abs/1412.7062 (2014)
6. Liu, Z., Li, X., Luo, P., Loy, C.C., Tang, X.: Semantic image segmentation via deep parsing network. CoRR, arXiv:abs/1509.02634 (2015)

7. Zheng, S., et al.: Conditional random fields as recurrent neural networks. CoRR, arXiv:abs/1502.03240 (2015)
8. Chandra, S., Kokkinos, I.: Fast, exact and multi-scale inference for semantic image segmentation with deep Gaussian CRFS. CoRR, arXiv:abs/1603.08358 (2016)
9. Luc, P., Couprie, C., Chintala, S., Verbeek, J.: Semantic segmentation using adversarial networks. CoRR, arXiv:abs/1611.08408 (2016)
10. Kozinski, M., Simon, L., Jurie, F.: An adversarial regularisation for semi-supervised training of structured output neural networks. CoRR, arXiv:abs/1702.02382 (2017)
11. Ronneberger, O., Fischer, P., Brox, T.: U-net: convolutional networks for biomedical image segmentation. CoRR arXiv:abs/1505.04597 (2015)
12. Zhao, H., Shi, J., Qi, X., Wang, X., Jia, J.: Pyramid scene parsing network. CoRR, arXiv:abs/1612.01105 (2016)
13. He, K., Gkioxari, G., Dollár, P., Girshick, R.B.: Mask R-CNN. CoRR, arXiv:abs/1703.06870 (2017)
14. Girshick, R.B., Donahue, J., Darrell, T., Malik, J.: Rich feature hierarchies for accurate object detection and semantic segmentation. CoRR, arXiv:abs/1311.2524 (2013)
15. Girshick, R.B.: Fast R-CNN. CoRR, arXiv:abs/1504.08083 (2015)
16. Ren, S., He, K., Girshick, R.B., Sun, J.: Faster R-CNN: towards real-time object detection with region proposal networks. CoRR arXiv:abs/1506.01497 (2015)
17. Ding, J., Li, A., Hu, Z., Wang, L.: Accurate pulmonary nodule detection in computed tomography images using deep convolutional neural networks. CoRR, arXiv:abs/1706.04303 (2017)
18. Ji, S., Xu, W., Yang, M., Yu, K.: 3D convolutional neural networks for human action recognition. IEEE Trans. Pattern Anal. Mach. Intell. 35(1), 221–231 (2013)
19. Kamnitsas, K., et al.: Efficient multi-scale 3D CNN with fully connected CRF for accurate brain lesion segmentation. CoRR, arXiv:abs/1603.05959 (2016)
20. Jesson, A., Arbel, T.: Brain tumor segmentation using a 3D FCN with multi-scale loss. In: Crimi, A., Bakas, S., Kuijf, H., Menze, B., Reyes, M. (eds.) BrainLes 2017. LNCS, vol. 10670, pp. 392–402. Springer, Cham (2018). https://doi.org/10.1007/978-3-319-75238-9_34
21. Yu, L., Yang, X., Qin, J., Heng, P.-A.: 3D FractalNet: dense volumetric segmentation for cardiovascular MRI volumes. In: Zuluaga, M.A., Bhatia, K., Kainz, B., Moghari, M.H., Pace, D.F. (eds.) RAMBO/HVSMR -2016. LNCS, vol. 10129, pp. 103–110. Springer, Cham (2017). https://doi.org/10.1007/978-3-319-52280-7_10
22. Çiçek, Ö., Abdulkadir, A., Lienkamp, S.S., Brox, T., Ronneberger, O.: 3D U-Net: learning dense volumetric segmentation from sparse annotation. CoRR, arXiv:abs/1606.06650 (2016)
23. Li, J., Zhang, R., Shi, L., Wang, D.: Automatic whole-heart segmentation in congenital heart disease using deeply-supervised 3D FCN. In: Zuluaga, M.A., Bhatia, K., Kainz, B., Moghari, M.H., Pace, D.F. (eds.) RAMBO/HVSMR -2016. LNCS, vol. 10129, pp. 111–118. Springer, Cham (2017). https://doi.org/10.1007/978-3-319-52280-7_11

The 1st International Workshop on Knowledge Graph Management and Analysis

Research on Video Recommendation Algorithm Based on Knowledge Reasoning of Knowledge Graph

Zhihong Xu[1,2], Xing Zhao[1], Yongfeng Dong[1,2(✉)], Wenjie Yan[1], and Ziqi Yu[1]

[1] School of Artificial Intelligence, Hebei University of Technology, 300401 Tianjin, China
xuzhihong@scse.hebut.edu.cn, dongyf@hebut.edu.cn
[2] Hebei Province Key Laboratory of Big Data Computing, 300401 Tianjin, China

Abstract. Since collaborative filtering algorithm cannot make full use of video attribute information, mining implicit information of video, this paper proposes a video recommendation algorithm based on knowledge reasoning of knowledge graph. The algorithm uses a knowledge graph with powerful semantic processing capabilities and open interconnection capabilities. The ontology model is used to formalize the implicit semantics in the data. By using the knowledge inference method of the knowledge graph, the existing triplet information is used to establish new relationships between entities and assign corresponding weights to the paths. All the semantic information is embedded in the low-dimensional vector space, and the potential semantic similarity of the video is calculated by combining the path weights. And then, integrate semantic similarity into collaborative filtering for recommendation. Experiments show that this algorithm can make up for the deficiency that collaborative filtering algorithms cannot fully utilize the hidden information of video, and enhance the effectiveness of recommendation at the semantic level, what is more, the recommendation results are interpretable. To a certain extent, it can solve the problem of data sparseness.

Keywords: Collaborative filtering · Knowledge graph · Knowledge reasoning Semantic similarity · Path weight

1 Introduction

In the age of the information explosion, a lot of video software has developed rapidly and a lot of video data is produced every day. Therefore, users face the trouble with fragmentation of network video information and information overload, how to make the user experience more humane and to meet the needs of users becomes an urgent problem to be solved. Thus, the recommendation systems came into being. In the recommendation system, the most important problem is to better serve the user, conduct an in-depth analysis of the user's past behavior information, obtain the user's actual preference information, and finally recommend interested video to the user. In

© Springer Nature Switzerland AG 2018
L. H. U and H. Xie (Eds.): APWeb-WAIM 2018, LNCS 11268, pp. 145–158, 2018.
https://doi.org/10.1007/978-3-030-01298-4_14

the continuous exploration of many people, several excellent recommendation algorithms have emerged. The current mainstream recommendation algorithms include collaborative filtering recommendation algorithm (CF) [1], content-based recommendation algorithm [2], knowledge-based recommendation algorithm [3, 4], semantic-based recommendation algorithm [5], and social network-based recommendation [6]. Every recommendation algorithm has its own advantages and disadvantages. The collaborative filtering recommendation algorithm can handle unstructured data such as multimedia, the recommendation results are rich, but there are problems of data sparseness, cold start and scalability. Content-based recommendation algorithm is not constrained by scoring sparseness problems, but the diversity of recommended results is limited. The knowledge-based recommendation algorithm is not constrained by scoring sparseness and avoids the problems such as the cold start of new users and new items, however, knowledge base needs to be constructed and it is restricted by knowledge representation methods. Semantic-based recommendation algorithm can mine implicit semantic information to improve recommendation accuracy and diversity, but domain ontology needs to be built.

Based on the above analysis, the collaborative filtering recommendation algorithm fails to make full use of the semantic information of the video itself, which has a certain impact on the recommendation result. Existing experiments show that knowledge graph as structured semantic knowledge bases [7], can accurately locate and acquire knowledge in depth. As a representative of the semantic processing ability with powerful functions, it can specifically describe one or more relationships among entities. Using the knowledge inference of knowledge graph can further excavate the deeper relationship between entity relations and retrieve the missing entities and relationships, then enrich and expand the knowledge graph. So this paper attempts to use knowledge inference of knowledge graph to embed existing formalized information into low-dimensional vector space, excavate the implied information of items, and enrich the semantic network of extended items. This method can not only realize the integration of semantic information into collaborative filtering, but also can effectively achieve the deep acquisition of the implicit semantic information of items, thereby improving the effectiveness of the recommendation.

2 Related Works

For user-based collaborative filtering algorithm [8], the burden and costs will continue to increase as the number of users continues to increase. On the contrary, the item-based collaborative filtering recommendation algorithm does not rely on user information, only uses the information of the item itself. For any item, its information will not change over time. Therefore, the similarity between the target item and the item that the user has watched or has evaluated can be calculated in advance using the cosine similarity algorithm, the Pearson algorithm, or the modified cosine similarity algorithm and so on. According to the obtained similarity matrix, the score prediction is performed, so that the top K items with high ratings that the user may be interested in are recommended to the user. The amount of calculations in this process is greatly reduced, and to a certain extent, the cold start problem is solved. However, since this process does not consider the semantic information of the item itself, it has a certain influence on its recommendation effect.

At present, the commonly used knowledge graph inference algorithms [9–11] including the tensor decomposition method (Rescal algorithm), the conversion-based method (TransE algorithm), and the path-based inference method (PRA algorithm). Based on Rescal algorithm, we can achieve better results in inference, but there is a problem that the time complexity is too large. When the amount of data is too large, the reasoning effect is limited. The TransE algorithm is relatively simple and the computational complexity is also low, and the complex semantic relationship between entities and relationships can be established efficiently. In PRA algorithm, learning and reasoning based on the structural characteristics of the knowledge graph can realize multipath reasoning between any two entities.

Liu et al. proposed Path-based translation model(PTransE) algorithm [12–15] and the relationship between entities on both sides is bidirectional for the semantic information they contain [16, 17], the algorithm finds all paths between any two entities using path sorting algorithm (PRA), gets the random walk path p and the probability of the path within a limited number of hops, for example, triple (h, r_1, e_1) is in the knowledge graph, and triple (e_1, r_2, t) exists at the same time, the first triplet's tail entity e_1 and the second triple head entity e_1 is the same entity, then get the path of h to t $P(h, t) = \{r_1, r_2\}$, taking into account the reliability of the triple, the score of the triplet $(h, r_1 \circ r_2, t)$ not less than 0.01 is added to the new triple set. Use translation models to embed entities and relationships into the d-dimensional vector space and normalize them. Calculate the loss function value of the model in vector space and calculate the semantic similarity between any two entities. There are many ways to combine semantics, for example, additive model, multiplicative model, and recurrent neural network model, etc. Existing experiments have proved that the semantic combination of the additive model works best.

3 Collaborative Filtering Recommendation Algorithm Based on Knowledge Inference of Knowledge Graph

To make use of the item information stability of item-based collaborative filtering recommendation algorithm, and introduce the reasoning mechanism of knowledge graph, a collaborative filtering recommendation algorithm called PTransE_CF is proposed. The algorithm not only uses the user-item rating data, but also introduces the information of the item itself, and fully excavates its implied information. It is expected to solve the shortcomings of insufficient information utilization, improve the recommendation efficiency, and to a certain extent solve the problem of data sparseness.

3.1 Item-Based Collaborative Filtering Recommendation Algorithm

In a complete recommendation system, suppose there are m users $U = \{u_1, u_2 ..., u_m\}$, n items $V = \{v_1, v_2, ..., v_n\}$, and the user-item rating data is constructed into a common one. The user-item rating matrix $R_{m \times n}$ is shown in (1).

$$R = \begin{bmatrix} R_{11} & R_{12} & \cdots & R_{1n} \\ R_{21} & R_{22} & \cdots & R_{2n} \\ \vdots & \vdots & & \vdots \\ R_{m1} & R_{m2} & \cdots & R_{mn} \end{bmatrix} \tag{1}$$

R_{ij} represents the user u_i's rating of the item v_j, and the score data indicates that the user's likeness to the item. In item-based collaborative filtering recommendation algorithm, the first step is to calculate the similarity between items; the second step is to use Top-K to recommend according to the user's history feedback. This paper uses cosine similarity to calculate the similarity between the two objects(items). This method assumes that there are two items V_i, V_j which are n-dimensional vectors, taking the cosine angle between two vectors as the similarity between the two items, the formula is as (2).

$$W(V_i, V_j) = \cos(V_i, V_j) = \frac{V_i \times V_j}{\|V_i\| \times \|V_j\|} \tag{2}$$

Considering the impact of popular items on the recommendation effect, when a user watches only the current popular movies, there may be no similarity between these movies, the introduction of popular item weight factors [16, 17] are helpful to solve this problem. Then the weight factor of any two popular movies is as (3).

$$A(V_i, V_j) = \frac{|N(V_i) \cap N(V_j)|}{\sqrt{|N(V_i)||N(V_j)|}} \tag{3}$$

Among them, $|N(V_i)|$ indicates the number of users who like the item V_j, and $|N(V_i) \cap N(V_j)|$ indicates the number of users who like the item V_i and the item V_j at the same time. $A(V_i, V_j)$ is the weight factor.

The item similarity matrix of the final item-based collaborative filtering recommendation algorithm can be expressed as (4).

$$\text{sim}_{\text{itemCF}}(V_i, V_j) = A(V_i, V_j) * W(V_i, V_j) \tag{4}$$

3.2 The Recommendation Process Based on Knowledge Inference of Knowledge Graph

Knowledge Inference Algorithm Based on PTransE. In the knowledge graph, the two entities can be represented in the form of triples (entity 1, relation, entity 2).

Entities can form a powerful and directed semantic network through relationships. Nodes are made up of entities and relationships form edges. Entities connect with each other through various relationships form a huge network structure.

Fig. 1. The movie knowledge graph triple

For the film field, the movie entities contain features such as release time, director, actors, and types, each feature of movie is represented in the form of a triplet, the triplet is shown in Fig. 1.

The interconnections among triples constitute a network of knowledge graph. The semantic network formed by some movies and movie attributes of movielens is shown in Fig. 2.

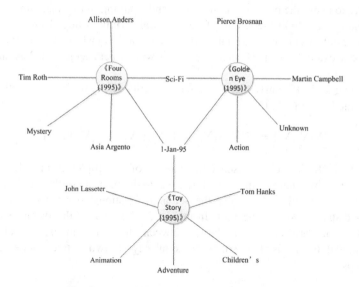

Fig. 2. Part of the knowledge graph of the film

From Fig. 2, we can see that the more relationship paths between two entities, the more similar attributes of two entities there are. That is, the greater similarity of the head entity of the triplet with the tail entity under the relation r, the greater the similarity between the two entities. Then Considering multi-paths and the weights on distribution of different relation paths, use the PTransE knowledge reasoning method combining path-based reasoning and transformation-based methods.

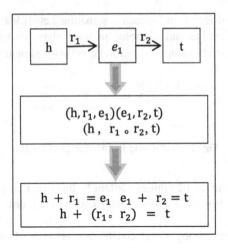

Fig. 3. PTransE model

In order to avoid the path getting too long, and resulting in too many duplicate data in the training data, limit the length of random walk within 4 hops(Here only get the even jump results according to actual needs). Using the PTransE model for training, the above model example of a PTransE algorithm with a path length of 2 is shown in Fig. 3.

According to the PTransE Algorithm, the similarity of any two movie entities V_i, V_j can be defined as formula (5).

$$W\left(V_i, V_j\right) = G\left(V_i, r, V_j\right) = E(V_i, r, V_i) + E\left(V_i, P, V_j\right) \qquad (5)$$

$E(h, r, t) = \|h + r - t\|$ represents the energy of the triplet formed by the two entities through the relationship r, which is equal to the measure of the similarity of the head entity to the tail entity under the semantic relationship r $d(h + r, t)$, d is the Euclidean distance. As can be seen from Fig. 2, the smaller the distance between two entities, the greater the similarity between the two entities. Therefore, the distance model (6) is used to solve the similarity of two entities under different relationships.

$$E(h, P, t) = \frac{1}{Z} \sum_{p \in P(h,t)} R(p|h, t) * 1.0/(E(h, p, t) + 1) \qquad (6)$$

P represents the collection of multi-path relationships between two entities. $R(r|h, t)$ represents the probability that the head entity h reaches the tail entity t through the relationship r. $Z = \sum_{p \in P(h,t)} R(p|h, t)$ is a normalization factor. A path from the head entity to the tail entity is $p = (r_1, \ldots, r_l) : e_0 \xrightarrow{r_1} e_1 \xrightarrow{r_2} e_3 \xrightarrow{r_3} \ldots \xrightarrow{r_l} e_l$, For any entity $m \in e_i$, the probability of walking along path p to m is (7).

$$R_p(m) = \sum_{n \in e_{i-1}(\cdot, m)} \frac{1}{|e(n, \cdot)|} R_p(n) \tag{7}$$

Where $e_i(n, \cdot)$ indicates that e_{i-1} is a direct successor. $R_p(n)$ indicates the probability that this path will travel to entity n and $R(p|h, t) = R_p(t)$. The above path from e_0 to e_l is: $p = r_1 \circ r_2 \circ \cdots \circ r_l$. For the relation p, use the formula (8) for semantic combination.

$$p = r_1 + r_1 + r_2 + \cdots + r_l \tag{8}$$

Based on the above ideas, PTransE algorithm can mine the hidden information, extend semantic network, and increase the accuracy of recommendations.

For any multipath triple (h, p, t), according to the PTransE translation model, Formula (9) is used to exercise the loss formula. Any two entities directly associated with loss formula training using formula (10).

$$L(p, r) = \sum_{(h, r', t)} [\gamma + E(p, r) - E(p, r')]_+ \tag{9}$$

$$L(h, r, t) = \sum_{(h, r, t) \in S(h', r, t') \in S'(h, r, t)} [\gamma + E(h, r, t) - E(h' + r + t')]_+ \tag{10}$$

h' and t' are erroneous triple vectors, they are negative training sample, the erroneous triples are derived from the following rules: reserve the head entity, replacing the tail entity, or keep the tail entity, replacing the head entity. [x] + is the hinge loss function, where [x] + represents keeping the original value when x is greater than 0, and taking 0 when it is less than 0. γ is a marginal parameter, and uses the idea of support vector machine (SVM) to maximize the distance between the correct triplet and the incorrect triplet.

Calculation of Fusion Semantic Similarity. According to the analysis, the user's rating of the already viewed movie has a dominant influence on the similarity between any two movies, and can reflect the user's interest better. Therefore, according to the "golden section" principle, the semantic similarity obtained from the knowledge inference of PTransE and the score similarity of the item-based collaborative filtering recommendation algorithm are fused at ratio 0.382:0.618. Then get the final list of similarities [18, 19]. The formula for similarity fusion is (11).

$$\_W(V_i, V_j) = 0.618 * sim\_itemCF(V_i, V_j) + 0.382 * W(V_i, V_j) \tag{11}$$

$sim_{itemCF}(V_i, V_j)$ is the similarity of ratings that obtained from item-based collaborative filtering recommendation algorithm, $W(V_i, V_j)$ is semantic similarity that obtained from PTransE algorithm. $\_W(V_i, V_j)$ is the similarity after fusion.

3.3   Algorithm Description

PTransE algorithm and CF algorithm are combined to make recommendation in movie domain, this algorithm is called PTransE_CF, the flow chart of the algorithm is shown in Fig. 4.

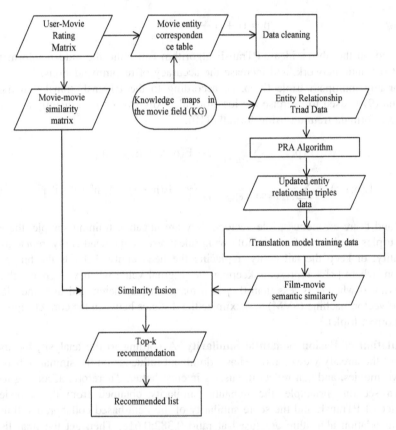

Fig. 4. The flow chart of PTransE_CF algorithm

The description of this algorithm is as follows:

Algorithm 1. PTransE_CF Algorithm

Input: Initial movie entity set E, Initial relationship set R, Triple set S, Boundary adjustment parameters γ, Learning rate δ, Low-dimensional space dimension d.

Output: Get the similarity between any two movie entities

/* initialization */

1. **While** $e_i \in E$

2. $e_i \leftarrow$ Uniform$((-6) / \sqrt{d}, 6 / \sqrt{d})$

                   //Embedding entity into a low-dimensional vector space

3.   Norm(e_i) // Normalized entity

4. **While** $r_i \in R$

5. $r_i \leftarrow$ Uniform$((-6) / \sqrt{d}, 6 / \sqrt{d})$

                   //Embedding relationship into a low-dimensional vector space

6.   Norm(r_i) // Normalized relationship

/* PTransE_CF algorithm process*/

7. **For each** $e_i \in E$

8.   Use PRA to calculate the path and weight of each entity pair

9.   Use additive combination of multipaths

10.   Using translation model PTransE to train loss function:

11. $L(h, r, t) = \sum_{(h,r,t) \in S}(h',r,t') \sum_{\in S'(h,r,t)}[\gamma + E(h, r, t) - E(h' + r + t')]_+$

12.   Update the triplet vector in the batch

13. **End for**

14. **For each** $e_i \in E$ // Calculate the similarity between any two entities

15.   **For each** $e_j \in E$

16.   Calculate the score similarity according to Item_CF algorithm

17.   Calculating Semantic Similarity According to PTransE Algorithm

18.   Fusion semantic similarity and score similarity

19.   **End for**

20. **End for**

21. **Return** Similarity matrix

---

4 Experiments and Analysis

4.1 Experiment Environment

The hardware environment of this experiment is: Dell dual-core server, 16 GB storage, 200 GB memory; the software environment is python 3.6.3

4.2 Data Sets

This experiment selected the MovieLens data set of the GroupLens study group. A total of 943 users generated 100,000 rating data on 1682 movies, each user rated not less than 10 movies. According to the rating data of these movies and the knowledge graph of the movie field, take the main 23 feature attributes of each movie, altogether 7911 triples were tested.

4.3 Evaluation Index

A recommendation algorithm can be evaluated from multiple perspectives, and different evaluation indicators can be used from different perspectives, the following evaluation indicators are used for this algorithm:

- Accuracy and Recall Rate: Accuracy is the ratio of the recommended movies to the actual number of recommended movies. Recall rate refers to the ratio of recommended correct movies (Including movies that the recommended user actually likes and those that are not recommended to the user and that the user does not like) to the total number of movies. Therefore, the higher the accuracy of the recommendation algorithm, the better recommendation effect is. Similarly, the higher the recall rate, the better recommended effect is.
- F1 Value: In the evaluation indicators, the contradiction between the accuracy rate and the recall rate often occurs. That is, when the accuracy rate is too high, the recall rate is relatively low, whereas if the recall rate is too high, the accuracy rate is relatively low. Therefore, we need to use the comprehensive evaluation index F-measure to measure the accuracy of the recommendation.

$$F - \text{measure} = \frac{(a^2 + 1)P * R}{a^2(P + R)} \tag{12}$$

This paper uses F1 to evaluate the recommendation effect, that is, take the parameter a = 1. The higher F1 is, the better the recommendation effect is.

For example, the confusion matrix for a movie is as Table 1.

Table 1. confusion matrix

	Users like	Users don't like
Recommended	TP	FP
Not recommended	FN	TN

According to the confusion matrix, the accuracy rate P, recall rate R, and F1 value are obtained as formula 13–15.

Accuracy P:

$$P = \frac{TP}{TP + FP} \tag{13}$$

Recall Rate R:

$$R = \frac{TP + TN}{TP + FP + FN + TN} \tag{14}$$

F1 value:

$$F = \frac{2 * P * R}{P + R} \tag{15}$$

4.4 Experiment Design and Analysis

A MovieLens 10 million user rating data is randomly divided into 80% training set data and 20% test set data. According to the video recommendation algorithm based on knowledge inference of knowledge graph proposed in this paper, each group of experiments was performed 5 times and the average was taken as the final result.

1. K films were randomly selected from 1682 films to find the influence of film numbers to F1.

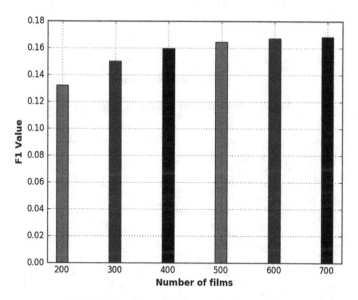

Fig. 5. F1 values at different movie numbers

As can be seen from Fig. 5, with the increase of the film numbers, the comprehensive evaluation value F1 is also increasing, so the algorithm can be applied to the environment with more items.

2. Select 200 dimensions as the embedded dimension of knowledge reasoning, take 40, 60, 80, 100, 120 as the neighbors number, and compared with several typical

algorithms in the collaborative filtering recommendation algorithm, each group of experiments was performed 5 times and finally averaged.

From Figs. 6, 7 and 8, it can be seen that with the change the neighbors number, the accuracy rate, recall rate and F1 value of various algorithms fluctuate in different degrees. In general, the algorithm of this paper has achieved a more ideal result. It can be concluded that the algorithm has certain advantages over other collaborative filtering recommendation algorithms in terms of accuracy, recall rate, and F1 value.

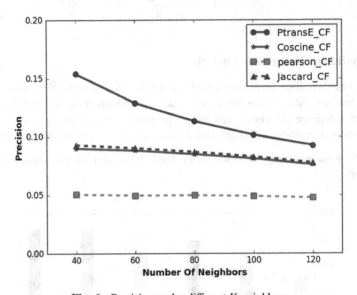

Fig. 6. Precision under different K neighbors

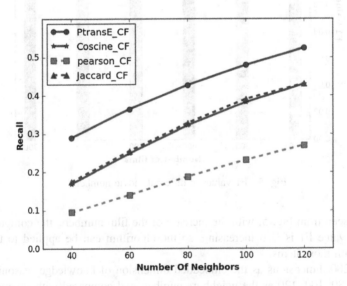

Fig. 7. Recall rates under different K neighbors

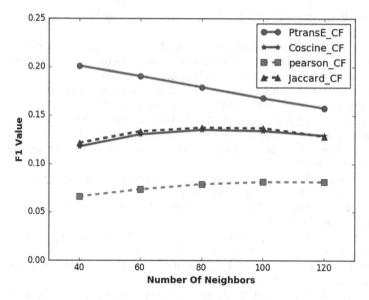

Fig. 8. F1 values under different K neighbors

5   Conclusion

This paper proposes a video recommendation algorithm based on knowledge inference of knowledge graph. It makes full use of the potential information of items, mines the hidden information of objects through knowledge reasoning algorithms, establishes multi-path relationships between entities, and can represent the attributes of items and the similarities between items more accurately. The algorithm not only increases the recommendation effect at the semantic level, but also has scalability, and solves the cold start problem of the collaborative filtering recommendation algorithm to some extent. Experiment results show that the algorithm has a certain improvement over other collaborative filtering recommendation algorithms. In the future research, we will try to apply the algorithm to cross-domain recommendations. At the same time, for a larger number of video recommendations, the efficiency of the algorithm needs to be improved.

Acknowledgements. This work of the paper is supported by National Natural Science Foundation of China (No.61702157), Science and Technology Support Program of Hebei Province of China (No.15210506), and Natural Science Foundation of Tianjin (No.16JCQNJC00400, No. 16JCYBJC15600).

References

1. Weng, X.L., Wang, Z.J.: Research progress of collaborative filtering recommendation algorithm. Comput. Eng. Appl. **54**(01), 25–31 (2018)

2. Ren, L.: Research on Key Technologies of Recommending System. East China Normal University (2012)
3. Zhou, X.T.: Internet commodity information classification and recommendation system based on knowledge base. In: The 29th China Database Conference Proceedings, DNBC 2012:4. China Computer Federation, Hefei, Anhui, China (2012)
4. Lu, W.J.: Design and implementation of gift recommendation system based on knowledge base. Beijing University of Posts and Telecommunications (2011)
5. Huang, Z.H., Zhang, J.W., Zhang, B., Yu, J., Yang, X., Huang, D.S.: A review of semantic recommendation algorithm research. Chin. J. Electron. **44**(09), 2262–2275 (2016)
6. Xing, X.: Research on personalized recommendation method of social network. Dalian Maritime University (2013)
7. Xu, Z.L., Sheng, Y.P., He, L.R., Wang, Y.F.: An overview of knowledge graph technology. J. Univ. Electron. Sci. Technol. China **45**(04), 589–606 (2016)
8. Sun, G.H., Liu, D.Q., Li, M.Y.: Summary of personalized recommendation algorithms. J. Sofeware **38**(07), 70–78 (2017)
9. Guan, S.P., K, X.L., Jia, Y.T., Wang, Y.Z., Cheng, X.Q.: Research progress on knowledge inference for knowledge graph. J. Softw. 1–29 (2018)
10. Liu, H., Li, Y., Duan, H., Liu, Y., Qin, Z.G.: An overview of knowledge graph construction technology. J. Comput. Res. Dev. **53**(03), 582–600 (2016)
11. Liu, Z., Sun M., Lin, Y., Xie, R.: Research progress of knowledge representation learning. J. Comput. Res. Dev. **53**(02), 247–261(2016)
12. Wu, Y., Chen, Q.M, Liu, H., He, C.B.: Collaborative filtering recommendation algorithm based on knowledge graph representation learning. Comput. Eng. **44**(02):226–232 + 263 (2018)
13. Liang, Y., Yu, H., Liu, Y., Liu, L., Yao, J.M.: Study on domain adaptation of translation model based on similarity of semantic distribution. J. Shandong Univ. (Sci. Ed.) **51**(07), 43–50(2016)
14. Liu, Y., Han, M.B., Jiang, L.Y., Liu, Y., Yu, J.: Relationship inference algorithm based on double-random random walk. Chin. J. Comput. **40**(06), 1275–1290 (2017)
15. CNKI Homepage: http://kns.cnki.net/kcms/detail/51.1196.TP.20180208.1714.068.html. Accessed 13 May 2018
16. He, R.N., Wang, C.K., Mcauley, J. L.: Translation-based recommendation. In: ACM RecSys (2017)
17. Liu, G.J.: Research on recommendation algorithm based on popular prediction. University of Electronic Science and Technology (2016)
18. Yin, L.T., Yu, Y., Lu, L., Ying, C.T., Guo, G.: Video recommendation algorithm based on fusion comment analysis and implicit semantic model. J. Comput. Appl. **35**(11), 3247–3251 (2015)
19. Zhang, Q.G., Zhang, X.X., Li, Z.Q.: Music depth recommendation algorithm based on attention mechanism. Comput. Appl. Res. **2019**(08), 1–6 (2018)

A Hybrid Framework for Query Processing and Data Analytics on Spark

Haokun Chen[1,3], Xiaowang Zhang[1,3(✉)], Jiahui Zhang[1,3], and Zhiyong Feng[2,3]

[1] School of Computer Science and Technology, Tianjin University,
Tianjin 300350, People's Republic of China
xiaowangzhang@tju.edu.cn
[2] School of Computer Software, Tianjin University,
Tianjin 300350, People's Republic of China
[3] Tianjin Key Laboratory of Cognitive Computing and Application,
Tianjin 300350, People's Republic of China

Abstract. In this paper, we propose a hybrid framework for query processing and data analytics over large-scale data on Spark, to support multi-paradigm process (incl. SQL, OLAP, data mining, machine learning etc.) in distributed environments. The framework features a three-layer data process module and a work flow module which controls the former. We will demonstrate the strength of our framework properly applying traffic scenarios in a real world.

Keywords: Query processing · Data analytics · OLAP · Work flow
Spark

1 Introduction

Adaptive data processing, as an advanced automatic data processing, can select models and parameters by a system itself when processing variable data from applications [6]. The core problem of adaptive data processing is to design a "smart" mechanism in generating dynamically optimal workflows for variable requirements. There are some existing approaches to generate workflows, which are mostly based on manual configurations, such as Apache Oozie [16]. They are often taxing and poor in reusability due to the limitation of single developer. It becomes interesting in generating workflows to support adaptive data processing.

Query processing and data analysis have become two useful techniques to organize, process, and analyze large amounts of data in order to obtain useful results or knowledge effectively such as hidden patterns, implicit correlations, future trends, customer preferences, valuable business information etc [2].

There are many techniques for query processing and data analysis. For instance, SQL represents queries over relational databases (e.g., MySQL) and OLAP (*online analytical processing*) [1] provides online analytical processing (e.g., Oracle OLAP [17,18] and IBM DB2 OLAP [4,19]). Taking advantage of big data processing and big data analysis, we could treat many complicated

© Springer Nature Switzerland AG 2018
L. H. U and H. Xie (Eds.): APWeb-WAIM 2018, LNCS 11268, pp. 159–173, 2018.
https://doi.org/10.1007/978-3-030-01298-4_15

tasks and then obtain more valuable information/knowledge [20] by applying those techniques together. For instance, OLAP provides rapid access to data (mostly relational data) for analysis to gain useful knowledge from data via efficient query processing tools (e.g., SQL). It is necessary to arrange workflows for organizing and scheduling those techniques of query processing and data analytics in handling more complicated tasks such as personalized recommendation [10]. For instance, we will recommend a possible place for a Taxi driver so that he/she might pick up some passenger. To this end, we require a workflow to schedule tools of querying the present time and location of the driver over the historical orders and then employ some clustering techniques to obtain *hot spots*. Finally, we apply some machine learning techniques to rank those *hot spots* close to the driver and recommend to him/her. The integrated techniques in a workflow is called *multi-paradigm* [15] *technique*, as a structured methodology for the systematic design of enterprise processes [12]. On the other hand, Apache Spark [13], built on HDFS of Hadoop, provides a high-performance computing architecture for big data with supporting query processing and big data analytics. However, as we investigated, there is few multi-paradigm technique for query processing and big data analysis. Moreover, there is no related tool automating these tasks. To complete the automation of the work flow is the goal in our future work. In order to achieve the goal of adaptive data processing, we must solve the workflow arrangement firstly, that is, the transfer between data query and data analysis. Not only we can analyze the data that is queried, we can query and analyze the analyzed results.

In this paper, we propose a hybrid framework for big data analysis on Spark. The framework features a three-layer data process module and a business process module which controls the former. Within this framework, we can support multi-paradigm data process in order to handle many complicated tasks (incl. SQL, OLAP, data mining, machine learning etc.), which are interoperated to process the analysis of various applications of big data (incl. data cube [3], intelligent prediction, and complex network etc.) respectively. Moreover, our proposed framework built on Spark can process large-scale data efficiently. Multi-paradigm data process ensures that our framework is in general, and workflow scheduling ensures the accuracy of these tasks.

2 Multi-paradigm Architecture

In this section, we present a multi-paradigm architecture of our framework in Fig. 1. This architecture consistes of four parts, namely, *storage management*, *resource scheduling*, *query processing* and *work flow*. In the following sections, we will introduce each part in detail.

2.1 Storage Management

The storage management in Fig. 2 contains two parts, namely, *physical storage* and *logical storage*. The rapid growth of data makes the physical storage of data

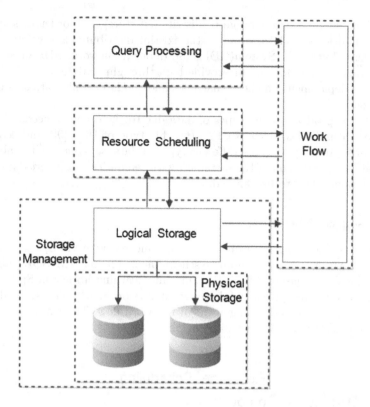

Fig. 1. A multi-paradigm architecture

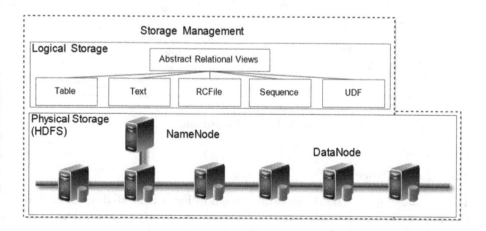

Fig. 2. Storage management.

change from single source storage to distributed storage. In order to solve the storage of multi-source data, we adopt the existing distributed file system such as Hadoop Distributed File System(HDFS)[8]. In our framework, HDFS is a highly fault-tolerant system and it can provide high throughput data access, which is suitable for deployment on cheap machines and the large-scale data sets on the application.

Besides, it products many types of data due to the different needs of applications, such as tables, texts, RCFile (the file type of Hive [9]) and sequence data. In order to support those different types of data, we compose the abstract relational view by designing the metadata with semantics to convert data types to the relational data to be handled.

2.2   Resource Scheduling

The resource scheduling is to manage computing resource via Spark shown in Fig. 3. In this framework, the resource scheduling can manage MySQL (querying over relational databases), MLlib [13] (machine learning library in Spark), and GraphX [7] (querying graph databases in Spark) on Spark. The scheduling of resources is based on the arrangement of the work flow. The work flow decide when and which the resources will be invoke.

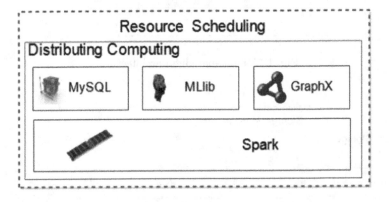

Fig. 3. Resource scheduling.

2.3   Query Processing

The module of query (task) processing shown in Fig. 4 is located on the top of the framework. This module consists of two layers, namely, *querying* and *DAP (Data Analysis Process)* tools , where the first layer presents queries by applying SQL or user defined functions and the second layer contains data analysis tools such as OLAP, DAP on machine learning and DAP on graph.

Our big data analysis and processing of the query language is based on the improvement of the fusion of SQL and HiveQL in multi-paradigm. First of all, we analyze the support of HiveQL and SQL respectively and count the amount of operations which can be supported by the traditional relational algebra model. On the basis of the relational algebra model, we add other necessary operators to construct an extension of the algebraic language model, which can fully support the operation of HiveQL and standard SQL.

For the operator with higher complexity, it is split into smaller sub operator or used other methods to optimize it. For the machine learning analysis, we count the commonly used analytical processing methods, such as classification and clustering, and define the abstract interfaces for the common machine learning analysis processing methods. For the graph analysis processing, we also count the commonly used analytical processing methods, such as the shortest path algorithm, and define the abstract interfaces for them.

In this module, the framework also relates to the implementation of the OLAP on the relational database and the machine learning and graph data processing tasks on the distributed framework. The traditional relational database query optimization method is no longer applicable to this situation. According to the different characteristics of relational storage management query engine and distributed file system of computing engine, we summarize the query information and optimize the performance. Firstly, we investigate the statistical index system used in traditional database and analyze the interaction between each index and the index in the system. Then, for each index in the index system of statistical information, we design efficient and accurate sampling methods to calculate the cost model in query optimization. According to the above statistics, we can also design a storage and maintenance programs which is easy to update and manage. And we may use the cost model in the traditional relational database to design a new cost model which can reflect the query cost of the mixed data.

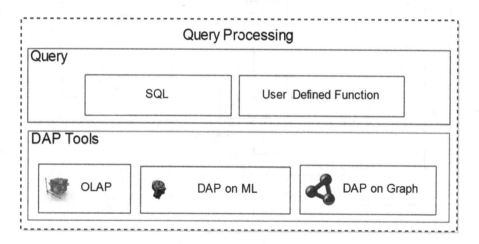

Fig. 4. Query processing.

2.4 Work Flow

The module of work flow is the most important to realize multi-paradigm process of queries (tasks). In this module, we need to do two things: developing a multi paradigm fusion analysis process orchestration language syntax and the complex business process scheduling method, which is also called work flow.

In the first part, we need to analyze the patterns and characteristics of service orchestration language in service oriented architecture design and design an

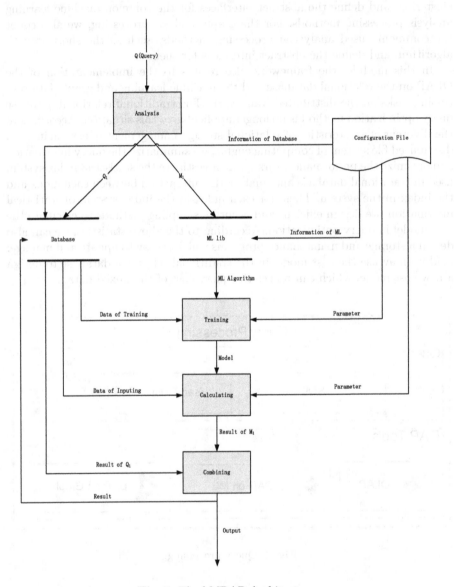

Fig. 5. The hMDAP Architecture

abstract model of the executable process. On the basis of the abstract model, we summarize the basic activities of complex business process analysis. Finally, we define the grammar of the business process. In the semantic, we need to research and analysis the meanings of basic business activities and define the start point, end point and the basic command.

In the second part, we need to study and analyze complex business processes in practical applications. Then, we build complex business process models and refine the way to exchange messages in public business processes. After that, we need to control the interaction of each part of the resources through the interaction sequence of messages, achieving a reasonable call for each resource service. We still need to investigate the applicability of existing object-oriented design patterns. For the analysis of complex business process integration model, we design data business processes. We refine the design patterns in complex business processes based on the advantages and principles of existing design patterns.

In the real world, the business process model is complex and it takes a lot of time to analyze. The Fig. 5 illustrates the details of the business process in our framework. In Fig. 5, when we get a query task, it will invoke the machine learning algorithms to mining the data which is acquired by quering the database. The results obtained by the maching learning process will be stored in the database, that will be utilized to query until getting the precise results. This module includes the following procedures:

- To design a united language (multi-paradigm process) for representing queries (e.g., SQL or user defined functions);
- To develop a scheduling method to manage tools of query processing and data analysis;
- To configure tools of query processing and data analysis.

3   Applications in Traffic

In this section, we present three applications of our framework in traffic. In the three applications, we employ MySQL as storage and SparkSQL and MLlib of Spark as the distributed computing framework. In the following subsections, we mainly design three work flows for the three applications in traffic respectively.

3.1   Order Analysis

The first query Q_O is to ask *how orders of express cars change in a month*. To realize Q_O, we design a work flow as follows:

1. representing this query Q_O via StandardSQL;
2. transforming OLAP to StandardSQL by employing Mondrian, which is an OLAP server with supporting the MDX [14] (multidimensional expressions) query language [20];
3. configuring and generating Cube by employing XML for further processing via MDX.

3.2 Taxi Driver

The second query Q_T is to ask *where is the best place to pick up passengers*. To realize Q_T, we design a work flow as follows:

1. representing this query Q_T via XML;
2. configuring XML;
3. designing a scheduling method in the following way:
   (a) querying by applying SparkSQL;
   (b) clustering via MLlib;
   (c) recommending by employing *collaborative filtering* [11] via MLlib.

3.3 Ride Driver

The third query O_R is to ask *where is the best place to pick up other passengers with a close destination*. To realize Q_R, we design a work flow as follows:

1. representing this query Q_R via XML;
2. configuring XML refer to Spark Streaming [13];
3. desiging a scheduling method in the following way:
   (a) querying by applying SparkSQL;
   (b) clustering via MLlib;
   (c) recommending by employing *collaborative filtering* via MLlib.

Note that the work flow of Q_R is slightly different from the work flow of Q_T in the configuring XML, where the work flow of Q_R requires a dynamic configuring via Spark Streaming while the work flow of Q_T is static in configuring.

4 Experiments and Evaluations

All experiments are carried out on a 4-site cluster connected by a gigabit Ethernet switch. Each node has one CPU with 6 cores of 2.2 GHz, 64 GB memory, and 1.2 TB disk. The surrounding of the cluster is as follows: Ubuntu 14.04.5 LTS, jdk 1.8.0.91, Hadoop 2.6.0, and Spark 2.0.0. The dateset in our experiments consists of 800,000 orders of July 2015 in Beijing provided by CAR Inc. (We state that the copyright of those datasets is completely owned by CAR Inc.)[1].

In the following, we will evaluate three applications stated in Sect. 3.

4.1 Order Analysis

OLAP will be used in our workflow. We can analyze orders from different dimensions in this way. The *data cube* we construct is shown in Fig. 6.

Q_O is to ask how orders of express cars change in a month. Q_O contains two parameters, namely, region and month. Given two regions of Beijing (i.e., South of Beijing and North of Beijing) and July, we demonstrate two cases of O_T per day and per week shown in Figs. 7 and 8, respectively.

By comparing Fig. 7(a) with Fig. 7(b) and Fig. 8(a) with Fig. 8(b), we find that South of Beijing is slightly different from North of Beijing in volume, where both regions of Beijing have the maximal orders in the middle of July.

---

[1] https://en.zuche.com/.

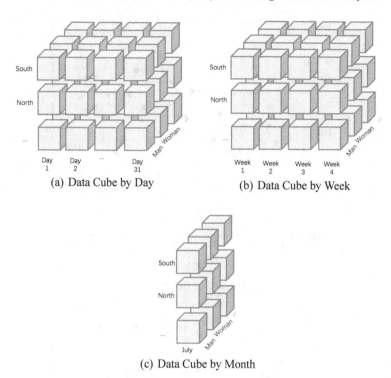

(a) Data Cube by Day

(b) Data Cube by Week

(c) Data Cube by Month

Fig. 6. The Cube Data of OLAP

4.2   Taxi Driver

The query O_T is to ask which is the best place to pick up passengers. Q_T consists of three parameters, namely, DriverId, date, and location. Given a DriverId 1019 and a fixed start location (home) [40.000, 130.000], we demonstrate three cases of O_T with three different dates 7:00:00, 12:00:00, and 18:00:00 in August 1, 2015 and top 3 positions recommended and their evaluation shown in Table 1) where the query execution time is in minute. Moreover, the recommendation is based on three distances (1 km, 3 km, 5 km), three spans of time (60 min, 30 min, 15 min), and two classifications of a week (weekend-nonweekend, per day) in total 18 cases.

We will get the corresponding hotspots through clustering in each case. If the position is a hot spot we make the corresponding weight is 1, otherwise the corresponding weight is empty, so that we get the weight under the condition that the weight is empty through the collaborative filtering algorithm. So that the highest weight hot spot is recommended to the user.

In Table 1, we can obtain the best place to pick up passengers for a given date. For instance, at 7:00:00, we recommend that the driver goes to the position labeled P56 [116.317847, 39.739821] based on analysis of historical orders at the same date.

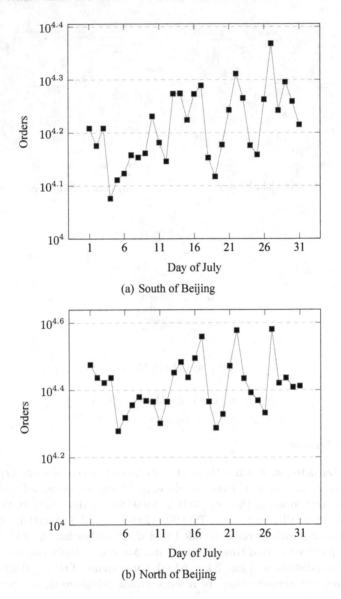

Fig. 7. Changes in the number of orders per day

4.3 Ride Driver

The query O_R is to ask which is the best place to pick up other passengers with a close destination. Note that the difference between O_R and O_T is that the position of the rider changes in real time. Every 5 min we receive a real-time position, that is, a hotspot recommendation every 5 min. Q_R also consists of

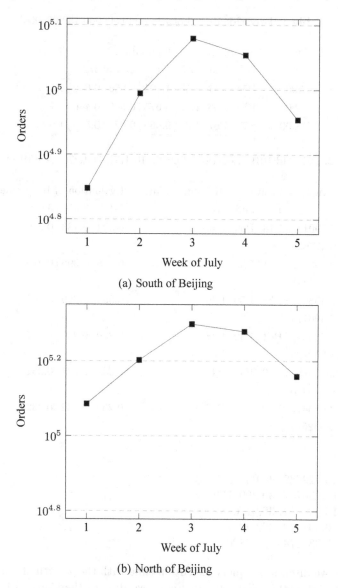

(a) South of Beijing

(b) North of Beijing

Fig. 8. Changes in the number of orders per week

three parameters, namely, DriverId, date, and location. Given a DriverId 1019, a start location [39.979000,116.477000], and a start date 2015-08-01 06:38:28, we obtain results of O_R shown in Table 2.

In Table 2, we can obtain five best positions to pick up other passengers with a close destination per 5 min as follows:

Table 1. DriverId: 1019, Location: [40.000000,130.000000]

Date	Positions			Evaluation			Time
	1st	2nd	3rd	1st	2nd	3rd	
7:00:00	P56	P11	P55	1.101	0.656	0.682	1.5
12:00:00	P55	P56	P11	0.675	0.675	0.486	1.7
18:00:00	P7	P67	P2	0.669	0.620	0.520	1.9

Table 2. DriverId:1019, Location:[39.979,116.477], Date:2015-08-01 06:38:28

Date	Position	Position of Recommendation			Evaluation of Recommendation		
		1st	2nd	3rd	1st	2nd	3rd
06:38	39.97901, 116.477633	P18	P50	P49	0.604	0.433	0.289
06:43	39.959325, 116.496318	P11	P50	P49	0.448	0.390	0.273
06:48	39.945829, 116.496174	P58	P50	P49	0.369	0.369	0.263
06:53	39.943506, 116.512703	P41	P49	P48	0.339	0.247	0.247
06:58	39.923588, 116.525495	P49	P48	P47	0.231	0.231	0.231
07:03	39.905547, 116.552516	P49	P48	P47	0.232	0.232	0.232

– P18: [116.422682, 40.051593];
– P11: [116.436737, 39.899177];
– P58: [116.590594, 39.911063];
– P41: [116.361342, 39.922777];
– P49: [116.334574, 39.748205].

Finally, we discuss the query execution time with the growth of dataset scale (range from 200,000 to 800,000). In Fig. 9, we discuss the change of the query execution time of DriverId 1019 and Location: [40.000000,130.000000] with the three different date (7:00:00, 12:00:00, and 18:00:00) with the growth of dataset scale.

In Fig. 10, we discuss two more drivers as follows: (1) DriverId: 2003, date: 2015-08-01 12:34:02, location: [39.885065,116.184283] and (2) DriverID 3247, date: 2015-08-01 19:28:10, location: [39.914624,116.403613]. As a result, we find that the query execution time is highly efficient (close to a linear time).

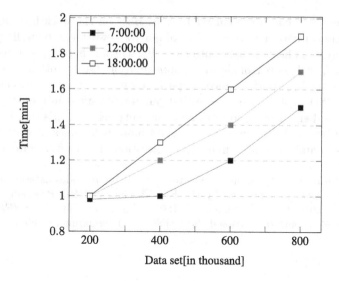

Fig. 9. DriverId: 1019

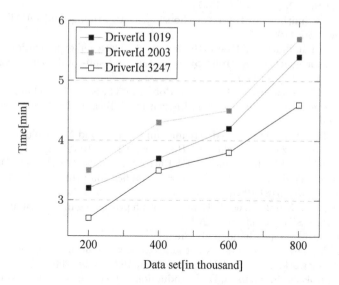

Fig. 10. DriverId:1019, 2003, 3247

5   Conclusion

In this paper, we proposed a hybrid framework for query processing and large-scale data analytical processing with supporting multi-paradigm process on Spark. We employ three applications under the framework. In the special cases, we can make special customization by the configuration file. For example, change the cube of dimension in OLAP query, change the model of cluster, which parameters are best suitable in the cluster models. Base on the work flow process we design, the multi-paradigm mechanism of our framework can be well applied to process many complicated tasks automatically, which is hardly processed by single technique, such as personalized recommendation. On the other hand, taking advantage of the high-performance of Spark, our proposal can process large-scale data in an efficient way. We believe that our framework is interesting to those researchers and engineers who work on processing big data in a multi-paradigm way. In the future work, we will investigate more techniques of work flows for many tasks in applications to make the generation of work flow automately.

Acknowledgements. We would like to thank CAR Inc provides datasets for science research. This work is supported by the Key Technology R&D Program of Tianjin (16YFZCGX00210), the the National Key R&D Program of China (2016YFB1000603, 2017YFC0908401), and the National Natural Science Foundation of China (61672377).

References

1. Chaudhuri, S., Dayal, U.: An overview of data warehousing and OLAP technology. SIGMOD Rec. **26**(1), 65–74 (1997)
2. Berson, A., Smith, S.J.: Data Warehousing, Data Mining, and OLAP. McGraw-Hill, New York (1997)
3. Gray, J., Chaudhuri, S., Bosworth, A., et al.: Data cube: a relational aggregation operator generalizing group-by, cross-tab, and sub-totals. Data Min. Knowl. Discov. **1**(1), 29–53 (1997)
4. Baragoin, C., Bercianos, J., Komel, J., Robinson, G., Sawa, R., Schuinder, E.: DB2 OLAP server theory and practices. International Technical Support Organization (2001)
5. Zaharia, M., et al.: Discretized streams: an efficient and fault-tolerant model for stream processing on large clusters. HotCloud **12**, 10–10 (2012)
6. Fernández-Delgado, M., Cernadas, E., Barro, S., Gomes Amorim, D.: Do we need hundreds of classifiers to solve real world classification problems. J. Mach. Learn. Res. **15**(1), 3133–3181 (2014)
7. Gonzalez, J.E., Xin, R.S., et al.: GraphX: graph processing in a distributed dataflow framework. OSDI **14**, 599–613 (2014)
8. Hadoop (2015). http://hadoop.apache.org/
9. Thusoo, A., Sarma, J.S., Jain, N., et al.: Hive: a warehousing solution over a mapreduce framework. Proc. VLDB Endow. **2**(2), 1626–1629 (2009)
10. Ricci, F., Rokach, L., Shapira, B.: Introduction to recommender systems handbook. In: Ricci, F., Rokach, L., Shapira, B., Kantor, P.B. (eds.) Recommender Systems Handbook, pp. 1–35. Springer, Boston, MA (2011). https://doi.org/10.1007/978-0-387-85820-3_1
11. Zhou, Y., Wilkinson, D., Schreiber, R., Pan, R.: Large-scale parallel collaborative filtering for the Netflix prize. In: Fleischer, R., Xu, J. (eds.) AAIM 2008. LNCS, vol. 5034, pp. 337–348. Springer, Heidelberg (2008). https://doi.org/10.1007/978-3-540-68880-8_32
12. zur Muehlen, M., Rosemann, M.: Multi-paradigm process management. In: Proceedings of CAISE 2004, pp. 169–175 (2004)
13. Meng, X., Bradley, J.K., Yavuz, B., et al.: Mllib: machine learning in apache spark. J. Mach. Learn. Res. **17**, 1–7 (2016)

14. Spofford, G.: MDX Solutions: With Microsoft SQL Server Analysis Services. Wiley, New York (2001)
15. Sheth, A.P., et al.: Supporting state-wide immunisation tracking using multi-paradigm workflow technology. In: Proceedings of VLDB 1996, pp. 263–273 (1996)
16. Oozie: Apache workflow scheduler for Hadoop. The Apache Software Foundation (September, 2010). http://oozie.apache.org/
17. Dodge, G., Gorman, T.: Oracle Data Warehousing. Wiley, New York (1998)
18. Schrader, M., Vlamis, D.: Oracle Essbase & Oracle OLAP. Peter Gbolagade Akintunde (2009)
19. Bontempo, C., Zagelow, G.: The IBM data warehouse architecture. Commun. ACM 41(9), 38–48 (1998)
20. Rouse, W.: What is big data analytics? TechTarget.com (2012). http://searchbusinessanalytics.techtarget.com/definition/big-data-analytics

Improving Network-Based Top-N Recommendation with Background Knowledge from Linked Open Data

Zhuoming Xu[✉], Chengwang Mao, Xiuli Wang, Wei Xu,
and Lifeng Ye

College of Computer and Information, Hohai University, Nanjing 210098, China
{zmxu, cwmao, xlwang, xwmr, lfye}@hhu.edu.cn

Abstract. The boom in Linked Open Data (LOD) has recently stimulated the research of a new generation of recommender systems—LOD-enabled recommender systems. ROUND (Random walk with restart on an Object-User Network towards personalized recommenDations) is a state-of-the-art method for network-based top-N recommendation. However, the ROUND method relies solely on the historical data (i.e., the ratings matrix) and does not take full advantage of background knowledge from LOD. This paper addresses the problem of improving network-based top-N recommendation using background knowledge from LOD by proposing an improved ROUND method called ROUND-APICSS. The core idea of ROUND-APICSS is that we exploit a knowledge graph constructed from LOD to calculate semantic similarities between the objects (items) involved in the recommender system, thereby improving the object-user heterogeneous network model and the random walk with restart model on the network. Our experimental results on real datasets suggest that the incorporation of background knowledge from LOD into the network-based top-N recommendation models can improve recommendation accuracy. The results also show the superiority of our ROUND-APICSS method over the ROUND method in terms of recommendation accuracy.

Keywords: Top-N recommendation
Linked open data enabled recommender systems · Knowledge graph
Object-user heterogeneous network · Random walk

1 Introduction

As personalized information service tools, recommender systems play an increasingly critical role in alleviating the information/choice overload that people face today. Advances in Web technology have been becoming the main driving force for the development of recommender systems technology. Recently, massive structured data have been published on the Web as freely accessible Linked Open Data (LOD) [1]. At the same time, various Web knowledge graphs have been built by extracting knowledge from the LOD cloud or large-scale knowledge bases [2]. For example, DBpedia [3], as a central interlinking hub for the LOD cloud, is a huge RDF dataset [4] extracted from Wikipedia; it is also a knowledge base (graph) describing millions of entities

© Springer Nature Switzerland AG 2018
L. H. U and H. Xie (Eds.): APWeb-WAIM 2018, LNCS 11268, pp. 174–187, 2018.
https://doi.org/10.1007/978-3-030-01298-4_16

(Web resources) that are classified in a consistent ontology. The boom in LOD has recently stimulated the research of a new generation of recommender systems—LOD-enabled recommender systems [5, 6]. The ultimate goal of an LOD-enabled recommender system, such as the sound and music recommender proposed by [7], is to provide the system with background knowledge about the domain of interest in the form of an application-specific knowledge graph.

To overcome the weaknesses of existing network-based recommendation methods, Gan and Jiang [8] proposed a novel method named ROUND (Random walk with restart on an Object-User Network towards personalized recommenDations). Extensive evaluations on historical data were performed over two different datasets (MovieLens and Netflix), and the results showed ROUND's superiority over such state-of-the-art approaches as non-negative matrix factorization (NMF) and singular value decomposition (SVD) in terms of not only recommendation accuracy and diversity but also retrieval performance [8]. However, according to our observations, the ROUND method relies solely on the historical data (i.e., the ratings matrix) and does not take full advantage of the background knowledge provided by LOD datasets or Web knowledge bases. One critical question that how to improve a network-based top-N recommendation method like ROUND with background knowledge from LOD, is still not systematically explored in existing studies, to the best of our knowledge.

In this work, we address the problem of improving network-based top-N recommendation using background knowledge from LOD by proposing an improved ROUND method. The core ideas of our improvement to the ROUND method includes: (i) calculating semantic similarities between the objects (items) involved in the recommender system by exploiting a knowledge graph constructed from LOD and using an LOD-based semantic similarity measure—our previously proposed APICSS measure [9]; (ii) using the calculated semantic similarities to improve the object-user heterogeneous network model and the random walk with restart model on the network in the recommendation method. We abbreviate our method as ROUND-APICSS. We have conducted experimental evaluation to validate the effectiveness of our proposed ROUND-APICSS method and to demonstrate its performance strength of top-N recommendation compared to the original ROUND method.

The remainder of this paper is organized as follows. In Sect. 2, we elaborate on network-based top-N recommendation methods including the ROUND method and our proposed ROUND-APICSS method. Section 3 presents our experimental evaluation results and discussions. Finally, we conclude our work in Sect. 4.

2 Network-Based Top-N Recommendation Methods

2.1 The ROUND Method

The ROUND method [8] is a state-of-the-art network-based top-N recommendation method which consists of two fundamental models: (i) the object-user heterogeneous network model that integrates relationships among objects[1], relationships among users,

---

[1] The term "object" in the ROUND method refers to the "item" in a recommender system.

and relationships between objects and users; *(ii)* the random walk with restart model on the heterogeneous network, as illustrated in Fig. 1.

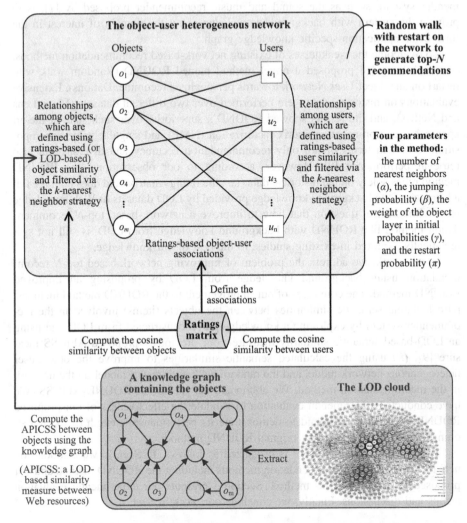

Fig. 1. The models of the ROUND method and our improved method.

Construction of the Object-User Heterogeneous Network. The object-user hetero-geneous network [8], as shown in Fig. 1, is composed of an *object layer* that contains all the objects involved in the recommender system and the relationships among the objects, a *user layer* that contains all the involved users and the relationships among the users, and an *interconnection layer* that embodies the associations between objects and users. In Gan and Jiang's work [8], the object-user heterogeneous network was pre-sented by a 5-tuple $\mathbf{H} = (\Omega, \Psi, \mathbf{A}, \mathbf{O}, \mathbf{U})$, where $\Omega = \{o_1, o_2, \ldots, o_m\}$ is the set of all objects, $\Psi = \{u_1, u_2, \ldots, u_n\}$ the set of all users, $\mathbf{A} = (a_{ij})_{m \times n}$ the adjacency matrix of the interconnection layer, $\mathbf{O} = (o_{ij})_{m \times m}$ the weight matrix of the object layer, and $\mathbf{U} = (u_{ij})_{n \times n}$ the weight matrix of the user layer.

As described in [8], $\mathbf{A} = (a_{ij})_{m \times n}$ is a binary matrix, where $a_{ij} = 1$ if object i is preferred by user j and 0 otherwise, and it is defined by re-assigning 1 to ratings no less than 3.0 and 0 to all other cases. The weight matrix $\mathbf{O} = (o_{ij})_{m \times m}$ is defined using ratings-based object similarities and filtered via the k-nearest neighbor strategy as described later. The object similarity matrix $\mathbf{S} = (s_{ij})_{m \times m}$ is calculated using the cosine measure as defined by Eq. (1) [10].

$$s_{ij} = \text{Cosine}(o_i, o_j) = \frac{\sum_{k \in U(o_i) \cap U(o_j)} a_{ik} a_{jk}}{\sqrt{\sum_{k \in U(o_i)} a_{ik}^2} \sqrt{\sum_{k \in U(o_j)} a_{jk}^2}} \tag{1}$$

where $U(o_i)$ and $U(o_j)$ represent the set of users who prefer object i and object j, respectively. For each object, a number of the weakest relationships are removed by sorting each column of the object similarity matrix $\mathbf{S}$ and resetting similarities to 0 for objects ranked lower than parameter α. The filtered object similarity matrix is denoted as $\mathbf{O} = (o_{ij})_{m \times m}$, which represents a nearest neighbor network of objects. After the filtration, $o_{ji} = o_{ij}$ is set for any pair (i, j) where $o_{ij} \neq 0$ but $o_{ji} = 0$, in order to ensure $\mathbf{O}$ is symmetric. The weight matrix of the user layer $\mathbf{U} = (u_{ij})_{n \times n}$, which represents a nearest neighbor network of users, is constructed in the same way as that of the object layer.

Random Walk with Restart on the Heterogeneous Network. The basic ideas of the random walk with restart model on the object-user heterogeneous network are as follows. As described in [8] and shown in Fig. 1, given a query user (i.e., the active user), the random walker starts a journey on the object-user heterogeneous network with some *initial probability*, with γ being the weight of the object layer in the initial probability and $1 - \gamma$ that of the user layer in the initial probability. In each step of the journey, the walker may choose to *restart* a new journey with probability π or proceed in the current journey with probability $1 - \pi$. When proceeding, the walker may choose to *jump* between the object layer and the user layer with probability β, or wander in either the object layer or the user layer with probability $1 - \beta$ by moving to one of its direct neighbors. The probability that the walker stays in each node of the object-user heterogeneous network will finally reach a *steady-state* after many steps. The steady-state probabilities indicate a measure of the strength of associations between the query user and candidate objects, which can then be used to generate top-N recommendations for the user.

Formally, given an object-user heterogeneous network $\mathbf{H} = (\Omega, \Psi, \mathbf{A}, \mathbf{O}, \mathbf{U})$, the transition matrix $\hat{\mathbf{A}} = (\hat{a}_{ij})_{m \times n}$ is constructed by applying row-normalization to adjacency matrix $\mathbf{A} = (a_{ij})_{m \times n}$, where $\hat{a}_{ij}$ represents the transition probability from object i to user j and is computed using Eq. (2) [8].

$$\hat{a}_{ij} = \begin{cases} a_{ij} / \sum_{j=1}^{n} a_{ij}, & \text{if } \sum_{j=1}^{n} a_{ij} \neq 0 \\ 0, & \text{otherwise} \end{cases} \tag{2}$$

Three other transition matrices are constructed in a similar way [8]: matrix $\hat{\mathbf{O}} = (\hat{o}_{ij})_{m \times m}$ with $\hat{o}_{ij} = o_{ij}/\sum_{j=1}^{m} o_{ij}$ ($\hat{o}_{ij} = 0$ if $\sum_{j=1}^{m} o_{ij} = 0$) representing the transition probability from object i to object j, matrix $\hat{\mathbf{U}} = (\hat{u}_{ij})_{n \times n}$ with $\hat{u}_{ij} = u_{ij}/\sum_{j=1}^{n} u_{ij}$ ($\hat{u}_{ij} = 0$ if $\sum_{j=1}^{n} u_{ij} = 0$) representing the transition probability from user i to user j, and matrix $\bar{\mathbf{A}}^{T} = (\bar{a}_{ij}^{T})_{n \times m}$ with $\bar{a}_{ij}^{T} = a_{ji}/\sum_{j=1}^{m} a_{ji}$ ($\bar{a}_{ij}^{T} = 0$ if $\sum_{j=1}^{m} a_{ji} = 0$) representing the transition probability from user i to object j.

Matrix $\mathbf{X}$ is then constructed using the above four transition matrices and parameter β (jumping probability), as defined by Eq. (3) [8].

$$\mathbf{X} = \begin{pmatrix} (1 - \beta)\hat{\mathbf{O}} & \beta\hat{\mathbf{A}} \\ \beta\bar{\mathbf{A}}^{T} & (1 - \beta)\hat{\mathbf{U}} \end{pmatrix} \tag{3}$$

The transition matrix for the network, $\mathbf{W} = (w_{ij})_{(m+n) \times (m+n)}$, is therefore constructed by applying row-normalization to $\mathbf{X}$, where w_{ij} represents the transition probability from node i to node j (Note that i and j can be either objects or users).

The initial probability matrix $\mathbf{P}^{(0)} = (\mathbf{p}_1^{(0)}, \mathbf{p}_2^{(0)}, \ldots, \mathbf{p}_n^{(0)})$ can then be defined by Eq. (4) [8].

$$\mathbf{P}^{(0)} = \begin{pmatrix} \gamma\bar{\mathbf{A}} \\ (1 - \gamma)\bar{\mathbf{U}} \end{pmatrix} \tag{4}$$

where $\bar{\mathbf{A}}$ is the column-normalization of $\mathbf{A}$, and $\bar{\mathbf{U}}$ the column-normalization of $\mathbf{U}$.

Let $\mathbf{P}^{(t)} = \left(p_{ij}^{(t)}\right)_{(m+n) \times n}$ contains probabilities that the walker stays at each node at time t, the iterative formula can be defined by Eq. (5) [8].

$$\mathbf{P}^{(t+1)} = (1 - \pi)\mathbf{W}^{T}\mathbf{P}^{(t)} + \pi\mathbf{P}^{(0)} \tag{5}$$

The iteration is repeated until the steady-state probabilities are obtained. Each user has a steady-state probability for each object (item). The final top-N recommendation process is as follows. First, all objects are ranked according to the steady-state probabilities, and the top-N objects that have not been consumed by the querying user are then generated according to the ranking result.

As for the parameters in the method, as described above and shown in Fig. 1, there are four parameters in the ROUND method [8]: the number of nearest neighbors (α), the jumping probability (β), the weight of the object layer in initial probabilities (γ), and the restart probability (π). Existing studies have shown that the random walk model is parameter-insensitive [11]. Gan and Jiang [8] also demonstrated this characteristic through the parameter robustness experiments, and these parameters were therefore set to reasonable default values: $\alpha = 100$ (for the smaller MovieLens dataset) or 200 (for the larger Netflix dataset), $\beta = 0.5$, $\gamma = 0.5$, and $\pi = 0.8$.

2.2   Our Proposed ROUND-APICSS Method

Our method, called the improved ROUND method with APICSS (abbreviated as ROUND-APICSS), is an improvement on the ROUND method. In ROUND, both the construction of the object-user heterogeneous network and the four transition matrices rely solely on the historical data (i.e., the ratings matrix). For instance, the relationships among objects in the object layer of the network are defined using ratings-based object similarity (the cosine similarity). This method of relying solely on ratings does not take full advantage of the background knowledge about the objects provided by various Linked Open Data (LOD) datasets or Web knowledge bases like DBpedia. Therefore, the core idea of our improvement to the ROUND method is to use background knowledge from the LOD to calculate the similarity between objects, thereby defining the relationships among objects in the object layer of the object-user heterogeneous network. In order to achieve this goal, our ROUND-APICSS method needs to complete the following three core steps in the process of constructing the heterogeneous network, as shown in Fig. 1:

- Constructing a knowledge graph containing the objects involved in the recom-mender system by extracting relevant knowledge from the LOD cloud;
- Calculating semantic similarities between the objects by exploiting the constructed knowledge graph and using an LOD-based semantic similarity measure such as our previously proposed APICSS measure [9];
- Using the calculated semantic similarities to redefine the object similarity matrix and thus improve the object-user heterogeneous network, thereby reconstructing the transition matrix of the object layer and improving the random walk model.

Constructing a Knowledge Graph Containing the Involved Objects. Various Web knowledge graphs have been built in the recent years, and many application-specific knowledge graphs have been built by extracting knowledge from the LOD cloud or large-scale knowledge graphs (such as DBpedia) [2]. In the emerging research field of LOD-enabled recommender systems [6], some work such as [7] has already proposed approaches to providing recommender systems with background knowledge in form of a knowledge graph [5]. Here we can adapt these approaches for constructing the knowledge graph that contains the objects (items) in the recommender system. More specifically, the approach for constructing the knowledge graph consists of two steps:

- *Object linking:* This step approaches the task of linking the objects (items) with the corresponding resources (entities) in the LOD knowledge bases like DBpedia. It takes as input the list of objects (items) in the recommender system and any dataset in the LOD cloud, and returns the mappings between the objects and the resources URIs [12].

- *Subgraph extraction:* This step approaches the task of extracting from the LOD dataset a descriptive and informative subgraph for each object (item), and then merging the extracted subgraphs to obtain a specific knowledge graph. It takes as input the list of resources URIs returned by the object linking step, and returns for each object a set of RDF triples [4] that ontologically describe these objects.

Each RDF triple is in the form of (subject, predicate, object), where the predicate is an object property defined in a formal ontology, whereas both the subjects and the objects are either the objects (items) in the recommender system or other Web resources in the LOD.

The resulting sets of RDF triples can be represented as a knowledge graph containing the objects involved in the recommender system, as shown in the lower left part of Fig. 1 and formally defined in Definition 1.

Definition 1 (Knowledge Graph). A knowledge graph KG is a labeled, directed multi-graph, defined as $<R, P, T>$, such that $R = \{r_1, r_2, \ldots, r_{|R|}\}$ is a set of resources (nodes) with each resource representing either an object in the recommender system or a Web resource in the LOD, $P = \{p_1, p_2, \ldots, p_{|P|}\}$ is a set of properties (directed edges) with each property representing an ontological object property, and $T = \{t_1, t_2, \ldots, t_{|T|}\}$ is a set of RDF triples like $\langle r_1, p_1, r_2 \rangle$, where $p_1 \in P$ is a property that expresses a relation between two resources $r_1, r_2 \in R$.

Calculating Semantic Similarities between Objects Using the Knowledge Graph. Semantic similarities between the objects can then be calculated by exploiting the constructed knowledge graph (Definition 1) and using an LOD-based semantic similarity measure such as the partitioned information content (PIC)-based semantic similarity (PICSS), a *symmetric* similarity measure proposed by Meymandpour and Davis [13] and our improved *asymmetric* similarity measure called the asymmetric PIC-based semantic similarity (APICSS) [9]. The reason for this improvement is that recent studies [14, 15] have shown that asymmetric similarity measures are more effective than symmetric similarity measures in solving recommendation and search problems. Based on several important concepts proposed by Meymandpour and Davis [13], including the features of a resource in LOD, the information content of features in LOD, and the partitioned information content (PIC) of resources in LOD, our APICSS employs our proposed two new concepts: the proportion of common PIC between two resources in the PIC of a resource (abbreviated as CPIC), and the PIC difference between two resources (abbreviated as DPIC), and calculates asymmetric similarity between two resources in LOD. Here we can adapt the APICSS measure for calculating semantic similarity between the objects contained in both the knowledge graph (Definition 1) and the recommender system.

More specifically and formally, resources can be defined in relation to their neighbor resources in the knowledge graph KG, that is, a resource $r \in R$ is defined as a set of its features F_r. According to Shannon's information theory, the information content (IC) of feature $f \in F_r$ is defined as Eqs. (6) and (7) [13]:

$$IC(f) = \log\left(\frac{1}{\pi(f)}\right) = -\log(\pi(f)) \tag{6}$$

$$\pi(f) = \frac{\varphi(f)}{N} \tag{7}$$

where the logarithm in Eq. (6) is usually to the base two, $\pi(f)$ is the probability of feature f, symbol $\varphi(f)$ is the frequency of the feature f, and N is the total number of resources in the knowledge graph.

According to the additivity property (i.e., the multiplication law of probability), the partitioned information content (PIC) of a resource r is defined as the sum of the IC values of its features, as defined by Eq. (8) [13].

$$PIC(r) = \sum_{\forall f_i \in F_r} IC(f_i) = \sum_{\forall f_i \in F_r} -\log\left(\frac{\varphi(f_i)}{N}\right) \tag{8}$$

Based on the concept of PIC, we define two new concepts, CPIC and DPIC, in Definitions 2 and 3 [9], respectively.

Definition 2 (Proportion of Common PIC between Two Resources in the PIC of a Resource). Given two resources $r, s \in R$, with their sets of features F_r and F_s, respectively. $F_r \cap F_s$ is the common features between r and s. The proportion of common PIC between the two resources in the PIC of r and that in the PIC of s are defined as Eqs. (9) and (10), respectively.

$$CPIC(r, s) = \frac{PIC(F_r \cap F_s)}{PIC(F_r)} \tag{9}$$

$$CPIC(s, r) = \frac{PIC(F_s \cap F_r)}{PIC(F_s)} \tag{10}$$

The value of CPIC always lies in the range [0, 1]. CPIC is an asymmetric measure, that is, $CPIC(r, s) \neq CPIC(s, r)$.

Definition 3 (PIC Difference between Two Resources). Given two resources $r, s \in R$, with their sets of features F_r and F_s, respectively. $F_r - F_s$ is the distinctive features of r relative to s, whereas $F_s - F_r$ is the distinctive features of s relative to r. The PIC differences between the two resources are defined as Eq. (11).

$$DPIC(r, s) = DPIC(s, r) = \frac{1}{PIC(F_r - F_s) + PIC(F_s - F_r) + 1} \tag{11}$$

The value of DPIC always lies in the range (0, 1]. DPIC is a symmetric measure.

Therefore, we use CPIC and DPIC to calculate APICSS measure between two resources $r, s \in R$ by using Eqs. (12) and (13).

$$\begin{aligned} APICSS(r, s) &= CPIC(r, s) \cdot DPIC(r, s) \\ &= \frac{PIC(F_r \cap F_s)}{PIC(F_r)} \cdot \frac{1}{PIC(F_r - F_s) + PIC(F_s - F_r) + 1} \end{aligned} \tag{12}$$

$$APICSS(s,r) = CPIC(s,r) \cdot DPIC(s,r)$$
$$= \frac{PIC(F_s \cap F_r)}{PIC(F_s)} \cdot \frac{1}{PIC(F_r - F_s) + PIC(F_s - F_r) + 1} \qquad (13)$$

The value of APICSS always lies in the range [0, 1]. APICSS is obviously an asymmetric measure, that is, $APICSS(r,s) \neq APICSS(s,r)$.

Improving the Heterogeneous Network with Calculated Semantic Similarities. Once the semantic similarities between the objects are obtained, we can define the object similarity matrix by using the semantic similarities (the APICSS measure) instead of the ratings-based similarities (the cosine measure as defined by Eq. (1)) in the ROUND method. Formally, the object similarity matrix $\mathbf{S} = (s_{ij})_{m \times m}$ is now calculated using Eq. (14).

$$s_{ij} = APICSS(o_i, o_j) \qquad (14)$$

After that, a number of the weakest relationships for each object are filtered out by sorting each column of the object similarity matrix $\mathbf{S}$ and resetting similarities to 0 for objects ranked lower than parameter α. In this way, we obtain a filtered object similarity matrix $\mathbf{O} = (o_{ij})_{m \times m}$. After the filtration, for any pair (i,j) where $o_{ij} \neq 0$ but $o_{ji} = 0$, the matrix element o_{ji} reverts to the original value of the element, which means that the filtered object similarity matrix in our method is asymmetric. This way, we have improved the object-user heterogeneous network model. It is noteworthy that the change in the network model also improves the random walk with restart model on the network, as the transition matrix $\hat{\mathbf{O}} = (\hat{o}_{ij})_{m \times m}$ has been reconstructed by using the new object similarity matrix $\mathbf{O}$.

Other aspects of our ROUND-APICSS method, including parameter settings, remain the same as the ROUND method.

3   Experimental Evaluation

3.1   Experimental Design

We have conducted experimental evaluation to validate the effectiveness of our proposed ROUND-APICSS method and to demonstrate its performance strength of top-N recommendation compared to the original ROUND method. The experimental design is described as follows.

Experimental Datasets. Our experiment used three real datasets, the MovieLens 100k dataset, the DBpedia 2016-04 release, and the DBpedia-MovieLens 100k dataset, as follows.

The MovieLens 100k Dataset. This benchmark dataset [16] (cf. https://grouplens. org/datasets/movielens/) contains 100,000 ratings from 943 users on 1,682 movies. Each user has rated at least 20 movies in the dataset. All the ratings are in the scale 1–5. During our experiment, we converted the ratings to binary ratings by re-assigning 1 as "relevant" to ratings no less than 3.0 and 0 as "not-relevant" to all other cases, as in Gan and Jiang's experiments [8]. This dataset was divided into two portions in our experiment:

– *Training data (80%):* This portion of the data was used to generate the list of recommended objects for each user contained in the testing data.
– *Testing data (20%):* This portion of the data was used as ground-truth data to test the recommendation accuracy of the two top-*N* recommendation methods.

The DBpedia 2016-04 Release. The English version of DBpedia dataset [3] (cf. http://wiki.dbpedia.org/dbpedia-version-2016-04) provides most of the movies in the MovieLens 100k dataset with a wealth of background knowledge in the form of RDF triples described with ontological properties (object properties).

The DBpedia-MovieLens 100k Datasets. This dataset was created by Meymandpour and Davis [13]. It contains the mappings from MovieLens object (movie) identifiers to DBpedia entity URIs. In fact, 1,569 movies out of the 1,682 movies in the MovieLens 100k dataset have been mapped to DBpedia entity URIs. These mappings can be used directly (i.e., there's no need to perform the object linking step) to construct the knowledge graph from the DBpedia dataset.

The Object-User Heterogeneous Network. The network was constructed from the above experimental datasets using the method described earlier. The constructed network contains 1,569 object nodes and 47,917 relationship edges among the object nodes in the object layer, 943 user nodes and 28,791 relationship edges among the user nodes in the user layer, and 65,726 object-user association edges in the interconnection layer.

The Knowledge Graph. During our experiment, 102,748 RDF triples were extracted from the DBpedia dataset to construct the knowledge graph according to the movie identifiers to DBpedia entity URIs mappings. The RDF triples use 35 object properties, as list in Table 1, to describe the 1,569 movies contained in the network.

Parameter Setting. During our experiment, the four parameters were set to the same default values as those in the experiment of Gan and Jiang [8] except for parameter α because our experimental dataset MovieLens 100k is smaller than the MovieLens dataset used in Gan and Jiang's experiments, that is, we set $\alpha = 40$, $\beta = 0.5$, $\gamma = 0.5$, and $\pi = 0.8$.

Accuracy Metrics. Three popular accuracy metrics, precision, recall and F1-measure [6, 17], were used to evaluate recommendation accuracy in the experiment. For a given user u, its precision, recall and F1-measure are defined as Eqs. (15)–(17), respectively.

$$Precision_u(N) = 100 \cdot \frac{|S_u(N) \cap GT_u|}{|S_u(N)|} \tag{15}$$

$$Recall_u(N) = 100 \cdot \frac{|S_u(N) \cap GT_u|}{|GT_u|} \tag{16}$$

Table 1. All ontological properties involved in the constructed knowledge graph.

Object properties	Object properties
http://dbpedia.org/ontology/artist	http://dbpedia.org/ontology/producer
http://dbpedia.org/ontology/author	http://dbpedia.org/ontology/soundRecording
http://dbpedia.org/ontology/basedOn	http://dbpedia.org/ontology/starring
http://dbpedia.org/ontology/cinematography	http://dbpedia.org/ontology/thumbnail
http://dbpedia.org/ontology/company	http://dbpedia.org/ontology/type
http://dbpedia.org/ontology/country	http://dbpedia.org/ontology/wikiPageDisambiguates
http://dbpedia.org/ontology/creator	http://dbpedia.org/ontology/wikiPageExternalLink
http://dbpedia.org/ontology/director	http://dbpedia.org/ontology/wikiPageRedirects
http://dbpedia.org/ontology/distributor	http://dbpedia.org/ontology/writer
http://dbpedia.org/ontology/editing	http://dbpedia.org/property/pictureFormat
http://dbpedia.org/ontology/editor	http://purl.org/dc/terms/subject
http://dbpedia.org/ontology/executiveProducer	http://www.w3.org/1999/02/22-rdf-syntax-ns#type
http://dbpedia.org/ontology/format	http://www.w3.org/2000/01/rdf-schema#seeAlso
http://dbpedia.org/ontology/genre	http://www.w3.org/2002/07/owl#differentFrom
http://dbpedia.org/ontology/language	http://www.w3.org/2002/07/owl#sameAs
http://dbpedia.org/ontology/musicComposer	http://www.w3.org/ns/prov#wasDerivedFrom
http://dbpedia.org/ontology/narrator	http://xmlns.com/foaf/0.1/#term_homepage
http://dbpedia.org/ontology/network	

$$F1_u(N) = \frac{2 \cdot Precision_u(N) \cdot Recall_u(N)}{Precision_u(N) + Recall_u(N)} \tag{17}$$

where $S_u(N)$ represents the list of recommended objects for user u, and the length of the recommendation list is $|S_u(N)| = N$. GT_u represents the true set of relevant items (ground-truth positives) that are consumed by the user. Finally, the experimental results show the Average Precision@N, Average Recall@N and Average F1@N for all the users in the testing data.

3.2 Experimental Results and Discussion

The top-N recommendation accuracies in terms of Average Precision@N, Average Recall@N and Average F1@N were yielded using the experimental datasets. Figures 2, 3 and 4 show the accuracy comparisons between the two top-N recommendation methods, ROUND and ROUND-APICSS, in terms of the three accuracy metrics, respectively, with N being 5, 10, 15, 20, 25, and 30.

As shown in the Figures, our ROUND-APICSS method outperforms the ROUND method in terms of all the three accuracy metrics in all the cases of different N values. More specifically, the maximum increase, 4.03%, in Average Precision@N occurs in the case of $N = 20$, whereas average increase in Average Precision@N reaches 3.66%. The maximum increase, 4.50%, in Average Recall@N occurs in the case of $N = 5$, whereas average increase in Average Recall@N reaches 3.20%. The maximum increase, 4.65%, in Average F1@N occurs in the case of $N = 5$, whereas average increase in Average F1@N reaches 3.33%. The experimental results suggest that the incorporation of the background knowledge from Linked Open Data into the network-based top-N recommendation model can improve recommendation accuracy.

It is noteworthy that Gan and Jiang's experimental results on two real datasets (MovieLens and Netflix) have shown the effectiveness of the ROUND method and its superiority over such state-of-the-art approaches as NMF and SVD in terms of not only recommendation accuracy and diversity but also retrieval performance [8]. Therefore,

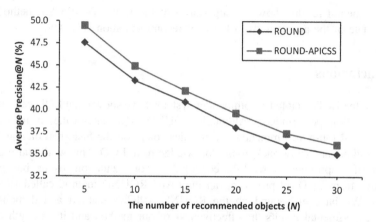

Fig. 2. Comparison of the two recommendation methods in terms of Average Precision@N.

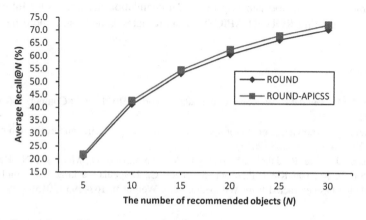

Fig. 3. Comparison of the two recommendation methods in terms of Average Recall@N.

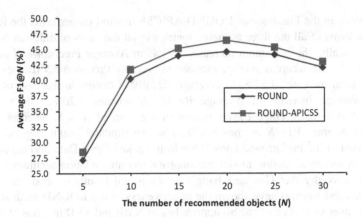

Fig. 4. Comparison of the two recommendation methods in terms of Average F1@N.

our experimental results show the superiority of the ROUND-APICSS method over existing state-of-the-art methods in terms of recommendation accuracy.

4   Conclusions

Studies on the LOD-enabled recommender systems have shown that LOD can improve recommendation performance. Although ROUND is a state-of-the-art method for network-based top-N recommendation, it relies solely on the historical data and does not take full advantage of background knowledge from LOD. This work addresses the problem of improving network-based top-N recommendation using background knowledge from LOD by proposing an improved ROUND method called ROUND-APICSS. We have elaborated on the ROUND-APICSS method and demonstrated through experimental results the effectiveness of our method and its strength of recommendation accuracy compared to the ROUND method. Our work also shows that the incorporation of background knowledge from LOD into a network-based top-N recommendation method can improve recommendation accuracy. Our future work focuses on evaluating ROUND-APICSS on more datasets and using more metrics.

References

1. Heath, T.: Linked data - Welcome to the data network. IEEE Internet Comput. **15**(6), 70–73 (2011)
2. Paulheim, H.: Knowledge graph refinement: a survey of approaches and evaluation methods. Sem. Web **8**(3), 489–508 (2017)
3. Lehmann, J., Isele, R., Jakob, M., Jentzsch, A., Kontokostas, D., Mendes, P.N., Hellmann, S., Morsey, M., van Kleef, P., Auer, S., Bizer, C.: DBpedia - a large-scale, multilingual knowledge base extracted from Wikipedia. Sem. Web **6**(2), 167–195 (2015)

4. Cyganiak, R., Wood, D., Lanthaler, M. (eds.): RDF 1.1 Concepts and Abstract Syntax. W3C Recommendation 25 February 2014. http://www.w3.org/TR/rdf11-concepts/. Accessed 20 May 2018

5. Di Noia, T., Ostuni, V.C.: Recommender systems and linked open data. In: Faber, W., Paschke, A. (eds.) Reasoning Web 2015. Lecture Notes in Computer Science (LNCS), vol. 9203, pp. 88–113. Springer, Cham (2015)

6. Di Noia, T., Cantador, I., Ostuni, V.C.: Linked open data-enabled recommender systems: ESWC 2014 challenge on book recommendation. In: Presutti, V., Stankovic, M., Cambria, E., et al. (eds.) Semantic Web Evaluation Challenge - SemWebEval 2014 at ESWC 2014, Revised Selected Papers. Communications in Computer and Information Science (CCIS), vol. 475, pp. 129–143. Springer, Cham (2014)

7. Oramas, S., Ostuni, V.C., Di Noia, T., Serra, X., Di Sciascio, E.: Sound and music recommendation with knowledge graphs. ACM Trans. Intell. Syst. Technol. (TIST) 8(2), Article No. 21 (2016)

8. Gan, M., Jiang, R.: ROUND: walking on an object–user heterogeneous network for personalized recommendations. Expert Syst. Appl. 42(22), 8791–8804 (2015)

9. Mao, C., Xu, Z., Wang, X.: Asymmetric item-item similarity measure for linked open data enabled collaborative filtering. In: Proceedings of 14th Web Information Systems and Applications Conference, WISA 2017, pp. 228–233. IEEE (2017)

10. Aggarwal, C.C.: Neighborhood-based collaborative filtering. In: Aggarwal, C.C. (ed.) Recommender Systems: The Textbook, pp. 29–69. Springer, Cham (2016)

11. Backstrom, L., Leskovec, J.: Supervised random walks: predicting and recommending links in social networks. In: Proceedings of the Forth International Conference on Web Search and Web Data Mining, WSDM 2011, pp. 635–644. ACM (2011)

12. Berners-Lee, T., Fielding, R., Masinter, L.: Uniform Resource Identifier (URI): Generic Syntax. IETF RFC 3986, the Internet Society, January 2005. http://tools.ietf.org/html/rfc3986. Accessed 20 May 2018

13. Meymandpour, R., Davis, J.G.: A semantic similarity measure for linked data: an information content-based approach. Knowl.-Based Syst. 109, 276–293 (2016)

14. Pirasteh, P., Hwang, D., Jung, J.J.: Exploiting matrix factorization to asymmetric user similarities in recommendation systems. Knowl.-Based Syst. 83, 51–57 (2015)

15. Garg, A., Enright, C.G., Madden, M.G.: On asymmetric similarity search. In: Proceedings of International Conference on Machine Learning and Applications, pp. 649–654. IEEE (2015)

16. Herlocker, J.L., Konstan, J.A., Terveen, L.G., Riedl, J.T.: Evaluating collaborative filtering recommender systems. ACM Trans. Inf. Syst. 22(1), 5–53 (2004)

17. Aggarwal, C.C.: Evaluating recommender systems. In: Aggarwal, C.C. (ed.) Recommender Systems: The Textbook, pp. 225–254. Springer, Cham (2016)

A Web-Based Theme-Related Word Set Construction Algorithm

Yingkai Wu[1], Yukun Li[1,2(✉)], and Gang Hao[1,3]

[1] Tianjin University of Technology, 300384 Tianjin, China
{wuyingkai521, liyukun_tjut, ganghao2018}@163.com
[2] Tianjin Key Laboratory of Intelligence Computing and Novel Software Technology, Tianjin, China
[3] Key Laboratory of Computer Vision and System, Ministry of Education Tianjin, Tianjin, China

Abstract. Constructing theme-related word set is a basic work for establishing theme-oriented information retrieval systems. Nowadays, most of previous studies focus on identifying representative words of a specific document, and few studies pay attention to constructing a word set related to a theme. By analyzing existing keywords extraction methods, this paper proposes a method to automatically construct theme-related word set based on the primary theme-related word set given by domain experts and the well-known websites related to the theme. As the first step, the method uses existing information extraction techniques to obtain the documents from the websites and every document's keyword set. Then it calculates the correlation degree between the known theme-related word set and the document keyword set, further gets a word set of the document related to the theme based on the document-theme relevance, and merges the word set to the theme-related word set. By using the method, the theme-related word set is supplemented by iteration based on the documents gotten from the theme-related websites. Because there is little research work focusing on this problem and no relevant experimental data set, this paper uses the proposed method to construct theme-related word sets towards two themes "electricity" and "college entrance examination", and we invite domain experts to evaluate the word sets. The results show that a relatively complete theme-related word set can be obtained based on this method, which shows the feasibility of our methods.

Keywords: Theme-related · Word set · Keyword · Relevance

1 Introduction

Internet is increasingly becoming an important data source with the arrival of the information explosion era. But most of the information on the Internet is unstructured text information, and the information on different themes is scattered on many different nodes of the Internet, which has brought difficulties to the effective use of Internet information. It is a fundamental research issue that how to find a method, which can

© Springer Nature Switzerland AG 2018
L. H. U and H. Xie (Eds.): APWeb-WAIM 2018, LNCS 11268, pp. 188–200, 2018.
https://doi.org/10.1007/978-3-030-01298-4_17

classify Internet information according to theme. For example, if a high school student wants to gain information of computer science majors in many schools, he has to go to many different websites to find relevant information; If a decision-making person of an electric power system wants to know about power related policy, the latest technology, and dynamic information of domestic industry, he also needs to go to many distinct sites to search for relevant information. In order to solve this problem, researchers put forward the concept, technology and methods of vertical search [1, 2]. However there still exist some problems that cause the searching results incomplete. The deeper cause is failure to find a high quality set of words related to the theme, which is the basis for finding the data sources related to the theme and sorting the searching results. What's more, there is no relevant work focus on the theme related word set construction.

For a given theme, there must be a corresponding set of theme-related words in the objective world. At the same time, due to the definition of subjectivity and uncertainty, it is very difficult to find a complete set of theme-related words. The goal of this paper is to propose a method for constructing the theme-related word sets and to make it as close as possible to a complete set of theme-related words.

This problem meets some challenges. The methods of existing document keyword extraction cannot be directly used to construct a related keyword collection of a certain theme. In addition, the evaluation of theme-related word sets is also a challenging issue. There is no experimental data set and benchmark for the evaluation of the methods to identify theme-related word set.

This paper studies this problem and proposes a web-based theme-related word set construction algorithm. The main contributions are as follows:

(1) A concept model of theme related word set is proposed, as well as the idea of constructing theme-related word set based on expert knowledge and large-scale Web data information;
(2) We put forward some methods of constructing the theme-related word set. Specifically, it includes: a method for calculating the relevance degree between Web documents and theme-related word set, and a method for supplementing theme-related word set based on the relevance degree;
(3) Because there are no relevant data sets for evaluation, we establish related experimental data sets and evaluation strategies based on the themes of "electricity" and "college entrance examination", and evaluate the method proposed in this paper based on the data set and its effectiveness is verified.

The rest of this paper is organized as follows: First, Sect. 2 describes the work related to this paper. Section 3 mainly introduces the theme-related word set construction algorithm, describes the related definitions and the specific construction process of theme-related word set. Section 4 uses the theme-related word set construction method proposed in this paper to construct the theme-related word set related to the fields of "electricity" and "college entrance examination", and validates the effectiveness of the method. In Sect. 5, we conclude our work.

2 Related Work

The current research work mainly focuses on how to extract keyword information that can represent the theme of a particular document, and keyword extraction methods are continuously deepened in both unsupervised and supervised directions.

The unsupervised keyword extraction methods usually take some evaluation indicators (such as TF-IDF, etc.) to rank the keywords, and then selects the top ones as keywords [3]. There are two main types of unsupervised keyword extraction methods: word frequency statistics and graph-based keyword extraction methods. The most representative of Word-frequency statistics-based keyword extraction method is TF-IDF algorithm proposed by Salton [4] with the central idea that if a word appears frequently in an article and rarely appears in other articles, it is considered that the word has a good distinguishing ability and can be used as a keyword of the article. There are many methods of keyword extraction based on graph. Earlier in 2004, Mihalcea proposed the TextRank method [5]. This method is based on the PageRank algorithm and uses the co-occurrence times of the edges instead of the out-degrees, achieving very good results. Xiaojun Wan and others proposed the ExpandRank method. This method adds a set of documents similar to the target document to help construct word graph, and solves the problem of lack of information in single-target document word graph [6]. Gollapalli et al. proposed the CiteTextRank method [7] with reference context in the scientific literature to enhance the information contained in the word graph. Rafiei et al. used theme information to propose a new method called TSAKE combining word graph and theme [8], but this method is computationally expensive. Florescu et al. added the position information of candidate words in the document to the PageRank model and proposed the PositionRank method [9]. Although the TF-IDF algorithm is simple, the actual extraction of keywords is very effective. And graph-based keyword extraction methods take into account the co-occurrence relationship between words and words, that is, the semantic relationship, they are usually better than word frequency extraction methods.

LDA [10] is an unsupervised machine learning technique. In natural language processing tasks, LDA is a document theme generation model, and it is also a three-level probabilistic model that contains vocabulary, document subject, and document three-tier structure. It can be used to identify topics hidden in large collections of documents or corpora [11], as well as to extract keywords.

The supervised keyword extraction methods usually consider the keyword extraction problem as a binary-class problem. Its main steps are as follows: First preprocess the text to get the candidate keywords; define and calculate the relevant features; then extract the keywords according to the set classifier [12]. Xie F et al. used the sequential pattern mining algorithm to enhance the candidate keyword generation effect, and used the experimental results to prove that the Naive Bayes algorithm can effectively extract keywords [13]. Joorabchi et al. used wikipedia to obtain external knowledge base features, and used a genetic algorithm to extract key words in combination with various features [14]. Sterckx L et al. used the decision tree to extract keywords [15], which is suitable for data sets with different user annotations. Yiqun Chen et al. used support vector machines to extract keywords by constructing a patent background knowledge base and defining relevant features [16]. Gollapalli SD et al. used word frequency, location, and other features and added expert knowledge features, and used conditional random field model [17] to extract keywords.

The above work is based on the purpose of extracting the keywords of the document. This paper analyzes the extraction methods of existing, and attempts to propose an algorithm that can construct a set of keywords related to a theme. The collection of theme related words obtained by using this method can be widely applied to many aspects such as retrieval, classification, and marking.

3   Theme-Related Word Set Construction Method

This section mainly introduces how to construct a theme-related word set. Section 3.1 gives the relevant definitions. Section 3.2 describes the pseudo-code representation of the construction method of the theme related word set. And next three sections show the process of the construction of the theme word set in detail.

3.1   Related Definitions

Definition 1: Theme (T). This paper defines a theme as a collection of objects related to a particular topic.

For example, the "college entrance examination" can be seen as a theme and its corresponding theme can be regarded as a collection of things or objects related to the college entrance examination. These objects can be all kinds of objects, including a university, major, admission score, related policies and more.

Definition 2: Theme-related word set (ST). Given a theme T, the theme-related word set ST is a set of two-tuples (W, Q), where W is a word and Q is the relevance degree of the word to the given theme T. And the theme-related word set ST must satisfy the following rules:

(1) The sum of relevance of all the words in the theme-related word set ST should be equal to 1.
(2) $Q_1 <= Q_2 <= Q_3 <= ... <= Q_n$.

For example, for a given theme "electricity", {(electricity, 0.2), (electric power, 0.2), (coal power, 0.2), (thermal power, 0.2), (wind power, 0.2)} can be seen as one of its related word set. It means all the relevance degrees of the words are 0.2.

Definition 3: The document keywords set (DK). The document keywords set DK is a set of two-tuples (K, Q), where K is a keyword of the document, and Q is the weight of this keyword to the document. The weight of keywords in DK of the document should satisfy $Q_1 <= Q_2 <= Q_3 <= ... <= Q_n$.

For example, let d be a document, $DK(d) = \{(K_1, Q_1), (K_2, Q_2), (K_3, Q_3), ..., (K_n, Q_n)\}$ is the keyword set of document d, there K_i is keyword, and Q_i is the relevance degree of the keyword K_i to the document.

Definition 4: Relevance of document and theme (R(DO,ST)). It represents the degree of relevance of the document DO with the theme-related word set ST.

For example, Let D' be a keyword set of document, T' be the theme "electric power", R(D',T') means the relevance degree of D' to the theme "electric power".

Definition 5: Theme-related word set of document (WS(D,T)). Let D be a keyword set of document and T be a theme, the WS(D,T) means a collection of keywords, which consists of the top N keywords according to their relevance degree to theme T, and N is directly proportional to the relevance of document and theme.

3.2   Theme-Related Word Set Construction Algorithm

The key idea of the algorithm proposed in this paper is: To a given theme, ask some experts on this theme give an initial word set and some websites related to the theme. By analyzing a large amount of articles of the given websites and continuously expanding the known theme-related word set, finally we can get a theme-related word set ST. And the specific algorithm is as follows:

Algorithm 1: Theme-related Word Set Construction Algorithm

Input: Initial theme-related word set ST_0, Theme-related website W_0;
Output: Theme-related word set ST.
1. Identify documents from the given websites W_0;
2. Identify keyword set and get DK of every document;
3. Merge all WS(D,T) which contained in DK and initial theme-related word set ST_0 and finally get theme-related word set ST;
4. **Return** Theme-related word set ST;

In Algorithm 1, first we should use information extraction techniques to identify documents from the given websites and get the collection of documents. Then, we identify keyword set of each document from the collection of documents and get DK of every document. Next, we merge all theme-related word set of document and theme-related word set, and expand theme-related word set continuously. Finally, we will get the theme-related word set.

Following, the Algorithm 1 will be described in detail in Sects. 3.3, 3.4 and 3.5.

3.3   Identifying Documents from Websites

According to theme-related word set construction algorithm, theme-related website W_0 is known. First, we should analyze the structure of webpages. The content of the webpage is given priority to with an article, mostly with navigation and some advertising etc. This paper mainly analyzes the content of the web page, and does not pay attention to other content in the webpage. Therefore, the algorithm of this paper uses existing crawler technology to filter other web content and only obtains the article as a document downloaded from webpages.

3.4   Identifying Keyword Set of Document

After getting all the documents in the theme related website W_0, next step is to extract the keywords for each document. The premise of extracting a keyword is to segment the document and remove the stop words. There are many methods for extracting keywords. This paper uses the statistics-based word frequency analysis method TF-IDF. The TF-IDF algorithm can be used to obtain the TF value, TF-IDF value, and IDF value of all keywords in each document. By ranked the words in each document in the descending order of TF-IDF values, then we can obtain the document keyword set DK. This paper uses the TF-IDF value of the keyword as the weight of the keyword in the document keyword set DK, where the keyword weight value satisfies $Q_1 <= Q_2 <= Q_3 <= ... <= Q_n$, and the keyword and the weight correspond one to one.

In this way, we can get $W(DK_n)$, which is a collection of DK from theme-related website. Then, the algorithm will expand the initial theme-related word set which given by experts.

3.5   Expanding Theme-Related Word Set

The related experts have given the initial theme-related word set ST_0. And in Sect. 3.4, we get a collection of $W(DK_n)$. Then it calculates the correlation degree between the known theme-related word set and the document keyword set, further gets a word set of the document related to the theme based on the document-theme relevance, and merges the word set to the theme-related word set. And the specific algorithm is as follows:

Algorithm 2: Expanding Theme-related Word Set Algorithm

Input: Initial theme-related word set ST_0, $W(DK_n)$ which is a collection of DK;
Output: Theme-related word set ST_n.
1. Initialize the input data;
2. **For** i = 1,2,3...,n **do**
3. $R_i(DK_i, ST_{i-1}) = N(DK_i, ST_{i-1}) / N(ST_{i-1})$;
         //$N(DK_i, ST_{i-1})$ means the number of keywords both DK_i and
            ST_{i-1} contained, and $N(ST_{i-1})$ means the number of key-
            words ST_{i-1} contained.
4. $N_{merge} = R_i(DK_i, ST_{i-1}) * N(DK_i)$;
         //$N(DK_i)$ means the number of keywords DK_i contained.
5. $WS(DK_i, ST_{i-1}) \leftarrow Top(N_{merge}, DK_i)$;
         //Get the top N_{merge} keywords of DK_i.
6. $ST_i \leftarrow WS(DK_i, ST_{i-1}) \cup ST_{i-1}$;
         //Merge the two keyword sets.
7. **End for**
8. **Return** ST_n;

As shown in Algorithm 2, this is an iterative process. And, the method that ST_{i-1} generates the theme-related word set ST_i can be expressed by the following formula:

$$\begin{cases} ST_0 = \{(W_1, Q_1), (W_2, Q_2), (W_3, Q_3), \ldots, (W_t, Q_t)\} \\ ST_i = ST_{i-1} \cup WS(DK_i, ST_{i-1}) \end{cases} \tag{1}$$

When the document keyword set DK_i merges with the theme-related word set ST_{i-1}, the keywords in both of them may be duplicated. This means that the relevance of the keyword to theme is high, otherwise it is low. Therefore, to ensure uniqueness of keywords in theme-related words set, the deduplication operation needs to be performed so that there are no duplicate words in the theme-related words. More importantly, while carrying out keyword deduplication, it is necessary to adjust the weights of keywords in the theme-related words set. To adjust the weights of keywords increases the proportion of weights of previously added keywords which have high relevance to the theme, and decreases the proportion of weights of keywords which have low relevance of the theme. The weight adjustment method from the ST_i to the ST_{i+1} is as follows:

$$\begin{cases} ST_i = \{(W_1, Q_1), (W_2, Q_2), (W_3, Q_3), \ldots, (W_t, Q_t)\} \\ ST_{i+1} = \{(W_1, Q_1), (W_2, Q_2), (W_3, Q_3), \ldots, (W_t, Q_t), (W_{t+1}, Q_{t+1}), \ldots, (W_p, Q_p)\} \end{cases}$$

$$\Downarrow$$

$$Q_i = \begin{cases} \frac{t}{t+N_{merge}} * Q_i & i \leq t \\ \frac{1}{t+N_{merge}} & t < i \leq p \end{cases}$$

$$\tag{2}$$

The keywords in ST are all unique, so when merging the $WS(DK_{i+1}, ST_i)$ and ST_i, both of them may have a certain keyword, it is necessary to remove duplicate keywords, and the corresponding weights should be combined. If the keyword ST_{i+1} (W_m) in the theme-related word set ST_{i+1} is the same as the keyword ST_{i+1} (W_n), where m <= t < n, then it is necessary to merge the keyword weights. Don not to add the keyword ST_{i+1} (W_n), and the weight of the recalculated keyword ST_{i+1} (W_m) is:

$$ST_{i+1}(Q_m) = Q_m + \frac{1}{t+N_{merge}} \tag{3}$$

After recalculating the weights of all the keywords of the theme-related word set ST_{i+1}, the two-tuples of theme-related words set need to be arranged in the descending order of the keyword weights.

According to the above method, when the keywords set of all documents in the theme-related website W_0 is merged into the original theme-related word ST_0, the theme-related word set ST_n will be generated.

4 Experiments and Analysis

There is no existing data set for evaluation of theme-related word set identification, and in order to verify the effectiveness of our methods, we decide to choose several themes and construct their theme-related words sets by our methods. In this paper we take the themes of "electric power" and "college entrance examination" as examples for evaluating our methods.

4.1 Experimental Data Set and Measures

We firstly ask related experts to give the initial theme-related word set of "electric power" and "college entrance examination". "Electric power" experts recommend that to use the ST_e = {(electricity, 0.2), (electric power, 0.2), (coal power, 0.2), (thermal power, 0.2), (wind power, 0.2)} as the initial theme-related word set, and recommend that to use "BeiJiXing Power Website", whose URL is "http://www.bjx.com.cn" as a theme-related website. Experts in the "college entrance examination" field recommend using ST_g = {(college entrance examination, 0.2), (voluntary, 0.2), (high school, 0.2), (university, 0.2), (score, 0.2)} as the initial theme-related word set, and recommend using the "GaoKao Website", whose URL is "http://www.gaokao.com" as a theme-related website.

Since the theme-related website W_0 is known, the data collected for the experiment is the document in its webpage. First, we enter the theme-related website and analyze the webpage structure. Then, we use the existing crawler technology to download the documents on the website to the local. Finally, the data is formatted and saved to the local MySQL database.

A total of 5,179 articles on "electricity" are integrated in the experiment. The data is stored in the table electricContent and contains 6 fields, namely Sid document number, SiteURL data source URL, DataTitle document title, DataURL document URL, DataTime document publication time, and DataContent document content. And we integrate 5066 theme on the "College Entrance Examination" that the data is stored in the table gaokaoContent and the table structure is same as electricContent.

For a given theme, there must be a corresponding theme-related word set in the objective world. But it is very difficult to find a complete theme-related word set. Using the method proposed this paper we can construct a theme-related word set. Then, by setting a threshold N_t, the keywords of the former N_t two-tuples of the theme-related word set ST_n were extracted to generate the maximum theme-related word set that make it as close as possible to a complete set of theme-related words.

In order to evaluate the experimental results, we use the keywords already labeled in the documents and manually labeled keywords as validation sets. And we use precision and recall to verify the validity of the experiment.

4.2 Experimental Results

The experiment uses the data set mentioned in Sect. 4.1. According to the theme-related words set construction algorithm proposed in this paper, the initial theme-related words set ST_0 and the theme-related website W_0 are known. For each theme, we use the method proposed in this paper and the LDA topic model to extract theme-related word sets. By calculating and comparing the precision and recall of the theme-related word sets obtained by the two methods, the effectiveness of the proposed method is verified.

According to the threshold value of N_t, the keywords of the former N_t were extracted to generate the maximum theme-related word set. And set the threshold from 1 to 70 and the interval to 5, and calculate the recall and precision of the corresponding keyword set. The following figures are available:

Figures 1 and 2 are comparison charts of the recall and precision of the electric power(ELEP) theme. Figures 3 and 4 are comparison charts of the recall and precision of the college entrance examination(CEE) theme. As can be seen from figures above, the abscissa indicates the threshold, and the ordinate indicates recall or precision.

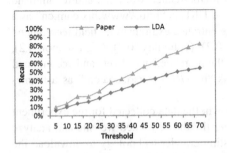

Fig. 1. Recall of ELEP

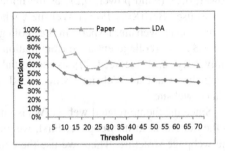

Fig. 2. Precision of ELEP

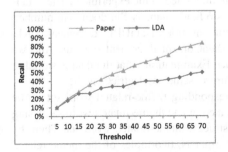

Fig. 3. Recall of CEE

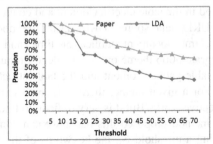

Fig. 4. Precision of CEE

As the set threshold increase, the recall continues to increase and the precision continues to decline. We can see from Figs. 2 and 4, when threshold is higher than a certain value, the precision tends to be stable and no longer declines. And the precision of our method is always higher than that of LDA. The power theme precision approaches 60% with our method and precision approaches 40% with LDA method after the threshold is higher than 30. The college entrance examination theme precision approaches 60% with our method and precision approaches 40% with LDA when the threshold is higher than 50. And we can see from Figs. 1 and 3, As the set threshold

Table 1. Electric power maximum theme-related word set

R	Key	R	Key	R	Key	R	Key	R	Key
1	Electricity	11	New energy	21	Coke	31	Charging	41	Clean
2	Energy sources	12	Technology	22	Kilowatt	32	Environment	42	Power station
3	Electric power	13	Battery	23	Group	33	Reform	43	Nuclear
4	Coal	14	Industry	24	Environmental protection	34	Productivity	44	Machine set
5	Volt	15	Motor vehicle	25	Propulsion	35	Power battery	45	Cost
6	Power grid	16	Domain	26	Kilowatt-hour	36	Price	46	Nation
7	Market	17	China	27	Natgas	37	System	47	Recycle
8	Electricity generation	18	Engineering	28	Wind	38	Service	48	Electric quantity
9	Company	19	Garbage	29	Nuclear electricity	39	Renewable energy sources	49	City
10	Trade	20	Emission	30	Pollution	40	Charcoal	50	Investment

Table 2. College entrance examination maximum theme-related word set

R	Key	R	Key	R	Key	R	Key	R	Key
1	College entrance examination	11	Reform	21	Evaluate	31	Schoolmate	41	College student
2	Student	12	Education	22	Study	32	Comprehensive	42	Parents
3	Recruit students	13	Awarded marks	23	Choice	33	China	43	Middle school
4	Examinee	14	Teacher	24	Journalist	34	Children	44	Senior high school
5	Examination	15	Child	25	English	35	College	45	Letter Of admission
6	University	16	Examine	26	Employment	36	Peking university	46	Academy
7	Colleges and universities	17	Freshman	27	Grade	37	Ministry of education	47	Tutor
8	Enroll	18	Subject	28	Remote	38	Rural	48	Check points
9	School	19	Patriarch	29	Plan	39	Graduate	49	Policy
10	Major	20	Autonomous enrolment	30	College aspiration	40	Engineering	50	Information

increase, the recall continues to increase and the recall of our method is always higher than that of LDA. In turn, we set the threshold to 50. The maximum theme-related word set obtained based on the theme-related word set construction algorithm is shown in the following table:

In Tables 1 and 2 above, R indicates the rank of the keyword weight in the theme-related word set, Key represents the corresponding keyword.

4.3   Experimental Analysis and Discussion

We can see from figures above, the experimental results of the power theme and college entrance examination theme are closely related to the threshold size. If the threshold is too small, the number of keywords in the theme-related words is small, and it is difficult to ensure the completeness of the theme-related word sets. On the contrary, the number of keywords in the theme-related words is too much. If the threshold is large, the completeness of theme-related word sets can be guaranteed, there will be a large number of keywords with low relevance to the theme, which reduces the representation of theme-related word sets. Moreover, as the threshold increases, the precision of the keyword set tends to be stable, and increasing the threshold blindly has little significance. From the above experiments, we find that setting a threshold value of 50 can ensure that the precision of theme-related word sets is the best, and it can also avoid adding a large number of keywords with low relevance to theme-related word sets.

We use the power theme and the college entrance examination theme to generate the top 50 keywords of the theme-related words set as corresponding maximum theme-related word sets. From the results, all the keywords in the power-related theme related words are related to the theme of "electricity". And the keywords of the comprehensive theme-related words in the college entrance examination are related to the theme of the college entrance examination and meet the expected expectations. We present the theme-related word sets to experts in the respective fields and have been approved by experts.

In the theme-related words of electricity, keywords such as "Electricity", "Energy Sources", "Electric Power", "Coal", "Volt", and "Power Grid" that are closely related to the theme of "Electric Power" are in front of the theme-related words. Keywords such as "Investment", "City", "Recycle", "Nation", "Cost", "Machine Set", etc. are related to the theme of "Electric Power" but do not particularly highlight theme. These keywords are behind the theme-related words set. In the theme headings of the college entrance examination, the words "College Entrance Examination", "Student", "Recruit Students", "Examinee" and "University" which are closely related to the theme of the "College Entrance Examination" are in front of the theme-related words. The keywords "Information", "Policy", "Parents" and other keywords are related to the theme of "College Entrance Examination" but they have no prominent theme. These keywords are behind the theme-related words. This is because in the theme-related words set, the keywords are generated according to the size of the keyword weights. The more relevant the keyword is to the theme, the higher the position is.

We find that in terms of the Electric Power theme-related words, the keywords "China", "Reform" and "Nation" are not highly related, and "China" and "Nation" appear in various fields. It seems that they should not appear in the Electric Power theme word set. Similarly, in the College Entrance Examination theme word set, the keyword "Journalist" and other words are not highly related to the "College Entrance Examination" theme and should not appear in the College Entrance Examination theme word set. For the above problem, we look for the documents containing relevant keywords in the data and find that for the theme of "Electric Power", many of the documents in the theme-related websites are related to the relevant Electric Power policies issued by the state, which necessarily contains "China" and "Reform" and

"Nation", and the weights calculated by the TF-IDF algorithm are high. Similarly, for the "College Entrance Examination" theme, many of the documents in the theme-related websites are all local Journalists interviewing people or events related to the college entrance examination, and most of them contain "some Journalists". The weights calculated by the TF-IDF algorithm are very high. So it will eventually appear in the theme-related word set.

Based on the above analysis, this method can also be improved in the following two aspects: (1) set the dynamic threshold. For different themes, the theme-related words set construction algorithm uniformly sets a threshold for 50. However, when the threshold is 50, not all themes get the best theme-related words set. For example, in the "power" theme in the experiment, when the threshold is set to 30, the precision can stabilize at 60%. Therefore, the problem for setting the value of threshold for the given theme still needs further study. (2) Semantic analysis of documents to extract keywords. In the theme-related words set construction algorithm proposed in this paper, the word extraction of the document adopts the TF-IDF algorithm based on word frequency, without considering the context of the document and the related semantic environment, sometimes the extracted keywords may not be accurate enough. For the next work, we will consider using semantic analysis-based keyword extraction methods.

5   Conclusion

This paper points out that in the field of keyword extraction research, there is little studies on extracting a theme-related keywords collection, and most studies are limited to the keyword extraction of documents. By analyzing existing keywords extraction methods, this paper proposes a method to automatically construct theme-related word set based on the primary theme-related word set given by domain experts and the well-known website related to the theme. This paper uses this algorithm to experiment on the themes of "electricity" and "college entrance examination", and successfully obtains related word sets of the power theme and the college entrance examination, which show the feasibility of our methods.

Acknowledgement. This research was supported by the Training plan of Tianjin University Innovation Team (No.TD13-5025), the Natural Science Foundation of Tianjin (No.15J-CYBJC46500) and the Major Project of Tianjin Smart Manufacturing (No.15ZXZNCX00050).

References

1. Chen, Y., Liu, Y., Zhou, K., Wang, M., Zhang, M., Ma, S.: Does vertical bring more satisfaction? Predicting search satisfaction in a heterogeneous environment. In: CIKM '15 Proceedings of the 24th ACM International on Conference on Information and Knowledge Management, pp. 1581–1590. Melbourne, Australia (2015). https://doi.org/10.1145/2806416.2806473

2. Zhou, K., Cummins, R., Lalmas, M., Jose, J.M.: Which vertical search engines are relevant? In: WWW '13 Proceedings of the 22nd international conference on World Wide Web, pp. 1557–1568. Rio de Janeiro, Brazil (2013). https://doi.org/10.1145/2488388.2488524

3. Bokaetf, M.H., Sameti, H., Liu, Y.: Unsupervised approach to extract summary keywords in meeting domain. In: Signal Processing Conference, pp. 1406–1410. IEEE, Nice, France (2015). https://doi.org/10.1109/eusipco.2015.7362615

4. Hofmann, K., Tsagkias, M., Meij, E., Rijke, M.D.: A comparative study of features for keyphrase extraction in scientific literature. In: Proceedings of the 18th ACM Conference on Information And Knowledge Management, Hong Kong, China (2009)

5. Mihalcea, R., Tarau, P.: TextRank: bringing order into texts. EMNLP **4**, 404–411 (2004)

6. Wan, X., Xiao, J.: Single document keyphrase extraction using neighborhood knowledge, In AAAI'08 Proceedings of the 23rd national conference on Artificial intelligence, pp. 855–860. Chicago, Illinois (2008)

7. Gollapalli, S. D., Caragea, C.: Extracting keyphrases from research papers using citation networks. In: AAAI'14 Proceedings of the Twenty-Eighth AAAI Conference on Artificial Intelligence, pp. 1629–1635. Québec City, Québec, Canada (2014)

8. Rafiei-Asl, J., Nickabadi, A.: TSAKE: a topical and structural automatic keyphrase extractor. Appl. Soft Comput. **58**, 620–630 (2017). https://doi.org/10.1016/j.asoc.2017.05.014

9. Florescu, C., Caragea, C.: A position-biased pagerank algorithm for keyphrase extraction. In: Proceedings of the 31st American Association for Artificial Intelligence (AAAI 2017), San Francisco, California, USA (2017)

10. Blei, D.M., Ng, A.Y., Jordan, M.I.: Latent Dirichlet allocation. J. Mach. Learn. Res. **3**(4–5), 993–1022 (2003)

11. Nguyen, D.Q., Billingsley, R., Du, L., Johnson, M.: Improving topic models with latent feature word representations. Trans. Assoc. Comput. Linguist. **3**, 299–313 (2015)

12. Sfikas, G., Gatos, B., Nikou, C.: Semicca: A new semi-supervised probabilistic CCA model for keyword spotting. In: 2017 IEEE International Conference on Image Processing, pp. 1107–1111. Beijing, China (2017). https://doi.org/10.1109/icip.2017.8296453

13. Xie, F., Wu, X., Zhu, X.: Efficient sequential pattern mining with wildcards for keyphrase extraction. Knowl.-Based Syst. **115**, 27–39 (2017). https://doi.org/10.1016/j.knosys.2016.10.011

14. Joorabchi, A., Mahdi, A.E.: Automatic keyphrase annotation of scientific documents using Wikipedia and genetic algorithms. J. Inf. Sci. **39**(3), 410–426 (2013). https://doi.org/10.1177/0165551512472138

15. Sterckx, L., Caragea, C., Demeester, T., Develder, C.: Supervised keyphrase extraction as positive unlabeled learning. In: Proceedings of the 2016 Conference on Empirical Methods in Natural Language Processing, pp. 1924–1929. Austin, Texas, USA (2016). https://doi.org/10.18653/v1/d16-1198

16. Yiqun, C., Ruqi, Z., Weiheng, Z., Mengting, L., Jian, Y.: Mining patent knowledge for automatic keyword extraction. J. Comput. Res. Dev. **53**(8), 1740–1752 (2016)

17. Gollapalli, S.D., Li, X., Yang, P.: Incorporating expert knowledge into keyphrase extraction. In: Proceedings of the Thirty-First AAAI Conference on Artificial Intelligence (AAAI-17), San Francisco, California USA (2017)

Classifying Personal Photo Collections: An Event-Based Approach

Ming Geng[1], Yukun Li[1,2(✉)], and Fenglian Liu[1,3]

[1] Tianjin University of Technology, Tianjin 300384, China
[2] Tianjin Key Laboratory of Intelligence Computing and Novel Software Technology, Tianjin, China
{gengming1124,liyukun_tjut}@163.com
[3] Key Laboratory of Computer Vision and System, Ministry of Education Tianjin, Tianjin, China
lflian@tjut.edu.cn

Abstract. Due to the increasing number of photos taken by mobile phone, how to improve the efficiency of re-finding personal photos becomes an important research issue. Based on the survey of the current methods to re-find expected photos in mobile phones, this paper proposes a photo classification strategy based on specific events. We firstly discover the relationship of shooting time and shooting location to specific events, and propose a method to classify personal photos based on them. Because there is no existing data set for experiment, we collect several representative photo collections of different people as experimental data sets, and the experiments show the efficiency of our method. We implement an album system based on our method and compare it with other existing classification ones, and the results show that our strategy can improve the efficiency of re-finding photos.

Keywords: Specific events · Photo classification · Re-finding personal photos

1 Introduction

How to manage personal information based on knowledge is an important research issue. Personal information shows different types like documents, emails, photos and so on. In this paper we studied the problem by focusing on how to classify personal photos to make re-finding easy. With the wide use of mobile phone, people tend to use it to take photos for different purposes. Taking photos is not only a way to record good moments, but also a method to record important information for reusing later. For example, a person sometimes takes a photo to record the username and password of a website for reusing laer. So the number of photos taken by mobile phones is increasing day by day. How to effectively manage the photos and improve the re-finding efficiency are valuable research issues, and the classification of personal photos is a basic topic.

L. H. U and H. Xie (Eds.): APWeb-WAIM 2018, LNCS 11268, pp. 201–215, 2018.
https://doi.org/10.1007/978-3-030-01298-4_18

There are some challenges for people to study the problem. People have some personalized features, and the number of each person's photos is different. It is not easy to get a general method to classify personal photos. Because some personal photos have some privacy information, how to get a data set for experiment is also a hard problem.

This paper studied the problem and proposed a solution. The main contributions can be summarized as follows:

(1) We conducted a survey on this problem, and observed some features about personal photo classifying and re-finding. We propose the idea of re-finding personal photos by classifying them based on specific events. We also propose and define some relevant concepts.

(2) We studied the classification method based on specific events, and propose a photo classification method based on shooting time and location with high accuracy and low computational cost.

(3) Because there are no relevant data sets and benchmark, we established the experimental data sets based on the real personal photo collections and tested the proposed method from two aspects of the classification accuracy and the efficiency of re-finding personal photos. The results verify the effectiveness of our method.

2   Related Work

The existing classification methods of photos are divided into four categories. The first is based on geographical information, the second is face recognition and identity recognition; the third is based on annotations; the last one is based on events.

There are some works based on geographical information. Lo et al. [1] designed a clustering algorithm that clusters geo-tagged photos in accordance to thresholds of different scales. Papadopoulos et al. [2] proposed a way to automatically detect landmarks and events using visual features and tag similarity in a set of images with tags.

About face recognition and identity recognition, Kumar et al. [3] proposed a semi-supervised framework for recognizing faces. Brenner et al. [4] used sparse Markov Random Fields for face recognition, Xu et al. [5] and O'Hare et al. [6] implemented techniques to organize photos based on people's identity. This classification method mainly depends on the accuracy of image recognition technology. The image matching algorithm's time complexity is very high.

There are some works on annotations, Andrade et al. [7] focused on people annotation, location annotation and event annotation. Kim et al. [8] developed a system collecting and storing annotations about photos to manage and search for photos. Ulges et al. [9] and Zigkolis et al. [10] both presented an annotation framework to facilitate event classification. The method of annotation requires the image recognition technology to recognize the image content with high computational cost and time complexity, and the added annotations may not accurately express the photo semantics.

There are also some works to classify personal photos based on events. Bacha et al. [11] and Cao et al. [12] both use the connection between scenes and events to identify events. Because activities and events in our lives are structural, Bosselut et al. [13] presented a data-driven approach to learning event knowledge. Wang et al. [14] and Tang et al. [15] utilized metadata and visual features for event recognition. Yuan et al. [16] mined some features from GPS and visual cues for event recognition. There are a lot of events going on, and even the same event contains many scenes and objects. It is difficult to complete the scenes and objects statistics of all events. The same scene or object often corresponds to different events, such as hiking and beach, which may include the sky. Therefore, it is not accurate to use scenes and objects for image event recognition, and the time complexity and computational cost are very high.

According to our findings, these kinds of classification methods are only applicable in specific conditions. People taking a lot of photos in their resident locations, the classification based on geographical information is not effective to deal with this situation. For the classification methods based on face recognition and identity recognition, they only work well to the photos with human face. The annotations for classification may not accurately express the photo semantics. It is inaccurate to judge events by the scenes and objects, and the classification based on events is likely to separate photos taken in continuous time. The classification methods of photos should not only take the features of mobile phones into account, but also have high efficiency. Therefore, we need a general and efficient method to classify personal photos for re-finding.

3   Conceptual Model

3.1   Survey and Findings

We made a survey on the problem referring to 200 users, and have the following observations: (1) Current mobile phone albums use a flat storage structure to store the photos made by mobile phones, displaying thumbnails of all photos. However, the screen of the mobile phone is very small, and many photo thumbnails of the same event are all displayed. Then it is inconvenient for people to re-find photos, so it is necessary to have a suitable method for mobile phones classification and re-finding. (2) Most of the photos are made when people go out to take part in activities, and they often make dozens or even hundreds of photos, but these photos are not often re-used by people. People often reuse the photos recording some important information. Such photos are usually taken a small number at a time, and are mixed into those photos taken when people go out for fun or to take part in activities. (3) The main clues used by people to re-find a photo are the approximate shooting time and the specific event related to it. (4) Some mobile phone albums already have the function of classification based on image content. But this function destroys the memory info about time, and they often have a low performance for content analysis.

3.2   Definitions

Based on the survey and findings, this paper proposes a photo classification strategy based on specific events. To make our approach clear, we propose some concepts and give their definitions as below.

Definition 1. Specific event. A specific event can be taken as a series of actions taking place for the same target subject in a relatively continuous period of time and relatively neighboring location. For example, attending a wedding of a friend at noon one day is a specific event.

The concept of "specific event" is not the same as the concept of "event". An event may include two or more different specific events, such as the "wedding" event may include several different people's weddings. A specific person's wedding is regarded as a specific event.

Definition 2. Personal Photo Sequence. A personal photo sequence is list of personal photos ordered by their shooting time ascendingly.

Definition 3. Time Interval. Let P_i and P_j be any two photos, the time interval of them is the time difference between the shooting time of P_i and P_j. The unit is second (s). Because most mobile phones have the function of taking picture and the shooting time can be recorded automatically, to compute time interval of two given photos is not a hard thing.

Definition 4. Photo Distance. Let P_i and P_j be any two photos, the distance between the shooting location of P_i and P_j is the photo distance. The unit is meter (m). Now most mobile phones has GPS function and the location of shooting can be recorded easily, then to obtaining distance of two given photos is also not hard.

Definition 5. Rate of Location Change. For any two photos P_i and P_j, let their time intervals be T, their photo distance be L, and the rate of location change of P_i and P_j is denoted as V, then $V = L / T$. The unit of it is m/s.

4   The Classification Method

Considering the characteristics of specific events, we intend to classify photos by specific events based on shooting time and location. At present, photos taken by mobile phones contain some metadata such as shooting time and GPS (latitude, longitude, and altitude). It is not difficult to get the shooting time and location of the photos taken by mobile phone.

Table 1. Representative seven photos of personal photo sequence

Photos	Shooting time	Latitude(°)	Longitude(°)	Altitude(m)	Description of specific events
P_1	2017-01-29 08:58:34	39.13472	115.250954	0	A selfie on the train
P_2	2017-01-29 08:59:13	39.132011	115.239204	0	A selfie on the train
P_3	2017-02-02 09:31:59	39.060826	117.133378	53	Recording information
P_4	2017-02-02 11:33:34	39.060828	117.133378	54	The flowers on table
P_5	2017-02-02 11:33:40	39.060828	117.133378	54	The flowers on table
P_6	2017-02-04 22:23:05	39.051158	117.139953	0	The allergic eyes
P_7	2017-02-05 06:00:11	39.051156	117.139953	0	The allergic eyes

4.1 A Sample of Personal Photo Sequence

To make our method clear, seven photos taken continuously by one participant were selected as an example, as shown in Table 1. The P_1 and P_2 are the photos of two selfies made on the train. The P_3 is a photo for recording some information in the laboratory. The P_4 and P_5 are the photos of the flowers on the laboratory table. The P_6 and P_7 are two photos of eyes taken when the eyes were allergic. Although the time interval between P_6 and P_7 is long, the photos' content is same, and they belong to the same specific event. Therefore, the correct photo sets of the seven photos based on specific events are: $\{(P_1, P_2), P_3, (P_4, P_5), (P_6, P_7)\}$.

4.2 Specific Event-Based Classification Method

After manually classifying experimental training data sets, it was found that the shooting time and location of the same specific events were relatively continuous, and that the changes of shooting time and location within the same specific events were significantly different from the ones between different specific events. After observing some regulations about the relationship between specific events and the changing of shooting time and location, this paper proposes a method to classify photos based on shooting time and location. Algorithm 1 shows the process of our method.

Algorithm 1. The classification algorithm based on shooting time and location

---

Input: Personal photos, I_1, ... , I_n.
Output: Photo sets associated with specific event, S_1, ... , S_j.
1: Obtain the photos' shooting time and location
2: Arrange personal photos, I_1, ... , I_n, into personal photo sequence P_1, ... , P_n.
3: j = 1;
4: **For each** (i∈$[1,n-1]$)
5:     Calculate the time interval and the photo distance of two adjacent photos in the personal photo sequence T_i, as L_i;
6:     Calculate the rate of location change of two adjacent photos in the personal photo sequence V_i;
7:     If ($V_i > 2$) $L_i = L_i / V_i$;
8:     Else $L_i = L_i$;
9:     Normalize the time interval T_i and the photo distance L_i and the normalized values are $T_i{}'$ and Li';
10:     Calculate the spatio-temporal distance of two adjacent photos in the personal photo sequence ST_D_i;
11:     If ($ST\_D_i \leq BST\_D$) P_i, $P_{i+1} \in S_j$;
12:     Else $P_i \in S_j$, $P_{i+1} \in S_{j+1}$; j = j + 1;
13: **End For**
14: **Return** photo sets associated with specific event, S_1, ... , S_j.

As Algorithm 1 shown, we firstly obtain the photos' shooting time and location, and the location information is represented by a 3-ary (W, J, H), where the three parameters' meaning are latitude, longitude, and altitude. The unit of latitude and longitude is degree (°), and the unit of altitude is meter (m). Then, arrange the photos as a personal photo sequence, and afterwards calculate the time interval T and the photo distance L of any two adjacent photos. For the method to compute time interval is easy and obvious, we don't detailed it here

(1) Calculate photo distance

It is more accurate to consider three elements (W, J, H) than to only considering longitude and latitude for calculating distance. For example, there might be two photos with the same longitude and latitude, but different altitude. The commonly used method to calculate the geographical space distance is the spherical model, which regards the earth as a standard sphere, and the distance between two points on the sphere is the arc length. Any position point can be converted into three-dimensional coordinates of the sphere. The two locations are given (W_1, J_1, H_1) and (W_2, J_2, H_2). The steps to compute the distance of the two location is as below.

a. To translate H_1 and H_2 into the distance of the location to the core of the earth, as shown with the formula (1).

$$\begin{cases} h_1 = H_1 + 6371229m(the\ earth's\ radius) \\ h_2 = H_2 + 6371229m(the\ earth's\ radius) \end{cases} \quad (1)$$

b. Convert longitude J_1, J_2 and latitude W_1, W_2 into radians j_1, j_2, w_1 and w_2, as shown with the formula (2).

$$\begin{cases} w_1 = W_1 * \pi/180 \\ w_2 = W_2 * \pi/180 \end{cases}, \quad \begin{cases} j_1 = J_1 * \pi/180 \\ j_2 = J_2 * \pi/180 \end{cases} \quad (2)$$

c. Convert the position information of the two photos into three-dimensional coordinates of the sphere, as shown by formula (3).

$$\begin{cases} X_1 = h_1 * \cos(w_1) * \cos(j_1) \\ Y_1 = h_1 * \cos(w_1) * \sin(j_1) \\ Z_1 = h_1 * \sin(w_1) \end{cases}, \quad \begin{cases} X_2 = h_2 * \cos(w_2) * \cos(j_2) \\ Y_2 = h_2 * \cos(w_2) * \sin(j_2) \\ Z_2 = h_2 * \sin(w_2) \end{cases} \quad (3)$$

d. The formula of the distance L of the two locations is as formula (4), by which we can obtain the distance of two give locations.

$$L = \sqrt{(X_2 - X_1)^2 + (Y_2 - Y_1)^2 + (Z_2 - Z_1)^2} \quad (4)$$

(2) The fusion of time and distance

We consider the effect of time interval and photo distance on specific events. Before we do that, we first have to think about a situation. Some specific events happened by means of running or transportation, such as the $\{(P_1, P_2)\}$ in the sample of Sect. 4.1, who are made on the train, and the rate of location change is fast, so the photo distance is relatively large. For example, the time interval between P_1 and P_2 is just 39 s, the photo distance is 1057.3 m, and the rate of location change is 27 m/s. It may lead to wrong effect to the fusion of time and distance. Considering actual situations, when the speed is faster than 2 m/s, it can generally be concluded that the object has moved by means of running or vehicles. Therefore, after calculating the rate of location change V of two adjacent photos, if $V > 2$ m/s, the photo distance L will be reduced in proportion $L = L / V$.

Then, we consider the fusion of time and distance. Because the units are inconsistent, we must normalize them at first. The normalization method used is the Min-Max Normalization method. The formula is as follows:

$$X = \frac{x - \min}{\max - \min} \tag{5}$$

The x is the value before normalization, and the X is the value after normalization. Find the maximum and minimum values from the sequence of time intervals and photo distances. Then use the normalization method to convert the time interval T to T' and the photo distance L to L' between [0, 1].

After that we use time and distance to calculate spatio-temporal distance. Some photos are taken at the same location, but at large time intervals. These photos belong to different specific events, such as P_3 and P_4. Some photos with large changes in location but small changes in time belong to the same specific event, such as P_1 and P_2. So the weight of time and location for classification are likely to be different. We set weights to the time and location for calculating the spatio-temporal distance ST_D. We detail how to select the weight in Sect. 5.2.

$$ST\_D = \lambda * T' + (1-\lambda) * L' \tag{6}$$

(3) Choose the best spatio-temporal distance as a classifier

When the best weights are selected, calculate the accuracy of classification by different spatio-temporal distances. When the accuracy is the highest, the corresponding spatio-temporal distance are the best spatio-temporal distance BST_D. Take the best spatio-temporal distance BST_D as a classifier.

Then calculate the spatio-temporal distance ST_D of two adjacent photos in the personal photo sequence. Compare ST_D and BST_D, if ST_D is not bigger than BST_D, these two photos belong to the same specific event; if ST_D is bigger than BST_D, these two photos belong to different specific events.

5 Evaluation

5.1 The Training Data Sets

The object of this study is to classify personal photo collections taken by mobile phones, but there are no relevant public data sets. The public photo data sets don't contain all photos taken by a user with the mobile phone, such as Flickr. Only the photos that the user wants to share out, and the time interval of photos may be long apart. During this time interval, users actually took a lot of other photos with their mobile phones, but didn't share them. In addition, photos shared by users may not be the photos taken by their own mobile phones. But our classification method is for personal photo collections taken by mobile phones, so the public photo data sets cannot be used for experiment and evaluation.

To ensure the validity of our classification method, we collected several personal photo collections. Considering time-length of photo collections may affect experimental results, six representative photo collections with different time-length were selected as the training data sets. The six personal photo collections were manually classified based on specific events, and the information showing in Table 2.

Table 2. Representative users' information

Users	Sex	Age	Profession	Number of photos	Number of specific events	Time-length of photo cluster
User1	Female	24	Graduate student	491	148	10 months
User2	Female	25	Graduate student	634	178	12 months
User3	Male	24	Graduate student	646	220	7 months
User4	Male	25	Graduate student	209	53	3 months
User5	Female	23	Graduate student	886	265	16 months
User6	Male	25	Graduate student	1260	300	18 months

5.2   Experiments

(1)   Determine the weight in the spatio-temporal distance ST_D

We experimented with six training data sets, that is, a total of 4126 photos. When we calculated the spatio-temporal distances of these photos, we found that most of spatio-temporal distances of the two adjacent photos in the same specific events are less than 0.001, and most of spatio-temporal distances between different specific events are greater than 0.01. Therefore, the range of spatio-temporal distances we experimentally selected for classification is $0.001 \sim 0.01$. We selected different values of the weight λ to calculate the spatio-temporal distance. The value of λ is chosen from 0.1 to 0.9 with a step size of 0.1. Then we delineate the line charts of accuracy that select different spatio-temporal distances as the standard to classify photos when the weight λ takes different values.

As can be seen from the Fig. 1, when the accuracy is the highest in training data set, the weight λ is 0.7. Therefore, the spatio-temporal distances is calculated as $ST\_D = 0.7* T'+0.3* L'$.

(2)   Determine the best spatio-temporal distance BST_D

From the Fig. 1, it can be seen that the data set has the highest sum of accuracy when classify photos by the spatio-temporal distance of 0.002. So the best spatio-temporal distance BST_D is selected as 0.002.

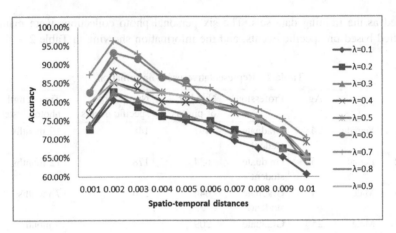

Fig. 1. The accuracy of classification by spatio-temporal distances with the weight λ taking different values

5.3   The Test Data Sets

Three users are selected randomly to test whether the method is effective, and it is also tested whether the album classifying photos based on this method can increase the speed of re-finding photos. The three users' information is as following Table 3.

Table 3. The three test users' information

Users	Sex	Age	Profession	Number of photos	Number of specific events	Time length of photo cluster
User7	Female	24	Graduate student	866	202	12 months
User8	Male	25	Graduate student	653	224	16 months
User9	Female	24	Graduate student	524	186	8 months

5.4   Evaluations of the Classification Method

In order to verify the accuracy of the method, we use the three test users' photo collections to classify photos with the spatio-temporal distance 0.002. Accuracy is a measure widely used in the field of information retrieval and statistics to evaluate the quality of results. We use the accuracy of the classification results to evaluate the methods. The accuracy is shown in Fig. 2. Since there is no relevant method of classifying based on specific events, we can't conduct comparative experiments with other methods in terms of classification accuracy.

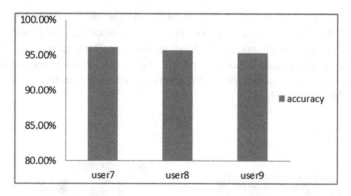

Fig. 2. The accuracy of classification by three users with the spatio-temporal distance 0.002

As shown by the Fig. 2, the accuracy of the classification based on shooting time and location is even over 95%. The biggest advantage of the proposed method is the time complexity is low, closing to O(n).

The reasons why the accuracy of the experimental results is less than 100% are that a few specific events' time interval is really small and the change in location is little. For example, one person takes a photo of the related knowledge of the papers, and then takes a selfie next minute in the laboratory. Or photos with long time intervals and little change in location, but still are the same specific event, such as $\{(P_6, P_7)\}$ in the sample of Sect. 4.1. Such situations can't be identified by this method. So this method has some limitations.

5.5   Evaluations of the Specific Event-Based Album

To verify whether the strategy classifying photos based on specific events can improve the efficiency of re-finding photos, we used the Android Studio software and the Recyclerview framework to complete the photo album based on this classification approach. In order to compare with the current photo albums in re-finding photos, the layout of design album imitated the current photo albums' structure. But the outermost layer of the album we designed is the representative photos of specific events, not all the photos. Each photo represents a specific event. Clicking the photo will jump to another page with all photos of this specific event. Currently, the characteristics of representative photos have not been studied yet, so we selected the first photos taken in specific events as the representative photos.

At present, there are a few mobile phones contains the photos classification by geographic information, events or face recognition. But there are some photo album applications including relatively complete photo classification, from which we have chosen one application with faster classification rate, and higher classification accuracy: ``time album''. We compared the speed of re-finding photos in four ways storing photos taken by mobile phones: the classification based on specific events, the classification based on geographic information, the classification based on events and the current photo album. Fig. 3. shows the albums in four ways storing photos taken by mobile phone. Fig. 3.(a) shows the layout of the current photo album. Fig. 3.(b) and (c) respectively show the classification based on geographic information and the classification based on events in the time album. Fig. 3.(d) shows the album we designed classifying photos based on specific events. We let three test users use the four ways to re-find photos, and record the time spent on each searching. The re-finding results are shown in Table 4.

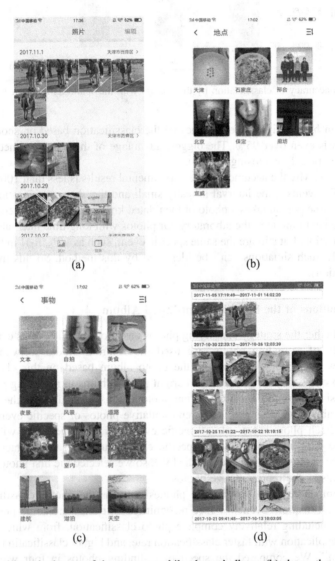

(a)

(b)

(c)

(d)

Fig. 3. (a) shows the layout of the current mobile phones' album; (b) shows the classification based on geographic information in the time album; (c) shows the classification based on events in the time album; (d) shows outermost layer in the album we designed with each photo representing a specific event

Table 4. The users' re-finding time

Users	Photos	Current album	Based on geographic information	Based on events	Based on specific events
User7	$Photo_{a1}$	6.93 s	13.56 s	8.66 s	3.29 s
User7	$Photo_{a2}$	9.82 s	31.24 s	24.13 s	4.33 s
User7	$Photo_{a3}$	18.03 s	26.1 s	15.74 s	23.5 s
User7	$Photo_{a4}$	19 s	13.62 s	25.74 s	6.63 s
User7	$Photo_{a5}$	28.12 s	40.45 s	24.56 s	10.24 s
User8	$Photo_{b1}$	25.6 s	33.23 s	40.69 s	10.47 s
User8	$Photo_{b2}$	32.96 s	23.65 s	16.78 s	8.51 s
User8	$Photo_{b3}$	22 s	45.29 s	19.56 s	9.5 s
User8	$Photo_{b4}$	15.1 s	19.64 s	20.77 s	12.3 s
User8	$Photo_{b5}$	26.76 s	15.96 s	19.34 s	12.85 s
User9	$Photo_{c1}$	36.94 s	48.7 s	39.5 s	28.61 s
User9	$Photo_{c2}$	13.95 s	9.52 s	16.54 s	6.31 s
User9	$Photo_{c3}$	26.64 s	24.33 s	18.1 s	11.17 s
User9	$Photo_{c4}$	21.7 s	17.87 s	24.33 s	15.73 s
User9	$Photo_{c5}$	12.3 s	9 s	8.51 s	6.7 s

As can be seen from the results of finding photos, our classification method used to photo albums can greatly improve the speed of finding photos. But there are one photo's re-finding time with our method is longer than that of other methods. By analysis, we find there are three main reasons:

(1) During the experiment, each re-finding process used the classification based on specific events firstly. Because of the person's own memory function, they would have some vague memory about the photos' information. It would be favorable when users used the other three ways to re-find the same photos.
(2) This is a new classification method. Users are not accustomed to using it, but they are skilled at using their own mobile phone's original album or other existing photo classification methods.
(3) The outermost representative photos selected by this album are not representative enough, and the user does not recall the specific event by this photo.

Even though there are some unfavorable factors to using this new classification to re-find photos, the results show that the classification approach based on specific events can greatly improve the speed of re-finding photos.

Thus, this classification strategy for albums can greatly improve the efficiency of re-finding photos. Because this approach can greatly reduce the time taken for sliding the screen up and down to re-find a photo, and the photos people often reuse are those ones recording information, which almost are on the outermost layer of the album. What's more, this classification strategy conforms to people's memory habits. Thus it saves time for re-finding.

6 Conclusions

This paper proposes a photo classification strategy which classifies photos based on specific events and meets the human memory habits. We also propose a classification method based on shooting time and location with high accuracy and low computational cost. It can be used for classifying photos and greatly improve the efficiency of re-finding photos. However, the accuracy of classification results cannot reach 100%, and a few photos' re-finding time is longer than using the original photo album. The next step we plan to consider combining the shooting time, location and the photos' semantic content to find a more accurate method classifying photos based on specific events. Besides, we'll do more research about the classification strategy to improve the re-finding efficiency.

Acknowledgements. This research was supported by Natural Science Foundation of Tianjin (No.15JCYBJC46500), the Training plan of Tianjin University Innovation Team (No.TD13-5025), the Major Project of Tianjin Smart Manufacturing (No.15ZXZNCX00050).

References

1. Lo, S.H.: Design and implementation the incremental clustering algorithm for geotagged photos on a map-enabled photo web service. In: Tenth International Conference on Mobile Data Management: Systems, Services and MIDDLEWARE, pp. 666–671. IEEE (2009). https://doi.org/10.1109/mdm.2009.114
2. Papadopoulos, S., Zigkolis, C., Kompatsiaris, Y., Vakali, A.: Cluster-based landmark and event detection for tagged photo collections. IEEE Multimed. **18**(1), 52–63 (2011). https://doi.org/10.1109/MMUL.2010.68
3. Kumar, V., Namboodiri, A., Jawahar, C.V.: Semi-supervised annotation of faces in image collection. Signal Image Video Process. **12**(1), 1–9 (2017). https://doi.org/10.1007/s11760-017-1140-5
4. Brenner, M., Izquierdo, E.: Joint people recognition across photo collections using sparse Markov random fields. In: Gurrin, C., Hopfgartner, F., Hurst, W., Johansen, H., Lee, H., O'Connor, N. (eds.) MMM 2014. LNCS, vol. 8325, pp. 340–352. Springer, Cham (2014). https://doi.org/10.1007/978-3-319-04114-8_29
5. Xu, Y., Peng, F., Yuan, Y., Wang, Y.: Face album: towards automatic photo management based on person identity on mobile phones. In: IEEE International Conference on Acoustics, Speech and Signal Processing (2017). https://doi.org/10.1109/icassp.2017.7952713
6. O'Hare, N., Smeaton, A.F.: Context-aware person identification in personal photo collections. IEEE Trans. Multimed. **11**(2), 220–228 (2009). https://doi.org/10.1109/tmm.2008.2009679
7. Andrade, D.O.S.D., Maia, L.F., Figueirêdo, H.F.D., Viana, W., Trinta, F., Baptista, C.D.S.: Photo annotation: a survey. Multimed. Tools Appl. 1–35 (2016)
8. Kim, J., Lee, S., Won, J.S., Moon, Y.S.: Photo cube: an automatic management and search for photos using mobile smartphones. In: IEEE Nineth International Conference on Dependable, Autonomic and Secure Computing, vol. 19, pp. 1228–1234. IEEE (2012). https://doi.org/10.1109/dasc.2011.199

9. Ulges, A., Worring, M., Breuel, T.: Learning visual contexts for image annotation from flickr groups. IEEE Trans. Multimed. **13**(2), 330–341 (2011). https://doi.org/10.1109/tmm.2010. 2101051

10. Zigkolis, C., Papadopoulos, S., Filippou, G., Kompatsiaris, Y., Vakali, A.: Collaborative event annotation in tagged photo collections. Multimed. Tools Appl. **70**(1), 89–118 (2014). https://doi.org/10.1007/s11042-012-1154-5

11. Bacha, S., Allili, M.S., Benblidia, N.: Event recognition in photo albums using probabilistic graphical model and feature relevance. In: International Conference on Pattern Recognition, pp. 2819–2823. IEEE (2017). https://doi.org/10.1109/icpr.2016.7900063

12. Cao, L., Luo, J., Kautz, H., Huang, T.S.: Image annotation within the context of personal photo collections using hierarchical event and scene models. IEEE Trans. Multimed. **11**(2), 208–219 (2009). https://doi.org/10.1109/tmm.2008.2009693

13. Bosselut, A., Chen, J., Warren, D., Hajishirzi, H., Choi, Y.: Learning prototypical event structure from photo albums. In: Meeting of the Association for Computational Linguistics, pp. 1769–1779 (2016). https://doi.org/10.18653/v1/p16-1167

14. Wang, Y., Lin, Z., Shen, X., Mech, R., Miller, G., Cottrell, G.W.: Recognizing and curating photo albums via event-specific image importance (2017)

15. Tang, F., Tretter, D.R., Willis, C.: Event classification for personal photo collections. In: IEEE International Conference on Acoustics, Speech and Signal Processing, pp. 877–880. IEEE (2011). https://doi.org/10.1109/icassp.2011.5946544

16. Yuan, J., Luo, J., Wu, Y.: Mining compositional features from GPS and visual cues for event recognition in photo collections. IEEE Trans. Multimed. **12**(7), 705–716 (2010). https://doi. org/10.1109/tmm.2010.2051868

Data Management and Mining on MOOCs

A Learning Analytic Model for Smart Classroom

Qunbo Wang$^{(\boxtimes)}$, Wenjun Wu$^{(\boxtimes)}$, and Yuxing Qi$^{(\boxtimes)}$

State Key Lab of Software Development Environment, Department of Computer Science and Engineering, Beihang University, Beijing, China
wangqunbo@nlsde.buaa.edu.cn

Abstract. With the popularity of Smart Classroom, it is necessary to study corresponding learning analytic methods to assist instructors. However, little research has investigated analyzing hidden state in class, which is an important analysis work. Therefore, focusing on the interactive learning through individual Pad devices, we propose a Learning Analytic Model to analyze hidden state with students' sequential behaviors that automatically recorded by devices. The model segments the class' process into multiple phases and construct a Hidden Markov Model (HMM) to infer students' state. In addition, a web page is developed to show students' behaviors and related analysis results intuitively. The experiment shows our model can fine-grained analyze and feedback the learning state of students in the smart classroom, which effectively help instructors improve teaching methods.

Keywords: Smart classroom · Interactive learning · Learning analytic model
HMM

1 Introduction

Smart Classroom is a popular classroom based on the multi-agent system paradigm. It takes the Ambient Intelligence (AmI) into educational process, which applies advantages of ubiquitous computing, augmented reality and mobile computing [1]. These technologies can assist to analyzing students' learning process, which can make Smart Classroom more intelligent. Research efforts have been made to investigate analysis methods for the behavior sequence in Smart Classrooms [2, 3], feedback methods of learning environment [4, 5] and multimodal learning analysis [6, 7]. They analyze learning behaviors using some basic methods of data mining, such as frequent pattern mining and pattern clustering. However, little attention has been paid to analyzing class' phase change and students' learning state, which are demanded by instructors to deeply understand students' learning process.

There is a Smart Classroom using Electronic Interactive Whiteboard (IWB), individual Pad devices, monitoring equipment, etc. in which corresponding interactive behaviors of instructors and students are automatically recorded. In order to analyzing hidden state in this Smart Classroom, we propose a new learning analytic model in this paper. For a class, this model segments the activity sequences of all the students into multiple phases. Then, it evaluates the quality of students' interactive behaviors and

© Springer Nature Switzerland AG 2018
L. H. U and H. Xie (Eds.): APWeb-WAIM 2018, LNCS 11268, pp. 219–229, 2018.
https://doi.org/10.1007/978-3-030-01298-4_19

constructs a HMM to infer the corresponding learning state of every student. After getting class' phase and students' learning state, a web page is developed to show analysis results and students' behaviors in this class intuitively. The experimental result shows our model can effectively help instructors understand class' phase change and students' learning state.

The organization of the paper is as follows: Sect. 2 briefly summarizes the related work of learning analysis in Smart Classrooms. Section 3 shows our Smart Classroom and dataset. Section 4 describes our Learning Analytic Model. Then, we show our experiment in Sect. 5. Finally, in Sect. 6 some conclusions and future work are exposed.

2   Related Work

For behavior analysis in the scenario of Smart Classroom, there are already many researches that mainly include mining frequent behavior patterns, statistical analysis of measurable behavioral information and analyzing multi-types behaviors.

Traditional methods focus on mining frequent behavior patterns in Smart Classroom. Some analyze difference of significant behavior patterns between student groups with different engagement levels [2]. Other authors cluster system-student sequential moment-to-moment interactive trajectories to capture the distinct characteristics of students' personalized learning experience [3]. These efforts often mine frequent behavior patterns and cluster behavior patterns to analyze features of students' behaviors. They cannot support continuously tracking every student cognitive state in a smart-room environment, which are really demanded by instructors.

Recently, some Smart Classrooms deploy teaching tools that can analyze class' situation. For example, Mavrikis, etc. design tools that can assist teachers in classroom settings where students are using Exploratory Learning Environments (ELEs). These tools can visualize classroom state and notify class events to facilitate teachers in focusing their attention across the whole class and inform their interventions [4]. These studies only take simple statistical analysis on measurable factors in class. It is necessary to study more accurate analysis methods.

In addition, some Smart Classrooms can record multi-types behaviors that are also investigated by some efforts. Andrade, etc. study multimodal learning analytics within an Embodied Interaction Learning Environment to understand learning trajectories of students [6]. Olney, etc. propose proportion models for Imbalanced Classes to assess the dialogic properties of classroom discourse automatically [7]. However, most Smart Classrooms only record sequential interactive behaviors with individual Pad devices of each student. We need to study this common scenario deeply.

For sequential interactive behaviors of each student in this Smart Classroom, our model segments the activity sequences of all the students into multiple phases that help instructors know the variation of class' status. And it can make fine-grained analysis of students' interactive learning state that help instructors track the students' learning state. In addition, we develop a web page that shows interactive behaviors and learning state in class intuitively.

3 Smart Classroom Scenario and Data Collection

A smart-classroom environment, enables an instructor and his students to participant in multiple kinds of interactions through their electronic devices. In such a smart-classroom environment shown in Fig. 1, every student has his/her own tablet device, through which they can access interactive courseware to read textbooks and finish on-class exercises. The instructor in such a classroom has also an upfront electronic interactive whiteboard, on which he can show course slides, display the results of student exercises, and draw annotations for his teaching. In addition, the instructor can also assign on-class quizzes to students and check the correctness of their answers.

Fig. 1. A smart classroom scenario

Behind the scene, there is often an integrated smart-classroom system that connects all these electronic devices, manages the on-class courseware and coordinates group interactions among student-teacher. The system can capture interactive behaviors from both the instructor and the students. By working with the system vendor, we integrated a xAPI-standard [8] based agent that can automatically record their interactive behaviors. This agent captures every interaction made by the instructor and students, and formats each interaction event in the form of xAPI records, as shown in Fig. 2. Specifically, our dataset of this smart classroom has 297599 behavior records of 4399 students, including 120 courses of six schools.

4 The Learning Analytic Model

In this section, we introduce a HMM-based learning analytic model to track the learning state of every student. Based on this model, we implement a learning analytic tool to mine the xAPI dataset and visualize the learning behaviors of every student.

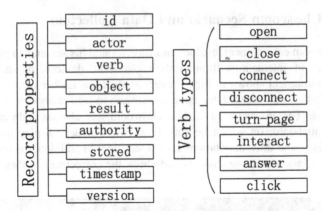

Fig. 2. The properties of a xAPI record

This tool consists of three major modules including an interaction activity parser, a HMM-based model and a graphical visualizer for rendering student behavior sequences. The next subsections explain each module in detail.

4.1 The Interaction Activity Sequence Classifier

This module parses the xAPI records generated by the smart-classroom system and segments the activity sequences of all the students into multiple phases based on the teaching-learning procedure in this class. We define three types of the phase in class shown in Table 1. In addition, both the TD and TI phases can be generally called as the non-NT phase.

Table 1. The notations of the phase type in class.

Notation	Description
NT	The instructor explains knowledge or discusses with students
TD	Every student connects or disconnects their tablets to the smart-classroom system
TI	Students use their individual Pad devices to read courseware or do exercises

During the entire process of a class in the smart-classroom, these phases often occur in an interleaved way. Such a class starts with a TD phase where everyone including the instructor connects their devices to the smart-classroom system. After that, for each knowledge topic, the class involves a NT phase and TD phase.

In order to analyze the phase type, the classifier separates the process of a class into small intervals with the time unit of 30 s, and can determines the phase type of these intervals by calculating the frequency of student activities. Specifically, the occurrence frequencies of students' activities in different class' phase are obviously different. We can know that most students don't interact with their devices during a NT phase. Therefore, a low occurrence frequency of student activities suggests a NT phase. On the contrary, a high occurrence frequency should indicate a non-NT phase. In order to determine the occurrence frequency threshold of interactive behaviors between NT phase and non-NT phase. We cluster the occurrence frequency of interactive behaviors in all time units of all courses and can get two clusters that correspond to NT and non-NT respectively. The boundary value between these two clusters can used to as the threshold value. Further, for a non-NT phase, we also cluster the occurrence frequency of debug behaviors in all time units of all courses and can get two clusters that correspond to TD and TI respectively. In addition, there would be some outliers in the clustering process as there always have individual students who don not follow instruction. Therefore, we should remove these outliers that can disturb classifying class' phases.

4.2   The HMM-Based Student Behavior Model

When the instructor explains knowledge or discusses with students (class is in NT phase), students' learning state are bad if they have behaviors in tablets. In addition, students don not need to learn when class is in TD phase. In summary, we do not need to analyze students' behaviors during NT phase and TD phase. However, during the TI phase, every student actively interacts with their tablet devices to read courseware or finish on-class quizzes. Through our observation of classroom video clips, students may behave differently in this phase. Most students with an attentive learning state work seriously with their devices whereas a handful of students touch their devices randomly or even play with them. It is necessary to infer every student's latent learning state in TI phase. Therefore, we introduce a HMM-based student behavior model that assess the quality of interactive behaviors and infers the corresponding hidden state in TI phase. This analysis process is shown in Fig. 3.

Firstly, we propose some rules to evaluate the behavior quality of students in TI phase. For example, students must take time to make a behavior. The behavior spending too little time is probably invalid. With these rules, we evaluate behavior quality into two statuses – high or low. Then, we construct a HMM to describe the process of student' learning state transition with behavior quality as observation value and use EM-algorithm to train the parameters of HMM, as shown in Fig. 4.

Different students have different probabilities because student' characteristics are unique. As shown in Fig. 4, each hidden state S_i of a student has corresponding behavior quality Q_i. The $P(S_0)$ is the probability that the initial state is attentive. Probability matrixes corresponding to the HMM are shown in Fig. 5. Among them, The $P(SC)$ is the probability of hidden state change, the $P(MH)$ is the probability that behavior in inattentive state is misjudged as high quality, and the $P(ML)$ is the opposite.

4.3   The Visualization of Student Interaction Sequences

In this module, we implement an AJAX based HTML interface to visualize the learning process in the classroom. Due to the limit of the page size, only a portion of such an interface shown in Fig. 6.

Class time	1'st time unit	2'st time unit	3'st time unit	4'st time unit	...
Student 1	Behavior sequence				
		(BQ) ↓ HMM *(LS)*	(BQ) ↓ HMM *(LS)*		...
Student 2	Behavior sequence				
		(BQ) ↓ HMM *(LS)*	(BQ) ↓ HMM *(LS)*		...
...	...	...	...	...	...
The class phase	NT/ TD	TI		NT/ TD	...

Note. BQ: behavior quality, LS: students' hidden state in class.

Fig. 3. The process of inferring hidden learning state. In the figure, each student has a behavior sequence and the class process is segmented into multiple phases. For each student in TI phase, we calculate the quality of behaviors in each time interval and use our HMM model to infer hidden learning state.

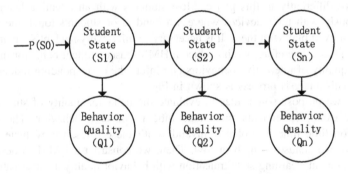

Fig. 4. The HMM-based student behavior model

The chart displays behavior sequences of all the students with the class time on the horizontal axis and student IDs on the vertical axis. For each student, a horizontal lane demonstrates the behavior sequence of a student interacting with his tablet in different icons, which defined in the legend section of the figure. In addition, we also render gray shades with different levels of darkness over those students whose learning state is inattentive. The degree of the shade darkness represents the level of a student's absent-mindedness in the class process. If a view clicks an icon of a student action, the corresponding details of this action will display. Across the chart, there are both solid lines and dashed lines to mark different phases in the teaching-learning process of the classroom and the major interaction activities of the instructor. The solid lines with different colors represent the starting point of different class phases – the green color denotes a NT phase, the red color denotes a TD phase, the yellow color denotes a TI

Initial state probability matrix π	
attentive	P(InitialState)
inattentive	1 - P(InitialState)

State transition probability matrix A		
	attentive	inattentive
attentive	1-P (SC)	P (SC)
inattentive	P (SC)	1 – P (SC)

Observation probability matrix B		
	High quality	Low quality
inattentive	1 - P (MH)	P (MH)
attentive	P (ML)	1 - P (ML)

Fig. 5. The probability matrixes of the model

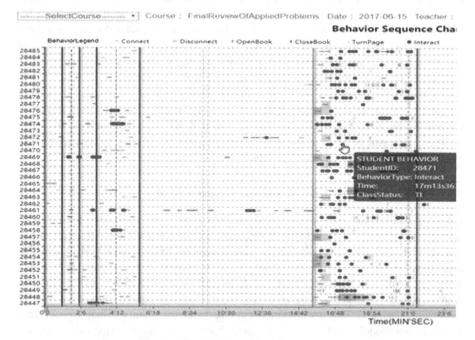

Fig. 6. Visualization of learning process

phase. The dashed lines with different colors represent the instructors' behavior activities - specifically, the green represents connecting IWB or opening book in IWB, the red represents disconnecting IWB or closing book in IWB, and the red represents paging in IWB.

5 Experiments

In this section, we use the Learning Analytic Model to analyze our dataset (Sect. 3) and evaluate the performance of the model. In Sect. 5.1, with the dataset, we cluster the behavior frequency of all time intervals and determine the threshold to distinguish different phases. Accordingly, Interaction Activity Sequence Classifier segments the class process into different phases. In Sect. 5.2, with evaluating the behavior quality in TI phase, we train the HMM-based Student Behavior Model and evaluate its performance. In Sect. 5.3, for a class, we make a video-checking experiment with the corresponding video clips to verify the visual interface.

5.1 Determine the Threshold Between Different Phases

We define the proportion of students who have behaviors in the classroom as BP, and distinguish NT and non-NT in each unit with BP. In fact, most of class time is spent explaining knowledge and discussing when students don not need to use tablets. Therefore, the number of time units is the most in NT that correspond to minimum BP value. At the same time, there always are individual students who do not listen carefully and do other things with the tablet in NT, resulting in many time units that correspond to the medium BP value. By contrast, there are not much time spent to interact with the tablet, so the number of time units is minimum in non-NT that correspond to the maximum BP value. After clustering all time units with BP value in our dataset, there are two obvious ranges – 0 to 0.17 and 0.17 to 1.0 shown in Fig. 7. Obviously, these ranges correspond to NT and non-NT. The boundary of BP value between them can be used to distinguish NT and non-NT. In a similar way, we can cluster the occurrence frequency of debug behaviors to distinguish TD and TI.

5.2 Evaluate the HMM-Based Student Behavior Model

After segmenting the class process into multiple phases and evaluating the behavior quality in TI phase, we evaluate the model using the last interval in activity sequence as test samples, and use other intervals as training samples to train the HMM by the EM-algorithm. We predict behavior quality of last interval and compare with actual behavior quality. At the same time, we define a baseline model for such prediction, which calculates the average value of a student's behavior quality along his activity sequence as student state to predict his future behavior quality. Then we compare results of our model with the base line model. Table 2 shows the average accuracy and AUC of both models on all the student sequences in our dataset. Clearly, our HMM-based model is more accurate. Comparing with other analysis methods for Smart Classroom [2–4, 6, 7], our method is more effective for instructors to understand the learning process of every students in class.

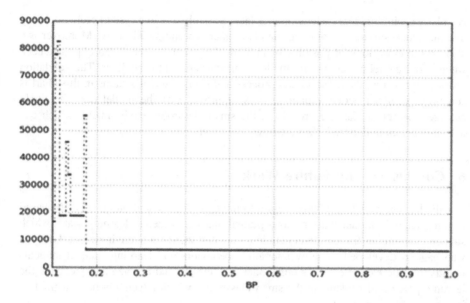

Fig. 7. The graph displays the cluster result of all time units with the BP value on the horizontal axis and corresponding number of time units on the vertical axis. Before 0.17 on the horizontal axis, it is obvious correspond to the NT phase when all students listen carefully or individual students do other things with the tablet. After 0.17, the number of time units in each BP value is small because there are not many practices in class and different practices correspond to different BP values.

Table 2. Comparison of results

Evaluation index	Baseline model	Our HMM-based model
Accuracy	0.749	0.849
AUC	0.727	0.836

5.3 A Video-Checking Experiment

We select *Final Review of Applied Problems* as the example class. For the visual results shown in Fig. 6, we can clearly compare the analysis results with the indoor video clips of this course. During the first five minutes of the class process, the class phase is marked as NT or TD because students were trying to connect their tablets to the smart-classroom server at the beginning of the class. From the 5th min to the 15th min, the class enters to a NT phase, when the instructor explains knowledge and discusses with students. After the 14th min, the chart of Fig. 3 shows the yellow solid vertical line, which indicates that the class entered a TI phase. At this stage, there are lots of interaction behaviors of students, who were working on their on-class quizzes through tablet devices. Among these activities, some icons are painted in a gray background, indicating that these students at this moment have low learning state. After the 21st min, Fig. 3 shows the green solid vertical line, which indicates that the class enters a

NT phase again. Suddenly a few minutes later, possibly due to a system problem, many disconnecting activities occur, and the class enters an ending TI phase. Moreover, we also took a close look at postures of individual students in the video clips to check whether our model correctly infer the learning states of all the students. The validation result confirms that the overall class process and every student's status in this chart is consistent with the actual situation. Thus, it proves that the model can effectively analyzes the actual learning process in the smart-classroom environment, which can help instructors understand class situation very well.

6   Conclusions and Future Work

Our model segments the class process into multiple phases, and builds a HMM-based learning analytic model that can fine-grained analyze students' learning state, which supports continuous tracking every student state in a smart-room environment. Lastly, a web page is developed to show interactive behaviors and learning state of students intuitively. The example shows our model can fine-grained analyze and feedback the learning process of students in the smart classroom, which effectively help instructors improve teaching methods.

In the future work, we will evaluate the proposed model with more datasets. In addition, we will consider other factors to evaluate behavior quality of students in class, such as the degree of concentration or knowledge mastery of students. Therefore, we will construct a multi-factorized HMM describing the student' state, and improve the EM algorithm to learn corresponding parameters.

Acknowledgement. This paper is supported by the NSFC (61532004), State Key Laboratory of Software Development Environment (Funding No. SKLSDE-2017ZX-03).

References

1. Aguilar, J., Valdiviezo, P., Cordero, J., et al.: Conceptual design of a smart classroom based on multiagent systems. In: International Conference on Artificial Intelligence Icai (2015).
2. Maldonado, R M., Yacef, K., Kay, J., et al.: Data mining in the classroom: discovering groups' strategies at a multi-tabletop environment. In: Educational data mining, pp. 121–128 (2013)
3. Shen, S., Chi, M.: Clustering student sequential trajectories using dynamic time warping. In: EDM, pp. 266–271 (2017)
4. Mutahi, J., Kinai, A., Bore, N., et al.: Studying engagement and performance with learning technology in an African classroom. In: International Learning Analytics & Knowledge Conference, pp. 148–152. ACM (2017)
5. Mavrikis, M., Gutierrezsantos, S., Poulovassilis, A., et al.: Design and evaluation of teacher assistance tools for exploratory learning environments. In: Learning Analytics and Knowledge, pp. 168–172 (2016)
6. Andrade, A.: Understanding student learning trajectories using multimodal learning analytics within an embodied-interaction learning environment. In: International Learning Analytics Knowledge Conference, pp. 70–79 (2017)

7. Olney, A.-M., Samei, B., Donnelly, P.-J., et al.: Assessing the dialogic properties of classroom discourse proportion models for imbalanced classes. In: EDM, pp. 162–167 (2017)

8. Ramirez, M.V., Collazos, C.A., Moreira, F., et al.: Relation between u-learning, connective learning, and standard xAPI: a systematic review. In: International Conference on Human Computer Interaction, pp. 1–4 (2017)

Understanding User Interests Acquisition in Personalized Online Course Recommendation

Xiao Li[1(✉)], Ting Wang[2], Huaimin Wang[3], and Jintao Tang[2]

[1] Information Center, National University of Defense Technology,
Changsha, China
xiaoli@nudt.edu.cn
[2] College of Computer, National University of Defense Technology,
Changsha, China
{tingwang, tangjintao}@nudt.edu.cn
[3] National Key Laboratory of Parallel and Distributed Processing,
National University of Defense Technology, Changsha, China
whm_w@163.com

Abstract. MOOCs have attracted a large number of learners with different education background all over the world. Despite its increasing popularity, MOOCs still suffer from the problem of high drop-out rate. One important reason may be due to the difficulty in understanding learning demand and user interests. To helper users find the most suitable courses, personalized course recommendation technology has become a hot research topic in e-learning and data mining community. One of the keys to the success of personalized course recommendation is a good user modeling method. Previous works in course recommendation often focus on developing user modeling methodology which learns latent user interests from historic learning data. Recently, interactive course recommendation has become more and more popular. In this paradigm, recommender systems can directly query user interests through survey tables or questionnaires and thus the learned interests may be more accurate. In this paper, we study the user interest acquisition problem based on the interactive course recommendation framework (ICRF). Under this framework, we systematically discuss different settings on querying user interests. To reduce performance-cost score, we propose the ICRF user interest acquisition algorithm that combines representative sampling and interest propagation algorithm to acquire user interests in a cost-effective way. With extensive experiments on real-world MOOC course enrollment datasets, we empirically demonstrate that our selective acquisition strategy is very effective and it can reduce the performance-cost score by 30.25% compared to the traditional aggressive acquisition strategies.

Keywords: Course recommendation · User modeling · Empirical study

© Springer Nature Switzerland AG 2018
L. H. U and H. Xie (Eds.): APWeb-WAIM 2018, LNCS 11268, pp. 230–242, 2018.
https://doi.org/10.1007/978-3-030-01298-4_20

1   Introduction

Since 2012, MOOCs (Massive open online courses) have attracted a large number of learners all over the world. Every day, millions of people with different education and demographics background study free online courses in popular MOOC platforms, such as Coursera[1] and Edx[2]. Compared with traditional university-level education, MOOCs dramatically speed up the way how college courses are delivered to normal crowd. From the view of higher education, MOOCs may have the potential to largely reduce the cost of higher education in a near future.

However, despite its increasing popularity, MOOCs still suffer from the problem of high drop-out rate (or low completion rate). According to the annual report of Coursera, the drop-out rate in Coursera's start-up stage ever reaches 91% [1]. Such a high drop-out rate not only means a waste of huge amount of human learning efforts but also brings negative opinions to MOOCs industry. Why does the vast majority of students fail to finish free online courses? One reason may be due to the difficulty in understanding learning demand and user interests. On the one hand, a good modeling of user interests can effectively help understand what kind of course students may like. On the other hand, a good modeling of learning motivation can also help estimate how likely a student will finish courses and thus avoids unnecessary course drop-out. However, due to the limitation of current MOOC platforms, it is usually difficult for users to explicit reveal their true interests and motivation.

To tackle this problem, personalized course recommendation technology has become a hot research topic in e-learning and data mining community. One of the key components in classic personalized course recommendation technology is user modeling which tries to learn the latent user interests from historic behavior data so that the decision module in recommendation system, i.e., a machine learning system, can estimate the likelihood of course preferences more accurately. Many works have been proposed, such as course enrollment pattern analysis [4, 13], profile text mining [5], and resource visiting log analysis [6] and so on. These methods have shown good recommendation performance. However, they are all passive as they do not actively query user interests from users but passively collect user behavior data to learn latent user interests. Due to the sparsity and noise problem, the modeling effect may not be satisfying.

Recently, the interactive course recommendation approaches have become more and more popular. Main-stream MOOC platforms, such as Coursera, have adopted such techniques to help students choose online courses (see Fig. 1). In these systems, before a user selects courses, the platform will show a list of survey tables (or questionnaires) to let her choose what topics she likes most. After the user submits the survey tables, the recommender system will integrate these explicit interest data with user demographics data together for better recommendation. For example, in Fig. 1, after a user selects the topics "Data Science" and "Machine Learning", the Coursera platform will recommend related courses, such as *Machine Learning Foundations* and *Probabilistic Machine Learning*. This recommendation mechanism has been shown to

---

[1] https://www.coursera.org/.

[2] https://www.edx.org/.

give more accurate courses to users. However, very few of research works have systematically studied the effect of human-system interaction in user interest acquisition.

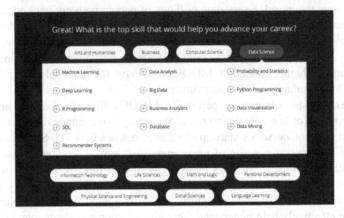

Fig. 1. The user interest survey table in Coursera's interactive course recommendation module.

In this paper, we study the user interest acquisition problem based on the interactive course recommendation framework (ICRF). We first give a formal model of this framework. After that, we systematically discuss different settings on integrating user interests into recommender system. We propose the ICRF user interest acquisition algorithm that selectively acquires interests from a small group of representative users and propagates the acquired interests to all other users. With extensive experiments on real-world MOOC course enrollment datasets, we find that selectively request users to show their interests before course selection do help improve the recommendation performance. On average, our strategy can significantly reduce the performance-cost score by 30.25% compared to the traditional aggressive acquisition strategy.

The rest of this paper is organized as follows. Section two reviews previous work in the field of course recommendation with a focus on user modeling methods. Section three describes the model of ICRF and proposes the ICRF user interest acquisition algorithm. Section four discusses the dataset, experimental setting and results. The last section concludes this paper and presents future works.

2   Related Works

In this section, we review literature of recommender system in the field of course recommendation domain with a focus on user modeling methods. Different from the movie or music recommendation, the course recommendation domain usually have complex context [1, 2, 7, 10].

Traditional data mining methods are first explored for user modeling in course recommendation. Aher et al. [4] use k-means clustering and Apriori association rule algorithm. Zhang et al. [13] develop a course recommendation system that implements distributed Apriori algorithm on computer clusters. These user modeling methods can reveal students' course enrollment pattern and show reasonable recommendation performance. However, they rely on analyzing the historical course enrollment behavior. Thus, traditional rule based methods may suffer from the cold-start problem when serving new users without any historical behavior [3].

To tackle the cold-start problem, several works try to explore contextual information, such as user profile, course prerequisite and other auxiliary information. Piao et al. [5] propose a similarity based user modeling method for MOOC course recommendation based on mining profile keywords such as job titles and skills. Jing et al. [6] propose a hybrid user modeling framework that mixes three item-based collaborative filtering scores based on user access behavior, user demographics and course prerequisite. Yu et al. [8] propose a method to mine discussion comments for user recommendation. Salehi et al. [11] propose a method combing sequential pattern mining and attribute based collaborative filtering. Social network based methods are also discussed in Zhang et al. [12]. These methods can be viewed as an extension of neighborhood-based collaborative filtering methods. They are all based on complex similarity metrics learned from contextual data. However, a major limitation of these methods is a lack of differentiable model that can be mathematically optimized to improve generalization ability.

Recently, latent factor models have become a popular research direction in the field of course recommendation. Due to their solid mathematical ground and good ability on modeling latent contextual behavior, these methods have shown better performance than item-based collaborative filtering methods. For example, Elbadrawy et al. [9] propose an ensemble of matrix factorization models each of which learns latent factors of students and courses at different levels of granularity. Sahebi et al. [14] propose a tensor factorization model for quiz recommendation. Their methods can model the temporal relationship between student knowledge and skill improvement.

Due to the effectiveness of latent factor model in course recommendation, we also use this type of recommendation methods in this paper. More specifically, we use the popular factorization machine [15] which learns the second-order interactions between features. This is especially useful for datasets with extreme sparsity, such as course recommendation datasets.

3 Interactive Course Recommendation Framework

In this section, we first describe the general model of the interactive course recommendation framework (ICRF). Second, we discuss various settings of user interest acquisition. Last, we propose the ICRF user interest acquisition algorithm to acquire user interest in a cost-effective way.

3.1 Framework Modeling

In this paper, we adopt the popular rating based recommendation approach. The goal of rating based recommendation is to learn a regression function that can accurately estimate user ratings on different courses.

Different from the traditional rating based recommendation, in ICRF, we explicitly model user interests. Specifically, we use U to denote the set of all users, C to denote the set of all courses, I to denote the domain of user interests and R to denote the rating interval (usually between zero and one)[3]. The framework can be defined as

$$f : (U, C, I) \rightarrow R \tag{1}$$

We use factorization machine (FM) [15] as the basic regression function. The regression function of FM of degree two in ICRF is defined as

$$\hat{f}(x; w, V) = w_0 + \sum_{i=1}^{n} w_i x_i + \sum_{i=1}^{n} \sum_{j=i+1}^{n} \langle v_i, v_j \rangle x_i x_j \tag{2}$$

where w and V are the model parameters and x is the feature vector that concatenates u, c and I_u as $\left[u^T, c^T, I_u^T\right]$. The most important part of FM is the inner product of v_i and v_j on the cross product of the i-th feature and the j-th feature. This has been proved to allow high quality estimation of pair-wise feature interaction under sparsity [15].

In a typical ICRF application scenario, the recommendation process usually contains four steps (as shown in Fig. 2). First, a user tags a set of topic options in the interest survey table. The survey tables usually contains many topic options which include, for example, academic topics, habit topics and professional career topics. Second, the system acquires interest data from the survey table. Third, the system conducts rating estimation on the triplet (u, c, I_u) for any $c \in C$. Finally, the estimated top rated courses will be shown as a ranking list to the user for further selection.

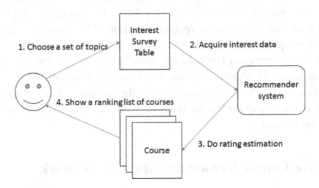

Fig. 2. The recommendation process in ICRF.

The most important part of ICRF is the second step (user interest acquisition). To represent user interests I in a computational way, we assume that any user interests can be modeled as a multivariate normal distribution of a k-dimensional random vector

$$I_u = \left[I_u^{(1)}, I_u^{(2)}, \cdots, I_u^{(k)}\right] \sim N\left(\mu, \sum\right) \tag{3}$$

---

[3] The ratings in real-world MOOC platforms can have different numerical intervals. To simplify our work, we assume those ratings can be normalized in the interval between zero and one.

where $I_u^{(i)}$ represents the degree of likelihood of user interests on the i-th application specific topic and μ and $\sum$ are the parameters.

3.2   User Interest Acquisition

Based on the above discussion, many interesting research questions can be discussed in ICRF. In this paper, we systematically study the user interest acquisition problem by answering how user and recommender system interact with each other. In real-world ICRF application, we can find four setting of user interest acquisition scenarios (see Table 1).

Table 1. Different setting on user interest acquisition.

System\Users	Cooperative answering	Reluctant answering
Aggressive acquisition	AC	AR
Selective acquisition	SC	SR

From the view of recommender system, there are two type of acquisition strategies including *aggressive acquisition* and *selective acquisition*.

Aggressive acquisition. The recommender system sends survey tables to every users. By doing so, the recommender system can build a user interests database for all users. Therefore, with the supplementary user interest information, the recommender system can model users better.

Selective acquisition. From the view of cost-sensitive learning, acquiring user interests is not free. Answering professional or privacy related questions can take users' time and even confuse them. To reduce the acquisition cost, it may be more realistic to intelligently acquire user interests from a partial group of representative users. Other user's interests can be deduced from these representative users. Thus, a significant amount of acquisition cost can be saved.

On the other side, from the view of users, there are two type of answering strategies including *cooperative answering* and *reluctant answering*.

Cooperative answering. When users receive survey tables either through emails or web portal, they are always willing to provide their course preferences and interests. For recommender systems, it means that users are very cooperative to provide personal interest data. Thus, the cost of sending survey tables will not be wasted.

Reluctant answering. A user may be reluctant to fill survey tables either because they are too busy or uncertain about questions. In such a case, the recommender system wastes amount of cost to send survey tables.

3.3   ICRF User Interest Acquisition Algorithm

Understanding the advantage and limitation of different settings of user interest acquisition is important for ICRF. In this paper, we propose the ICRF user interest acquisition algorithm[4] (see Algorithm 1) to empirically study this problem. As the SR setting (*selective acquisition + reluctant answering*) is the most complex, we will mainly discuss ICRF algorithm for SR in this section. Other settings can be easily modified from SR. The algorithm consists of an initialization part, a selective acquisition part and an interest propagation part.

In the initialization part, as it is very costly to collect true user interest online (step 1 and step 2 in Fig. 2), we have to resort simulation approaches. Specifically, we simulate the interest acquisition procedure by designing an oracle component[5] (i.e., oracles in active learning literature [18]) which can generate the true user interests. Specifically, for each user, we assume that her interests can be estimated based on the average topic distribution (learned by the LDA model [20]) on all her enrolled courses (step 1). Second, we define the probabilistic function e as a reluctant trigger which mimics the reluctant behavior (i.e., user will answer only when $e(u) > h$) for AR (*aggressive acquisition + reluctant answering*) and SR (*selective acquisition + reluctant answering*) settings (step 2).

Algorithm 1: ICRF user interest acquisition algorithm

Input: a training set $D^T=(U, C, R)$, the Gaussian parameters Σ, the cost per acquisition v, the damping factor d, number of clusters K, the reluctant threshold h, the nearest neighbor graph parameter k

Output: a new training set $\tilde{D}^T=(U, C, I, R)$ and the total cost V

Initialization:

1: Learn the topic distribution of each $c \in C$ in D^T by the LDA model. For each $u \in U$ in D^T, set $O(u) \leftarrow N(\bar{t}_{C_u}, \Sigma)$ where $\bar{t}_{C_u}$ is the average topic distribution on all courses enrolled by u.

2. Define the reluctant trigger function e.

Selective Acquisition:

3: Build a selective user pool $U^{(s)}$ by k-means clustering.

4: For each training instance $(u, c, r) \in D^T$ do

5:     $I_u \leftarrow$ If $u \in U^{(s)}$ and $e(u) > h$ Then $O(u)$ Else $\emptyset$

6:     $V \leftarrow V +$ If $u \in U^{(s)}$ Then v Else 0

7:     $\tilde{D}^T \leftarrow \tilde{D}^T \cup (u, c, I_u, r)$

Interest Propagation:

8: Repeat until converge

9:     For each training instance $u \in U - U^{(s)}$ do

10:        $I_u \leftarrow (1 - d) + d \times \sum_{v \in N_u} \frac{1}{|N_u|} I_v$ (N_u is the set of neighbour of u)

11:        Update (u, c, I_u, r) in $\tilde{D}^T$

---

[4] In the rest of this paper, the ICRF user interest acquisition algorithm and the ICRF algorithm will be used interchangeably.

[5] The oracle refers to a virtual agent that is omniscient to know true user interests.

In the selective acquisition part, we first adopt the representative sampling strategy to build the selective user pool $U^{(s)}$ (step 3). Specifically, we use k-means clustering to group users into different set and choose the users that are most closed to the clustering centroids as representative users. The recommender system with selective acquisition strategy will only query interest data for users in $U^{(s)}$. Second, for each user, we generate her user interests either as $O(u)$ or $\emptyset$ based on the in-pool condition and reluctant trigger condition (step 5). We also add an acquisition cost to the total cost V (step 6). The acquired user interests I_u will be added in a new training set $\tilde{D}^T$ (step 7).

In the interest propagation part (from step 8 to step 11), we propose a simple interest propagation algorithm to iteratively update interests for not-queried users (i.e., $u \in U - U^{(s)}$). The idea is borrowed from the famous PageRank [19] algorithm which has been applied to many domains, such as keyword extraction [16]. The interest propagation algorithm relies on the user nearest neighbor graph. We form the graph edges by connecting each pair of users if they are among each other's k nearest neighbors (measured by Euclidean distance on demographics data). The damping factor d plays the role of absorbing interest from other random users.

4 Experiments

In this section, we first describe the datasets used in our experiments. Second, we present the experimental configuration. Third, we conduct experiments to study the following research questions:

RQ1. Is the selective acquisition strategy better than the aggressive acquisition strategy in terms of the performance-cost score?

RQ2. How does the interest propagation method improve performance-cost score?

RQ3. Are the reluctant answering behavior harmful?

4.1 Dataset

The dataset used in this work is provided by a MOOC platform deployed at our university[6]. As there are usually no explicit rating data for MOOC courses, we use video watching progress as pseudo ratings. Generally speaking, the higher the ratings, the more likely students may finish all videos. Thus, the goal of course recommendation becomes recommending courses that students are likely to finish all videos.

This dataset contains 319,408 course enrollment events spanning from Sept 2014 to April 2015. During this period, 93,063 students have enrolled in 79 MOOC courses. For each student and each course, the dataset provides categorical features and text features (see Table 2). For example, an undergraduate student in College of Compute Science can be represented as a sparse 0/1 feature vector where only the feature index in *"Undergraduate"*, *"Student"* and *"College of Computer Science"* are one. For each

---

[6] The MOOC platform provides free online courses for both on-campus students and off-campus professional employees in China.

course, the dataset also provides title, introduction and syllabus text that can be used to learn topic distribution.

Table 2. Feature description.

Field	Feature	Example of feature value
User	Education level	Undergraduate/Master/PhD
	Career	Student/Engineer/Officer
	Company or college	College of Computer Science
Course	College	College of Computer Science
	Academic subject	Engineering/Science/Law/Art
	Course type	Audit/Exam
	Teachers	Bob/Alice/Jack
	Text	Words in title, introduction and syllabus

In our experiments, we split the full dataset into six groups by a time window of three months (see Table 3). Specifically, in each time window, the first two months of data is used as the training set and the rest one is used as the testing set. Thus, in total, six groups of datasets will be used in our experiments.

Table 3. Dataset split.

Dataset	I	II	III	IV	V	VI
Training	2014-09	2014-10	2014-11	2014-12	2015-01	2015-02
	2014-10	2014-11	2014-12	2015-01	2015-02	2015-03
Testing	2014-11	2014-12	2015-01	2015-02	2015-03	2015-04

4.2   Configuration

Experimental Protocols. For each group of dataset, we use the ICRF user interest acquisition algorithm to generate explicit user interests for users in both training and testing sets. For the selective acquisition strategy, we set the cluster number of k-means algorithm as 500. The k parameter to build the nearest neighbor graph is set as 100. Following the work of [16], we also set the damping factor d in the interest propagation algorithm as 0.85. When the oracle queries a user, we accumulate the total cost by one.

We use the famous libFM [17] library as the implementation of FM. The popular SGD (stochastic gradient descent) optimization method is used to train FM regression functions on the training set. In our pilot experiments, we find that setting the learning rate as 0.001 and the regularization parameter as 0.0001 for SGD can have reasonable performance. Thus, we use these hyper-parameters to train FM. The number of training iteration is restricted as 50.

To evaluate each strategy in a cost-sensitive way, we compare the performance-cost score (PCS) of different strategies. Intuitively, a good user interest acquisition strategy should not only have good recommendation performance (e.g., low RMSE score) but also have low acquisition cost (i.e., the lower the PCS score is, the better the strategy performs). Thus, we define a new evaluation metric by combining RMSE and cost acquisition as

$$PCS = \sqrt{RMSE \times Cost} \tag{4}$$

Implement of Other Settings. We summarize the differences of the four settings in Table 4. By default, the ICRF algorithm implements the SR settings. We can easily modified ICRF algorithm to implement the other settings by removing not used components.

Table 4. Differences of the four user interest acquisition settings.

Setting	Oracle	Selective acquisition	Interest propagation	Reluctant trigger
AC	Yes			
AR	Yes			Yes
SC	Yes	Yes	Yes	
SR	Yes	Yes	Yes	Yes

4.3 Aggressive Acquisition VS Selective Acquisition (RQ1)

Table 5 compares the PCS scores between AC (*aggressive acquisition + cooperative answering*) and SC (*selective acquisition + cooperative answering*). We can see that on all the six datasets, SC outperforms AC significantly. On average, the SC strategy can reduce the PCS score by 30.25% compared to AC. This means that in real-world interactive course recommendation application, the selective acquisition strategy that intelligently acquires interests from representative users and propagates their interest to other similar users is much better than the aggressive acquisition strategy that blindly acquires interest from all users.

Table 5. Comparing the PCS scores between AC and SC.

Dataset	I	II	III	IV	V	VI
AC	115.70	69.03	85.73	101.98	106.78	84.01
SC	92.17	47.03	66.54	62.94	65.71	58.65
Improvement	20.34%	31.87%	22.39%	38.29%	38.46%	30.19%

4.4 Effect of Interest Propagation (RQ2)

The selective acquisition strategy relies on the interest propagation (IP) algorithm to propagate interests to users not in the selected pools. How about the performance of SC without IP? To study this issue, we compare SC with two alternatives. For each user not in the selected pools, the *SC-IP* strategy assigns no interests while the *SC-IP +Center* strategy assigns the interests of cluster centers where the user belongs to.

From Table 6, we can see that on all the six datasets the SC strategy outperforms the other two variants. This demonstrates that propagating interests to similar users is important in the ICRF user interest acquisition algorithm.

Table 6. Compare PCS scores of SC, SC-IP and SC-IP+Center.

Dataset	I	II	III	IV	V	VI
SC	92.17	47.03	66.54	62.94	65.71	58.65
SC-IP	92.44	48.04	68.24	63.37	66.75	59.23
SC-IP+Center	93.34	48.77	69.02	67.61	68.89	60.81

4.5   Is Reluctant Answering Harmful (RQ3)

When users refuse to answer survey tables, the recommender system will waste amount of cost. How will be such reluctant behavior harmful to ICRF framework? To study this issue, we conduct experiments for SR (*selective acquisition + reluctant answering*) and AR (*aggressive acquisition + reluctant answering*) on the I and the V datasets where we find the smallest and biggest PCS improvement of SC over AC (both strategies do not consider reluctant behavior). The reluctant thresholds are set from 0.3 to 0.7.

Table 7 tabulates the PCS scores at different reluctant thresholds. We can see that with the increasing of reluctant thresholds, the PCS scores of both AR and SR continue to increase. This means that reluctant answering behavior do have negative impact on performance-cost score. However, the PCS scores of SR are always lower than AR in all cases. It means that the selective acquisition strategy is more robust than the aggressive acquisition strategy when reluctant users exist.

Table 7. Comparing PCS scores at different reluctant thresholds.

Dataset	Strategy	0.3	0.4	0.5	0.6	0.7
I	AR	137.9	140.0	149.3	148.1	158.7
	SR	96.5	111.4	124.5	136.4	147.4
V	AR	120.1	124.5	127.9	128.7	130.8
	SR	66.5	67.5	75.5	82.7	89.3

5   Conclusion

In this paper, we study the problem of user interest acquisition in interactive course recommendation. Different from traditional course recommendation methods where user modeling is conducted by mining historic user behavior data, in this paper, we study a new paradigm where recommender systems can directly query user interests through survey tables or questionnaires. We describe the formal model of ICRF (Interactive Course Recommendation Framework) and propose the ICRF user interest acquisition algorithm to reduce performance-cost score. Specifically, we use k-means clustering to choose representative users for querying oracles and propose an interest propagation algorithm to deduce interests for not selected users. With extensive experiments on real-world MOOCs enrollment datasets, we empirically demonstrate that our selective acquisition strategy is very effective and it outperforms the traditional aggressive acquisition strategy by 30.25% in terms of performance-cost score.

In our future work, we plan to study the ICRF framework by reinforcement learning where interest acquisition actions will be decided by a policy agent that learns the optimal policy from the environment to maximize expected recommendation reward.

Acknowledgments. We thank the reviewers for their helpful comments. This work is supported by the National Key Research and Development Program of China (2018YFB1004502), the National Natural Science Foundation of China (61702532) and the Key Program of National Natural Science Foundation of China (61532001, 61432020).

References

1. Koller, D.: MOOCs on the move: How coursera is disrupting the traditional classroom. Knowledge@Wharton Podcast (2012)
2. Parameswaran, A., Venetis, P., Garcia-Molina, H.: Recommendation systems with complex constraints: a course recommendation perspective. ACM Trans. Inf. Syst. **29**(4), 20:1–20:33 (2011)
3. Schein, A.I., Popescul, A., Ungar, L.H., Pennock, D.M.: Methods and metrics for cold-start recommendations. In: Proceedings of the 25th Annual International ACM SIGIR Conference on Research and Development in Information Retrieval, pp. 253–260. ACM, Tampere (2002)
4. Aher, S.B., Lobo, L.M.R.J.: Combination of machine learning algorithms for recommendation of courses in E-Learning System based on historical data. Knowledge-Based Syst. **51**, 1–14 (2013)
5. Piao, G., Breslin, J.G.: Analyzing MOOC entries of professionals on LinkedIn for user modeling and personalized MOOC recommendations. In: Proceedings of the 2016 Conference on User Modeling Adaptation and Personalization, pp. 291–292. ACM, Halifax (2016)
6. Jing, X., Tang, J.: Guess you like: course recommendation in MOOCs. In: 2017 Proceedings of the International Conference on Web Intelligence, pp. 783–789. ACM, Leipzig (2017)
7. Hou, Y., Zhou, P., Wang, T., Yu, L., Hu, Y., Wu, D.: Context-Aware Online Learning for Course Recommendation of MOOC Big Data. arXiv preprint (2016)
8. Yu, Y., Wang, H., Yin, G., Wang, T.: Reviewer recommendation for pull-requests in GitHub. Inf. Softw. Technol. **74**(C), 204–218 (2016)
9. Elbadrawy, A., Karypis, G.: Domain-aware grade prediction and top-n course recommendation. In: Proceedings of the 10th ACM Conference on Recommender Systems, pp. 183–190. ACM, Boston (2016)
10. Klašnja-Milicevic, A., Ivanovi, M., Nanopoulos, A.: Recommender Systems in e-Learning Environments: A Survey of the State-of-the-art and Possible Extensions. Artif. Intell. Rev. **44**(4), 571–604 (2015)

11. Salehi, M., Nakhai Kamalabadi, I., Ghaznavi Ghoushchi, M.B.: Personalized recommendation of learning material using sequential pattern mining and attribute based collaborative filtering. Educ. Inf. Technol. **19**(4), 713–735 (2014)
12. Zhang, Y., Wang, H., Yin, G., Wang, T., Yu, Y.: Social media in GitHub: the role of @-mention in assisting software development. Sci. China Inf. Sci. **60**(3),032102 (2017)
13. Zhang, H., Huang, T., Lv, Z., et al.: MCRS: a course recommendation system for MOOCs. Multimed. Tools Appl. **77**(6), 7051–7069 (2018)
14. Sahebi S., Lin Y., Brusilovsky P.: Tensor Factorization for Student Modeling and Performance Prediction in Unstructured Domain, pp. 502–506. Raleigh, NC, USA (2016)
15. Rendle S.: Factorization Machines, pp. 995–1000. Sydney, Australia (2010)
16. Mihalcea R., Tarau P.: Textrank: bringing order into text. In: Proceedings of the 2004 Conference On Empirical Methods In Natural Language Processing, pp. 404–411. (2004)
17. Rendle, S.: Factorization machines with libFM. ACM Trans. Intell. Syst. Technol. **3**(3), 1–22 (2012)
18. Settles B.: Active learning. Synth. Lect. Artif. Intell. Mach. Learn. **6**(1), 1–114 (2012)
19. Page L.: PageRank: Bringing Order to the Web. Stanford Digital Library Project (2002)
20. Blei, D., Ng, A., Jordan, M.: Latent dirichlet allocation. J. Mach. Learn. Res. **3**, 993–1022 (2003)

A Learning Analytics System for Cognition Analysis in Online Learning Community

Yinan Wu$^{(\boxtimes)}$ and Wenjun Wu$^{(\boxtimes)}$

State Key Lab of Software Development Environment,
Department of Computer Science and Engineering,
Beihang University, Beijing, China
wuyinan@nlsde.buaa.edu.cn

Abstract. While cognitive behaviors and social network structure in Online Learning Community (OLC) have been studied in the past, few research has proposed a model linking the two important factors to analyze students' cognitive learning gains, even though it has been widely acknowledged that interaction is a significant way for students to exchange knowledge and obtain learning gains. In this paper, for a better indication of cognitive gains, we introduce an analytic model to quantify the students' learning gains by using a redesigned taxonomy of cognitive behaviors while considering the flow of knowledge among students in discussion forums. And further, we implement a learning analytics system to streamline the data analysis pipeline of social network analysis, cognition classification and learning gain calculation and visualize the analytic results from multiple-level views including student, discussion thread and forum. We demonstrate the results on a MOOC course and confirm the effectiveness of our model. Our model and analytic system enable instructors and TAs to take active mediation among online discussions of students to improve their cognitive gains through OLCs.

Keywords: MOOC discussion forums · Social learning network
Cognitive learning gains · Taxonomy of cognition

1 Introduction

Massive Open Online Courses (MOOC) has attracted millions of registered users to learn over the Internet, which scaled distance education to a magical size that everyone can participate in courses developed by numerous universities and educational institutions [1]. However, many problems regarding MOOCs such as high dropout rate and low terminal efficiency remain unsolved. To address these problems, Online Learning Community (OLC) had been provided to increase the teacher-to-student ratios and face-to-face interaction [2]. In addition, the instructor and TAs are able to monitor the learning progress based on the posts [3]. Previous research efforts have qualitatively and quantitatively proved that students' participation in online discussion is positively correlated to their learning gains [4, 5]. For exploring the factors in forums that would influence students' learning gains, many methods are proposed from the two main aspects: content analysis and social structural analysis.

L. H. U and H. Xie (Eds.): APWeb-WAIM 2018, LNCS 11268, pp. 243–258, 2018.
https://doi.org/10.1007/978-3-030-01298-4_21

Currently, most researchers adopt content-related analysis, which focuses on students' posts observed in discussion forums, such as topic analysis, semantic analysis and emotional analysis. OLCs offer a new environment that provides computer supported collaborative learning activities, where social existence might reflect cognitive existence [6]. Besides, interpersonal relationships among students can provide cognitive and emotional support that ultimately benefit the learning process [7]. Online learning community as a social learning network (SLN), provides a place where teachers and students exchange information and share their ideas and insights into course topics through discussion threads. Thus, structural connectivity between each pair of students, is also an important factor of affecting students' learning gains [8]. However, the social structure of OLCs is not well applied on analyzing students' cognitive learning gains in previous investigations.

In this work, we aim to combine SLN structural analysis of OLCs and cognitive analysis to build a new analytic model to quantitatively assess the students' learning gains. Further on, considering the continued rising in popularity of OLCs [9], we parallelized the algorithm in our cognitive gains model and developed a Spark-based analytics system to streamline the large-scale data analysis from forum data collecting, data analysis to result visual presentation. In particular, we contribute to the existing literature by (1) redesigning an easy labeled coding scheme of cognitions based on Wang's framework [10] for precisely categorizing students' cognitive behaviors in OLCs; (2) developing a classification workflow including posts preprocessing, labeling, features extracting and modeling; (3) visualizing the learning gains from the perspectives of student, thread and course forum respectively.

The remainder of the paper is organized as follows: Related work section describes the related work and previous theoretical basis that we take advantage of in our research. Dataset section describes the datasets for the research of the paper, and in Methods section, we elaborate our redesigned coding scheme, post processing methods, cognitive gains model and the selection of latent parameters. Results section describes the implementation of our analytics system and experimental results of our analytic model. Conclusion section gives the conclusion and points out the future work.

2 Related Work

2.1 Social Network Analysis on OLCs

Online learning community provides students an online environment to interact with their learning partners. One of the issues about social learning pointed out by Putnik [11] is to analyze the students' interactions using techniques from the social network analysis field. The interactions between students in OLCs mainly includes asking questions, offering answers and giving opinions, and all of them promote the flow of knowledge exchange between students and bring them benefits. Brinton et al. [12] proposed a framework for modeling SLN efficiency of information exchange between students and evaluated students' gains in topic level. The study defines student's benefit from his/her learning benefits by asking questions and his/her teaching benefits by providing answers. The study has set up a good foundation for analyzing students' gains in social structure of OLCs.

However, it is not enough to express the gains from questions or answers in different levels of cognition. For example, novel questions would benefit students more than simple questions and reasoned answers would benefit students more than repeated answers. Such an observation motivates more fine-grained analysis method to be defined for classifying the gain sources in multiple cognitive levels.

2.2 Cognitive Analysis on OLCs

Investigating the online behaviors in cognition level in OLCs has been one major theme for research. Elouazizi [13] used linguistic cues to measure cognitive engagement in forum data and found evidence of low cognitive engagement. Wong [14] used content analysis to investigate the tendency of different levels of cognitive learning in OLCs. Wang [10] classified different behaviors into three cognition categories including active, constructive and interactive behaviors based on Chi's ICAP framework [15] and explored the relationships between each cognitive behavior and students' scores. Their finding imply that students' interactive discussion behaviors will benefit more than constructive behaviors, and constructive behaviors will benefit more than active behaviors.

These research results give our work strong theoretical support, especially Wong's study that provides a quantification basis for each cognition.

However, only the quantity of student's behaviors in each cognition is considered to affect the scores in Wong's study, but a student wouldn't obtain gains without information exchanges (i.e. no one replies his/her post) in the forum. Hence in our work, we take others' contribution into account to calculate students' cognitive gains.

3 Dataset

Data for this research comes from two different data sources. For testing the classification result of cognitive behaviors, we used the Stanford MOOC Posts dataset [16], which contains 29,604 posts from OLCs within the Education, Humanities and Medicine domain areas. It is convenient for testing because the opinion, question and answer classifications are labeled in the dataset.

But the Stanford dataset doesn't cover students' information required in our cognitive gains analysis. We have to collect the second dataset of an OLC from an online educational institution (referred as SETC), made available upon successfully fulfilling application requirements. The SETC dataset contains 15,536 posts and 128,202 answers in 1540 questions, covering 2,858 students in an OLC within the 420 high school courses.

To analyze students' cognitive gains over the social network, we select the course "Mathematic Thinking Method" from the second dataset, which includes the 1592 posts made by 403 students. "Mathematic Thinking Method" was designed as a 11-week course. For each week of class, students focused on a major topic, watched the video lecture, completed quizzes, and discussed in the forum. A screenshot of the course discussion forum is shown in Fig. 1.

Fig. 1. Screenshot of discussion forum in the course Mathematic Thinking Method

4 Methods

4.1 Taxonomy of Cognition

Among the taxonomy of cognition, Chi's ICAP framework [15] has been widely used to discriminate cognitive behaviors and interpret the learning result [10, 17]. The framework classified the learning behaviors into three categories, which is, active, constructive and interactive behaviors. The active behaviors infer to the degree that students engaged in the learning process. The constructive behaviors indicate how much students produce new information beyond the presented materials. And the interactive behaviors involve interaction and cooperation with partners. We choose this framework as the basis of our discourse classification algorithm and extend it into a three-level classification method.

For cleaning posts of futility in cognitive classification, we firstly distinguish on-task and off-task discourse in the dataset. Off-task posts refer to those discourses that talking about administrative issues about the course [10], or only have contents irrelevant to academics and nonsense words, such as greeting, self-introduction and emoticons. Within the on-task labeled posts, we drew on constructs of coding scheme on cognitive domain as articulated by Wang [10], based on Chi's ICAP (Active-Constructive-Interactive) framework [15]. For suiting our analysis and for convenience of labeling, we adjusted the coding scheme and determined abbreviated definitions of each category in Table 1.

Table 1. Cognition coding examples

Cognition level 1	Cognition level 2	Cognition level 3	Abbr.	Definition
Off-task			Off.	Talk about administrative issues about the course, contents irrelevant to academics and nonsense words
On-task	Active	Provide simple/repeat answer/opinion	Act. A/O	The student states an opinion or provides an answer without reasons, or just repeats the information that's already covered in the course material
		Ask simple question	Act. Q	The student proposes a question that just repeats the information given in the material without his/her own understanding
	Constructive	Provide reasoned answer/opinion	Con. A/O	The student supports his/her answer or opinion with evidence, e.g., giving examples, comparing or connecting with external resources
		Ask novel question	Con. Q	The student proposes a novel question based on his/her own understanding
	Interactive	Acknowledgement or expand on	Int.	The student acknowledges others' statements, or expand on them
		Defend and challenge		The student challenges others' ideas, or defends his/her own ideas, when there is a disagreement

4.2   Post Processing

4.2.1   Data Wrangling

As the first step in processing forum post data, we aim to extract the useful discourse information from every post by:

- Removing all punctuations except for question marks (probably a question), quotation marks (probably a repeat) and exclamation mark (probably an Interactive post)
- Converting all URLs and images to text notation (probably providing reasons)
- Removing all emails and emoticons
- Removing all stopwords from an aggressive stopword list
- Stemming all words
- Segmentation to get bag of words

After the data wrangling, we will get enough formatted information for feature extracting.

4.2.2 Labeling

In supervised learning, ground truth is significant for training, which requires expertise to distinguish each category. Hence we introduce distinctive features for each category high distinction that can be recognized via some simple features, as list in Table 2.

Table 2. Indicator of each cognitive behavior

Cognitive behavior	Indicator
Off-task	Could be indicated by administrative words, e.g., "homework", "submit", "download", and nonsense words, e.g. "hah", "lol"
Provide simple/repeat answer/opinion	Could be indicated by quotation marks, a short length of text, and a simple statement without details
Ask simple question	Could be indicated by question and quotation marks, topics about "what" and "where", and a short length of text
Provide reasoned answer/opinion	Could be indicated by longer length and cognitive action verbs, e.g., "propose", "imagine", as well as images and URLs
Ask novel question	Could be indicated by question marks, topics about "why" and "how", and a longer length of text
Interactive	Could be indicated by longer length and interactive action verbs, e.g., "agree", "disagree", as well as exclamation mark

Certainly, a post may contain multiple cognitive behaviors, but it makes no difference for analysis with two or more labels on one post because we calculate all behaviors distribution and put them in model, and for the sake of accuracy, we advise to do so.

4.2.3 Feature Extracting

On Cognition Level 1, we take the following features to make a distinction between on-task and off-task posts:

- The number of the top 30 linguistic words for content-related and non-content-related posts proposed by Cui [18]
- The number of words in the post

On Cognition Level 3, for identifying cognition related words, we adopt the action verbs for each level of revised Anderson and Krathwohl cognitive taxonomy [19] and extract the features below:

- The number of words in the post. Kovanović [20] pointed out the longer the message is, the higher the chances are for the message to display higher levels of cognitive presence
- The number of cognitive action verbs on each level
- The number of each punctuation mentioned in preprocessing section
- The number of text notation of URLs and images
- The number of the first person pronouns

- The number of the second person pronouns (probably an Interactive post)
- The number of upvotes (suggesting a gainful post)

Experimental results of Sect. 5 confirm that these feature choices can achieve more accurate classification on Cognition Level 3.

4.2.4 Classification Modeling

For binary classification of on-task and off-task, we designed 3 methods including SVM, Bayesian model and logistic regression model on the following three courses: Education How to Learn Math (DS1), Medicine Sci Write Fall2013 (DS2) from Stanford dataset and Mathematic Thinking Method (DS3) from SETC dataset. Table 3 shows that on the forum data of these courses, the logistic regression model can achieve the best classification effect of AUC over 0.8, which illustrates that the results are within a reasonable range for our further analysis.

Table 3. AUC of 3 datasets using 3 methods

Dataset	SVM	Bayesian model	Logistic regression
DS1	0.82	0.79	0.85
DS2	0.79	0.77	0.83
DS3	0.84	0.78	0.86

For predicting the on-task into categories in Cognition Level 3, considering a post may contain one or more cognitive behaviors, e.g., one asks a question immediately after he formulates his ideas or interacts with others in one post, we adopted 5 binary classifier using logistic regression model. On the forum data of the three courses, the average accuracy from 10-fold cross validation for each category has been made bold in Table 4.

Table 4. Average accuracy from 10-fold cross validation

Dataset	Active		Constructive		Interactive
	Q	A/O	Q	A/O	
DS1	0.78	0.79	0.82	0.85	0.74
DS2	0.76	0.75	0.79	0.83	0.73
DS3	0.82	0.80	0.83	0.87	0.78

4.3 Cognitive Gains Model

4.3.1 Model Design

When a student posts a comment or replies to someone in a discussion thread, his cognitive behavior influences both how much others gain from his post and what cognition level others can behave following his post in this thread. To analyze cognitive gains in OLC discourses, we introduce a new cognitive gain model to analyze student's cognitive gains in fineness of threads.

We define $f_{u,t,i}$ in (1) to be the probability of user u posting on cognitive behavior c in thread t, where $p \in P_{u,t}$ captures all the posts made by user u in thread t, $c_i \in C$ denotes cognition c_i in the set C of all the categories in Cognition Level 3 and $\varphi_{p,i} \in [0, 1]^{|P| \times |C|}$ is post-cognition distribution resulting from 6 binary classifier, giving the probability that the cognitive behavior in post p contains c_i.

$$f_{u,t,i} = \log\left(1 + \sum_{p \in P_{u,t}} \varphi_{p,i}\right) \tag{1}$$

To analyze cognitive gains of each user, we should know how much is the possibility that the spread of cognition from user u to user v in the course, that is, the probability of user u replying to user v if user v makes a post. Hence we set $r_{u,v}$ in (2) to quantify it.

$$r_{u,v} = \frac{\sum_t n_{u,v,t}}{N_u} \tag{2}$$

Since in some cases the number of times that user u replies to user v may be so small that the probability calculated in (2) would be unrepresentative, we introduce a heuristic definition instead: in the same thread t, if user u makes a post later than user v, we define that user u replies to user v, denoted by $n_{u,u,t}$. Let $N_u = \sum_t P_{u,t}$ be the total times that user u posts in the course.

By now, the total cognitive gains of user u in thread t can be modeled as:

$$G_{u,t} = \sum_{c_i \in C} g_{u,i} \log\left(1 + \sum_v r_{v,u} f_{v,t,i}\right) \tag{3}$$

Here, $r_{v,u} f_{v,t,i}$ captures the theoretic amount of cognitive response provided from user v to user u in thread t. And further on, we can get the gains of each user as $G_u = \sum_t G_{u,t}$ and sort out the threads by $G_t = \sum_u G_{u,t}$ where users obtain the most gains. We adopt a gain rate $g_{u,i}$ from the spread of others' cognitive behavior, considering that student u can obtain different degrees of cognitive gains from different cognitive behaviors c_i. The selection of gain rate is discussed in next section.

4.3.2   The Selection of Gain Rate

To evaluate the gain rate of each cognition, we take the test scores of students into two parts according to answering time: the scores before and after students participating the OLC's discussion. And then we use Huang's approach [21] to simplify the evaluation computing of gain rate.

Firstly, we model $s_{u,k,i}^{(T)}$ as a Bernoulli random variable, which is observation binary test score of the student u answering the question q_k at time instance T in terms of cognition c_i, where $c_i \in C$:

$$s_{u,k,i}^{(T)} \sim Ber\left(p_{u,k,i}^{(T)}\right) \tag{4}$$

$$p_{u,k,i}^{(T)} = \Phi\left(a_{k,i}c_{u,i}^{(T)} - b_{k,i}\right) \tag{5}$$

Here, Φ is the inverse logit link function $\Phi(z) = \int_{-\infty}^{z} \mathcal{N}(t)dt$, where $\mathcal{N}(t) = 1/\sqrt{2\pi}exp(-t^2/2)$ is the standard normal distribution. Let #C be the number of cognition in level 3, and the $c_{u,i}^{(T)}$ represents student u's latent cognition state of cognition c_i at time instance T. The $a_{k,i}$ and $b_{k,i}$ are properties of question q_k in terms of cognition c_i that can be estimated using Item Response Theory (IRT). For notational simplicity, we will omit the student index u and cognition index i in the following formulas, e.g., the quantities $s_{u,k,i}^{(T)}$ and $c_{u,i}^{(T)}$ are replaced by $s_k^{(T)}$ and $c^{(T)}$. Hence, the likelihood of an observation $s_k^{(T)}$ can be written as:

$$P\left(s_k^{(T)}|c^{(T)}\right) = p_k^{(T)s_k^{(T)}}\left(1 - p_k^{(T)}\right)^{1-s_k^{(T)}} \tag{6}$$

Then, we model the latent state transition between time instance $T - \tau$ and T as:

$$P\left(c^{(T)}|c^{(T-\tau)}, g\right) = \mathcal{N}\left(c^{(T)}|g^\tau c^{(T-\tau)}, \tau\sigma^2\right) \tag{7}$$

Where $\mathcal{N}(x|\mu, \sigma^2)$ represents a Gaussian distribution with mean μ and covariance σ^2. The g, shorthand for $g_{u,i}$, is student u's gain rate of the cognition c_i. The covariance σ^2, shorthand for $\sigma_{u,i}^2$ characterizes the uncertainty induced in the student u's cognition state transition by acting the c_i behavior.

Therefore, we can estimate the student's cognition state after participating in the discussion and gain rate parameter through the student's history answers of tests. And for simplifying the computing, we use the approximate result by the following methods:

$$P\left(c^{(T)}, g|s_{1:m_T}^{(1:T)}\right) \propto P(s_{1:m_T}^{(1:T)}|c^{(T)}, g) \cdot P\left(c^{(T)}, g\right) \tag{8}$$

Where m_T is the number of tests that student have done at time instance T.

We assume that the student's answers are independent of each other, the student previous answering will not impact on the current one. So the first item on the right side of (8) can be expressed as:

$$P\left(s_{1:m}^{(1:T')}|c^{(T')},g\right) = \prod_{T=1}^{T'}\prod_{k=1}^{m_T} P\left(s_k^{(T)}|c^{(T)},g\right)$$

$$= \prod_{T=1}^{T'}\prod_{k=1}^{m_T} \int P\left(s_k^{(T)}|c^{(T)},g\right) \cdot P\left(c^{(T)}|c^{(T')},g\right)dc^{(T)}$$

$$= \prod_{T=1}^{T'}\prod_{k=1}^{m_T} \int p_k^{(T)s_k^{(T)}}\left(1-p_k^{(T)}\right)^{1-s_k^{(T)}} \cdot \mathcal{N}\left(c^{(T)}|g^{T-T'}c^{(T')},(T-T')\sigma^2\right)dc^{(T)}$$

$$(9)$$

By using the Eq. (10) and definition of $\tilde{p}_k^{(T)}$ in (11), we can simplify (9) into (12):

$$\int \Phi(ax-b)\mathcal{N}(x|\mu,\Sigma) = \Phi\left(\frac{b-a\mu}{\sqrt{1+a^2\sigma^2}}\right) \tag{10}$$

$$\tilde{p}_k^{(T)} = \Phi\left(\frac{b_k - a_k g^{T-T'}c^{(T')}}{\sqrt{1+a_k^2(T-T')\sigma^2}}\right) \tag{11}$$

$$P\left(s_{1:m}^{(1:T')}|c^{(T')},g\right) \approx \prod_{T=1}^{T'}\prod_{k=1}^{m_T} \int \tilde{p}_k^{(T)s_k^{(T)}}\left(1-\tilde{p}_k^{(T)}\right)^{1-s_k^{(T)}} \tag{12}$$

Then, by putting log on both sides, we can get:

$$\log P\left(c^{(T')},g|s_{1:m_T}^{(1:T')}\right) = \log P\left(c^{(T')},g\right)$$
$$+ \sum_{T=1}^{T'}\sum_{k=1}^{m_T} s_k^{(T)}\log\tilde{p}_k^{(T)} + \left(1-s_k^{(T)}\right)\log\left(1-\tilde{p}_k^{(T)}\right) \tag{13}$$

With:

$$\log P\left(c^{(T')},g\right) = \log\mathcal{N}\left(c^{(T')}|^{(1)},\sigma^2\right) + \log g \tag{14}$$

Finally, our goal latent state variable g can be estimated by using BFGS-B algorithm [22] to maximize the objective function:

$$\underset{g}{maximize} \sum_{\mathcal{U}} \log P\left(c^{(T')},g|s_{1:m_T}^{(1:T')}\right) - \gamma\|g_1\|$$

Here, we define $\mathcal{U}$ as all the students that answer the tests between time instances 1 and T'. To prevent overfitting, we impose an ℓ_1-norm penalty on g. The stop condition is the difference between the two iterations is less than 0.0001.

4.4   System Design

We developed our analytics system as inspired by the Lambda architecture [23], and identify six functional phases from data aggregation to visualization (Fig. 2): (1) Data Aggregation, (2) Data Ingest, (3) Data Storage, (4) Streaming Analysis, (5) Batch Analysis and (6) Visualization. The OLCs data come from various of data sources provided by education institutions. We aggregate these data by using xAPI [24], an e-learning software specification software specification that allows learning content and learning systems to speak to each other in a manner that records and tracks all types of learning experiences. The aggregated data are temporarily stored in a Learning Record Store (LRS), hold by PostgreSQL in our system. Periodically, Sqoop (http://sqoop. apache.org/) imports raw data from PostgreSQL into Hadoop HDFS for permanent storage. For streaming analysis, we ingest high velocity forum data from relational database by using Flume (http://flume.apache.org/), and then forward to Spark Streaming for preprocessing. The preprocessed data store back into HDFS via Flume agent for further analysis, such as batch analysis. Apache Spark is chosen as the core component for batch analysis because it is efficient in iterative computing, provides various data sources supports, and can run on Hadoop YARN with multiple program-ming languages. In the batch analysis step, our analytic model is executed to generate the social network of the discussion forums and calculate the students' cognitive gains. At last, the analysis results are written back to HDFS for visualization and query.

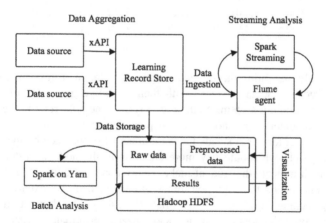

Fig. 2. Data flow in the Cognition Analysis System

5   Results

5.1   Student Cognitive Gains Analysis

We firstly select some typical students with their quantifiable data in the course to provide an intuitionistic view of students' behaviors. Table 5 lists the top 3 students who obtain the highest gains and the top 3 students who obtain the lowest gains in the course, along with the number of each cognition (the most probable cognition) in their posts and their total posts count (#Posts). An interesting observation is that some students with high posting quantity, e.g., No.32143, No. 7036 and No. 8149 student, have low cognitive gains, contrary to some previous research results which indicated that the more students posted, the higher learning gains they achieved [10]. For exploring the causes, we investigate their forum behaviors in the course. We found that the cognitive behaviors of those who have low gains are relatively concentrated in off-task and active cognition. Besides, most of the discussion threads for these students cease after their posts because no one makes comments on their thread posts, which means they receive almost no gains in these threads. These observations suggest that course instructors and TAs need to actively get involved in the discussion with these students in time.

Table 5. Representative students in the course Mathematic Thinking Method

Student ID	Gains	#Posts	Off-task	On-task				
				Act.		Con.		Int.
				Q	A/O	Q	A/O	
36856	40.38	49	0	3	6	20	9	11
27405	33.10	54	1	10	5	16	13	9
6547	27.04	42	0	9	0	10	18	5
32143	3.92	69	38	24	4	2	1	0
7036	4.01	57	22	20	12	3	0	0
8149	4.02	59	23	29	4	1	2	0

For intuitively presenting the students' cognitive gains in each course, we plot the structure of students' social network with their cognitive gains and other intuitive information, based on our system. As the Fig. 3 shows, the links refer to the interaction between the two students, and node's size depicts the relative gains of student in the course and node's color depth stands for the number of posts made by the student. By observing the graph, we should pay more attention to the node with deep color but small size, which indicate the student makes many posts but gains little.

To better understand the cognition trends of each student, we extend the graph in Fig. 3 and by clicking a node, a sunburst graph (Fig. 4) will be shown to represent the student's cognitive behaviors distribution and quizzes performance every week in the course. In the graph, the black arc around shows the number of correct quizzes out of all quizzes the student answered, and each sector indicates the proportion of each cognitive behavior in cognition level 3. Instructors and TAs could find problems according to the proportion of each student's cognitive behaviors.

5.2   Thread Cognition Analysis

Students obtain gains by posting in interested threads where they can get information from others. However, by inspecting the threads with low gains, we found that most of them have few participants or talk about off-task topics. It is for the reason of the first

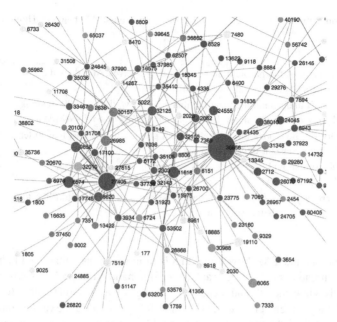

Fig. 3. Visualization of Students' cognitive gains in social network graph of course Mathematic Thinking Method

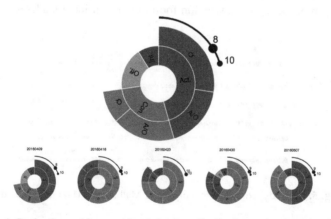

Fig. 4. A Student's weekly cognitive behaviors distribution and quizzes performance

post with insufficient details or asking a questions that no one is interested. Hence it is necessary for students to notice how to create or participate in threads that benefit them more and for teachers to track the trends of each thread and make timely adjustments.

To students, the Table 6 below with the top 5 threads where students obtain the most cognitive gains could give inspiration. From the table, we can see the threads with high gains usually start with a novel question with enough information (e.g., images), mostly followed by constructive answers or opinions, and interactive behaviors are also contributive.

Table 6. Top 5 threads with highest total gains in the course Mathematic Thinking Method

First post in thread	Total gains
Why does the angle of the problem about billiards not give? And does the last time the ball enters the box count a collision?	10.45
Excuse me, I don't understand the meaning that Xiao Ai walks up 1000 m along the vertical direction with the slope. How does she go? Is that it? ##IMAGE##	7.50
I would like to ask, how does the stroke theorem prove out, is it just that Euler has tried it through constant experimentation?	6.72
The stroke theorem should not just contain two singular point, right? A singularity should not work.	6.04
The problem about power supply can't always be symmetrical. I would like to ask you whether there is better solution?	5.58

To instructors and TAs, how to get involved in the threads that need help in time? We design a thread cognitive tree graph (Fig. 5) to handle this problem, where teachers can intuitively view each thread's cognitive changes via the circle colors (e.g., yellow refers to off-task, green refers to active, blue refers to constructive and red refers to interactive cognition). Instructors and TAs could intervene in a discussion thread when it is mostly covered by off-task and active cognitive behaviors, or a thread seems to be ignored. Also, teachers could pick out essential threads by their total gains and rank them on the top of the course discussion forum to benefit more students.

Fig. 5. Threads' cognition tree of course Mathematic Thinking Method

5.3   Course Forum Cognition Analysis

For overviewing the cognitive gains obtained by all students in the whole course forum, we draw a line chart to show the trends of total students' gains over time based on our system. As shown in Fig. 6, at the beginning and end of the course, the gains grow slower than the middle. It is due to the off-task and active cognition taking up a majority of posts at the beginning and less posts in vacations of the International Labor Day (May 1st) at the end. By intervening the preliminary posting patterns, such as increasing the number of higher level cognitive posts, stopping off-task topic threads and ranking gainful threads on top, instructors and TAs could increase the slope of

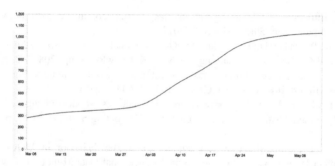

Fig. 6. Trends of students' gains in course Mathematic Thinking Method

students' gains. Also, the slope could indirectly reflect how deep students understand the lecture in every week, by which teachers could adjust the content or difficulty of the lecture.

6   Conclusions

In this paper, we propose a cognitive gains model on social learning network to represent students' cognitive benefits through interaction. With an easy-labeled coding scheme and common posting features defined in our model, teachers are able to conveniently and accurately classify students' postings into six cognitions (off-task, active question, active answer/opinion, constructive question, constructive answer/opinion and interactive cognition). The classification results on our OLC dataset show that the model is viable to estimate student's gains from conversation in cognitive level. Moreover, we have integrated the model into our learning analytics system for enabling instructors and TAs to assess the cognitive performance of students through three perspectives (student, thread and course).

Acknowledgement. This paper is supported by the NSFC (61532004), State Key Laboratory of Software Development Environment (Funding No. SKLSDE-2017ZX-03).

References

1. Anderson, A., et al.: Engaging with massive online courses. In: Proceedings of the 23rd International Conference on World Wide Web. ACM, p. 687–698 (2014)
2. Brinton, C.G., et al.: Learning about social learning in MOOCs: from statistical analysis to generative model. IEEE Trans. Learn. Technol. **7**(4), 346–359 (2014)
3. Stephens-Martinez, K., Hearst, M.A., Fox, A.: Monitoring MOOCs: which information sources do instructors value? In: Proceedings of the First ACM Conference on Learning@ scale Conference. 2014, ACM, p. 79–88
4. Barab, S.A., Duffy, T.: From practice fields to communities of practice. Theor. Found. Learn. Environ. **1**(1), 25–55 (2000)
5. Chi, M.T., et al.: Learning from human tutoring. Cogn. Sci. **25**(4), 471–533 (2001)

6. Shea, P., Bidjerano, T.: Learning presence as a moderator in the community of inquiry model. Comput. Educ. **59**(2), 316–326 (2012)
7. Oleksandra, P., Shane, D.: Untangling MOOC learner networks. In: Proceedings of the Sixth International Conference on Learning Analytics & Knowledge, pp. 208–212. ACM (2016)
8. Gillani, N., et al.: Structural limitations of learning in a crowd: communication vulnerability and information diffusion in MOOCs. Sci. Rep. **4**, 6447 (2014)
9. Russell, D.M., et al.: Will massive online open courses (moocs) change education? In: CHI'13 Extended Abstracts on Human Factors in Computing Systems. ACM, p. 2395–2398 (2013)
10. Wang, X., et al.: Investigating how student's cognitive behavior in MOOC discussion forums affect learning gains. In: International Educational Data Mining Society (2015)
11. Putnik, G., et al.: Analysing the correlation between social network analysis measures and performance of students in social network-based engineering education. Int. J. Technol. Des. Educ. **26**(3), 413–437 (2016)
12. Brinton, C.G., et al.: Social learning networks: efficiency optimization for MOOC forums, in computer communications. In: 35th Annual IEEE International Conference on IEEE INFOCOM 2016-The, pp. 1–9. IEEE (2016)
13. Elouazizi, N.: Point-of-view mining and cognitive presence in moocs: a (computational) linguistics perspective. In: Proceedings of the Empirical Methods in Natural Language Processing Workshop, pp. 32–37 (2014)
14. Wong, J.S., et al.: Analyzing MOOC discussion forum messages to identify cognitive learning information exchanges. Proc. Assoc. Inf. Sci. Technol. **52**(1), 1–10 (2015)
15. Chi, M.T.: Active-constructive-interactive: a conceptual framework for differentiating learning activities. Top. Cogn. Sci. **1**(1), 73–105 (2009)
16. The Stanford MOOCPosts Data Set. (2017) http://datastage.stanford.edu/StanfordMoocPosts/. Accessed 20 Aug 2017
17. Marzouk, Z., Rakovic, M., Winne, P.H.: Generating Learning Analytics to Improve Learners' Metacognitive Skills Using nStudy Trace Data and the ICAP Framework, in LAL@ LAK, pp. 11–16 (2016)
18. Cui, Y., Wise, A.F.: Identifying content-related threads in MOOC discussion forums. In: Proceedings of the Second (2015) ACM Conference on Learning@ Scale, pp. 299–303. ACM (2015)
19. Anderson, L.W., et al.: A Taxonomy for Learning, Teaching and Assessing: A Revision of Bloom's Taxonomy. New York: Longman Publishing; Artz, A.F., Armour-Thomas, E.: Development of a cognitive-metacognitive framework for protocol analysis of mathematical problem solving in small groups. Cogn. Instr. 2001 **9**(2), 137–175 (1992)
20. Kovanović, V., et al.: Towards automated content analysis of discussion transcripts: a cognitive presence case. In: Proceedings of the Sixth International Conference on Learning Analytics & Knowledge ACM, pp. 15–24 (2016)
21. Huang, J., Wu, W.: T-BMIRT: Estimating Representations of Student Knowledge and Educational Components in Online Education
22. Zhu, C., et al.: LBFGS-B: Fortran subroutines for large-scale bound constrained optimization. Report NAM-11, EECS Department, Northwestern University (1994)
23. Marz, N., Warren, J.: Big Data: Principles and Best Practices of Scalable Realtime Data Systems. Manning Publications Co. (2015)
24. xAPI: https://www.adlnet.gov/xAPI

A Semantic Role Mining and Learning Performance Prediction Method in MOOCs

Zhiqiang Liu[1,2(✉)] and Yan Zhang[1]

[1] School of Electronics Engineering and Computer Science, Peking University, Beijing, China
{lucien,zhy}@cis.pku.edu.cn
[2] Academy for Advanced Interdisciplinary Studies, Peking University, Beijing, China

Abstract. Massive Open Online Courses (MOOCs) have gained tremendous popularity in the last few years. Discussion forums are a common element in MOOCs. Previously, Numerous studies have been undertaken on the potential role that discussion forums play in education. However, the existing works don't effectively mine and analyze the rich text information of the forums associated with learner performances. In this work, we propose a hybrid method to mine learner roles based on MOOC discussion forums and to jointly evaluate the quality of learning with other learning activities. We pay more attention to extracting semantic features of posts and comments in forums, which help to promote the performance prediction. We evaluate the performance of our method based on the Coursera platform. Experiments show that our approach can improve the performance compared to existing works on these tasks.

Keywords: Role mining · Performance prediction · MOOCs
Discussion forums

1 Introduction

Massive Open Online Courses have become popular in the last few years. Thanks to MOOCs, millions of learners from all over the world have taken thousands of high-quality courses for free. Despite their rapid development and successes, there are still some problems within MOOCs. One prominent problem is the high dropout rates. Most of the students taking online courses do not complete the courses and drop out halfway. This could be a potential factor hindering the development of MOOCs. As a result, making the effective prediction of whether a student will drop out or whether a learner will obtain better learning performance is of great value for MOOC platforms [5].

To solve this problem, some methods have been proposed in recent years [15]. Previous research into predicting MOOC completion and performance has

© Springer Nature Switzerland AG 2018
L. H. U and H. Xie (Eds.): APWeb-WAIM 2018, LNCS 11268, pp. 259–269, 2018.
https://doi.org/10.1007/978-3-030-01298-4_22

focused on click-streams, demographics, and sentiment analysis in MOOC platforms [10]. Although inspiring results have been achieved by these methods [4], most of them still suffer from the insufficiency of discriminative activity features for learner behaviors. Besides, the sparsity of MOOC platforms data still exists. However, discussion forums play an important role in MOOCs, which have accumulated plenty of posts and their associated comments over time. Techniques used for the analysis of MOOC discussion forums can be characterized as content-related or activity-related. The content analysis aims to uncover the nature of forum contributions and activity analysis to reveal the features of user learning behaviors.

In detail, based on the existing work, we find out that learners performance, regarding whether he/she could get certificated eventually, can be predicted by looking into several features of their learning behaviors such as video click-stream behaviors, forum view behaviors and so on. Experiment results of the previous work indicate that these features can be trained to effectively estimate whether a learner is probably to complete the course successfully. Besides, these methods have the potential to partially evaluate the quality of both teaching and learning in practice. Meanwhile, role mining of contributors in discussion forums of online courses is an important part in MOOCs. We aim to jointly analyze the learner role in discussion forums to predict the learning performance more accurately.

In the performance prediction problem, the data we have are raw activity records of learners in the online course platform over a period of time. The prediction we need to make is whether these students drop out from the courses or whether they obtain good performance in the future, which is a regression problem. In this paper, we propose a hybrid method to mine learner roles based on MOOC discussion forums and to jointly evaluate the quality of learning with other learning activities. We pay more attention to extracting semantic features of posts and comments in forums, which help to promote the performance prediction. The framework of our hybrid prediction model is shown in Fig. 1.

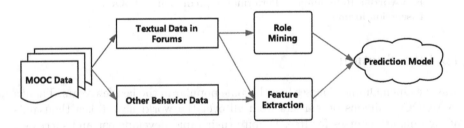

Fig. 1. The framework of our hybrid prediction model.

As shown in Fig. 1, after collecting raw MOOC data, our basic idea is to extract features including text content features and other behavior features. Then we will mine learner roles via textual data in forums and finish the final prediction task. In conclusion, the major contribution of this paper is listed below.

1. Unlike previous methods, we propose a hybrid method to mine learner roles based on MOOC discussion forums and to jointly evaluate the quality of learning with other learning activities.
2. To take full advantage of learners' limited activities, we employ the features including learners' temporal clickstream behaviors and posts view behaviors to predict the quality of learning more accurately.

The rest of this paper is organized as follows. In Sect. 2, we provide a brief review of the related work about of role mining and performance prediction in MOOCs. We describe the hybrid method in detail in Sect. 3. A variety of experimental results are shown in Sect. 4. Finally, we provide some conclusions and insights of this work in Sect. 5.

2 Related Work

2.1 Dropout Prediction and Learning Performance Prediction

Many scholars at home and abroad have studied on when the learners of MOOCs will drop out of the course. There are two kinds of analysis data in the current research: the data of the forums and the data of the clickstream behaviors. Here, several typical dropout studies are analyzed and presented [6,15,20].

Amnueypornsakul [2] used learners' clickstream data to predict whether or not a student would drop out of the course. Researchers formed a sequence of weekly learning behaviors for each learner. Then the researchers defined three learners: active, inactive, drop. The results showed that the accuracy rate was significantly improved when the inactive learners were excluded from the model construction, and when including inactive learners, the accuracy rate of modeling inactive learners as active learners was relatively higher.

Sinha [13] leveraged combined data of video clickstream and forum to form the action sequence to seek traits that were predictive of decreasing engagement over time. The results showed that dropout behavior was more affected by the learning behavior of recent weeks. And most of the dropout students started classes a few weeks after the beginning of the course. There were two possible explanations. One was that these dropout students have needs in specific information, and they ceased to attend classes after they met their needs. Or, the students who joined later had to give up due to the excessive material and work to keep up with the course. Taylor [14] used different machine learning approaches to predict the dropout students, including logistic regression, support vector machines, deep belief networks, decision trees, and hidden Markov models. Kloft [11] used clickstream data and machine learning algorithms to predict dropout behaviors.

Besides, In MOOCs, a similar problem to dropout prediction is completion prediction or learning performance prediction. In this problem, instead of predicting whether a student will drop out from a course, it aims to predict whether a student can complete a course and get the corresponding certificate. Some researchers used logistic regression to identify students who seem to be not able to complete the course [8,18]. Others used latent dynamic factor graph model for the prediction [21]. There are also some works that comprehensively considered activity records and completion of the course, and proposed the prediction problem suitable for their problem setting [16].

However, these existing work consider this problem as activity feature analysis without rich text semantic information. In Scott Crossley paper [4], they explore the potential for natural language processing (NLP) tools that include but also go beyond sentiment analysis to predict success in an educational data mining MOOC. The goal is to develop an automated model of MOOC success based on NLP variables such as text length, text cohesion, syntactic complexity, lexical sophistication, and writing quality that can be used to predict learning performance. In this work, we aim to propose a hybrid model jointly between learner role mining in MOOC discussion forums and learning activities to solve this problem.

2.2 Semantic Role Mining in MOOC Discussion Forums

Role modeling is of particular interest to characterize learners, such as peripheral participants or "lurker" or active advice-givers in the community. [9,17,21]. In a recent study [19], a combination of content analysis and machine learning was used to distinguish forum threads in which participants discuss the course content from those merely socializing or discussing organizational matters. Content analysis is also used to characterize forum users based on the types of contributions they make [3]. Social network analysis is commonly applied for communication-related analytic approaches. Social networks of users based on common discussion threads can serve to investigate the coherence of the underlying social network [7], detection of communication patterns and community support [12].

However fine-grained network modeling is required to adequately reflect and represent the concrete post/reply communication between participants. In discussion forums with nested threads, these relations can be observed directly from the thread structure [1]. However, in forums with a more linear thread structure, such as the Coursera forums investigated in this paper, the identification of direct communication between users requires content-analytic approaches such as discussion act tagging [3]. In our work, we combine role mining in MOOC discussion forums to our learning performance prediction task.

3　Role Mining and Performance Prediction

In this section, we will describe the hybrid model proposed. Firstly, we mine the important learner roles in MOOC discussion forums and then we combine it to other learner activities in order to predict the learning performance.

3.1   Role Mining

First of all, we mine the posts written by learners in MOOC forums to obtain learner role features, which can be seen as a text classification task for posts. Since there are not class tags in this data, we need to make use of data processing methods to annotate the original text data. The detail operation can be seen in Fig. 2.

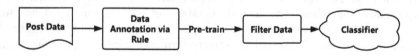

Fig. 2. The framework of data processing.

As shown in Fig. 2, after collecting raw MOOC post data, our basic idea is to label data via the sentiment keywords rules. Then we will use a simple classifier to filter this data. Finally, we will obtain the labeled data for the classification task.

We first describe how we use collective sentiment analysis to study learners' attitudes towards the course and mastery of knowledge based on forum posts. In our work, we define learner roles in MOOC discussion forums, which is reflected via the learner sentiment attitudes and the post intentions such as information-seeking or information-giving. As a result, we will explore the learner roles by results of post-classification. Firstly, the posts will be divided into 3 categories including *positive, negative* and *neutral*. And another text classification is defined as post intention identification, which mainly divided into 3 classes including *information-seeking, information-giving* and *other*.

Then we propose a text classification model based on attention mechanism to finish the post sentiment analysis and post intention identification. The model include the Bi-directional LSTM with self-attention mechanism. A Bi-directional LSTM consists of a forward LSTM and a backward LSTM. The forward LSTM reads each word w_i (i.e., from w_1 to w_i) in sequence as it is ordered, and generate the hidden states of each word as $\left(\overrightarrow{h_1}, ..., \overrightarrow{h_i} \right)$. For the backward LSTM, it processes each sentence in its reversed order w_i(i.e., from w_i to w_1) and form a sequence of hidden states $\left(\overleftarrow{h_1}, ..., \overleftarrow{h_i} \right)$. We calculate the hidden states $\overrightarrow{h_i}$ by following equations:

$$i_t = \delta(W_i x_t + G_i h_{t-1} + b_i)$$
$$\hat{C}_t = \tanh(X_c x_t + G_f h_{t-1} + b_f)$$
$$f_t = \delta(W_f x_t + G_f h_{t-1} + b_f)$$
$$C_t = i_t \cdot \hat{C}_t + f_t \cdot C_t$$
$$o_t = \delta(W_o x_t + G_o h_{t-1} + V_o C_t + b_o)$$
$$h_t = o_t \cdot \tanh(C_t)$$

where δ represents the sigmoid activation function; W_s, U_s and V_o are weight matrices; and b_s are bias vectors. There are three different gates (input, output, forget gates) for controlling memory cells and their visibility. The input gate can allow incoming signal to update the state of the memory cell or block it and the output gate can allow the state of the memory cell to have an effect on other neurons or prevent it. Moreover, the forget gate decides what information is going to be thrown away from the cell state.

In addition, Not all words contribute equally to the representation of the sentence meaning. Hence, we introduce attention mechanism to extract such words that are important to the meaning of the sentence and aggregate the representation of those informative words to form a sentence vector. Specifically,

$$u_i = \tanh(W_i h_i + b_i)$$
$$\alpha_i = \frac{\exp(u_i^T u_w)}{\sum_t \exp(u_t^T u_w)}$$
$$s = \sum_t \alpha_i h_i$$

That is, we first feed the word annotation hit through a one-layer MLP to get u_i as a hidden representation of hit, then we measure the importance of the word as the similarity of u_i with a word level context vector u_w and get a normalized importance weight α_i through a softmax function. After that, we compute the sentence vector s (we abuse the notation here) as a weighted sum of the word annotations based on the weights. The context vector u_w can be seen as a high-level representation of a fixed query "what is the informative word" over the words like that used in memory networks. The word context vector u_w is randomly initialized and jointly learned during the training process. The total model for classification is shown in Fig. 3.

In this way, we separately accomplish the sentiment role classification and post intention classification. The Table 1 shows the confusion matrix for post sentiment classification and the Table 2 shows the confusion matrix for post intention classification that is is to identify the post intentions as information-seeking, information-giving and other.

Through the above classification method, we can analyze the roles played by learners in forums. We define the role features as below.

- **Positive learning rate**: The percentage of positive posts in the total posts.
- **Negative learning rate**: The percentage of negative posts in the total posts.
- **Information-seeking learning rate**: The percentage of information-seeking posts in the total posts.
- **Information-giving learning rate**: The percentage of information-giving posts in the total posts.

Meanwhile, we extend the text statistic features as below Table 3. Through the above methods, we analyze the textual information proposed via learner in the forums and discover the features of the user's role. In next step, we will combine these role features to learner other behaviors in order to predict the learning performance.

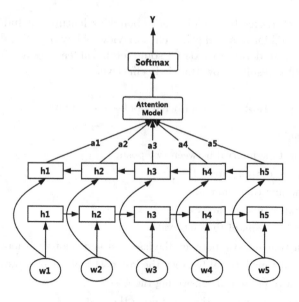

Fig. 3. The framework of Bi-directional LSTM with attention.

Table 1. Confusion matrix for post sentiment classification

Class	Precision	Recall	F1
Positive	0.82	0.80	0.81
Negative	0.79	0.78	0.80

Table 2. Confusion matrix for post intention classification

Class	Precision	Recall	F1
Information-seeking	0.73	0.72	0.72
Information-giving	0.75	0.76	0.75

Table 3. Text feature definitions

Feature	Meaning
l_{max}	Maximum post length
l_{min}	Minimum post length
l_{avg}	Average post length
c_{topic}	Count of topic keywords

3.2 Performance Prediction

Through the above mining of characters and features extraction, we get the text features of the learners in the forums. Next we will combine the other learners' behavior characteristics to model. These behavior features include leaner view features in MOOC forums and other click or view behaviors. The detail is listed below in Table 4. In detail, we extract some temporal features in MOOC forums. The experimental results show that it is effective.

Table 4. Forum behaviors feature definitions

Feature	Meaning
t_{mean}	Mean time interval between two posts
t_{max}	Maximum time interval
t_{min}	Minimum time interval
t_{var}	Variance of time interval
t_{sum}	Weighted sum of time interval
t_{last}	Last time interval between the last post and the post before it
t_{recent}	Recent time interval between the last post and the observation deadline
r_{mean}	Mean response time answering the post
r_{max}	Maximum response time answering the post
r_{min}	Minimum response time answering the post
r_{var}	Variance of response time answering the post

Finally, the hybrid features are extracted and we proposed to predict the learning performance scores based on the regression model. The total framework has been shown in Fig. 1.

4   Experiments

4.1   Data Preparation

In the autumn of 2013, Peking University released its first six courses on the Coursera[1] platform. In this work, we select two courses for experimental evidence, which are *Crowd and Network* and *MOOC Wendao*. We evaluate the performance of our method on the two courses. The whole dataset consists of over 6683 posts and comments and 1164 learners who participate in these posts and learn in MOOCs. In our experiment, we use the total posts for role mining and post intention prediction. Meanwhile, the learner set is divided into training data and test data. because this task is a regression task that the result value is between 0 and 1, We evaluate the performance of our proposed hybrid method based on **Root Mean Square Error**(RMSE) and the smaller the value, the better the performance of the model. The RMSE is defined as Formula 1.

$$RMSE = \sqrt{\frac{\sum_{i-1}^{n}(X_{obs,i} - X_{model,i})^2}{n}} \tag{1}$$

---

[1] https://www.coursera.org.

4.2   Experimental Results and Analysis

Table 5 shows the experimental results of some regression models and we find that the Gradient Boost Regression Method obtains best results.

Table 5. Experiment results for prediction learning performance

Method	RMSE
Lasso	0.422
Ringe	0.417
SVM	0.405
ElasticNet	0.383
Random Forest	0.387
Gradient BoostRegression	**0.375**

Feature contribution. We analyze how the different kinds of feature contribution to the model. The results are shown in Fig. 4. The forum behavior features have a predominant influence on the final results, and learner role feature is second important. Other features are also contributive but the difference is small. Even so, every kind of features to model positively. The result indicates our semantic role mining and learning performance prediction model is useful and effective. Especially, the semantic role mining based on MOOC discussion forums is necessary.

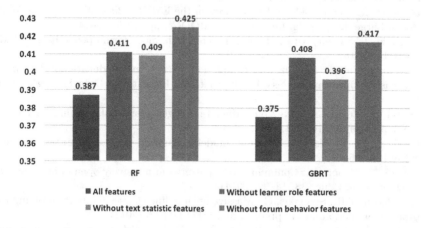

Fig. 4. Efficacy of different features.

In addition, we find that the number of people participating in the forum is only a small part of the total number of learners, but the forum data can help us effectively analyze the behavior of users and predict the learning performance. Therefore, it is of necessity for teachers to encourage more learners to participate in the forum discussion and it is of great significance in enhancing the participation of the courses.

5   Conclusions

In this paper, we present a hybrid method that mines learner role in MOOC discussion forums and predict the quality of learning. Experiments on the Coursera platform datasets show that the hybrid model outperforms other existing methods. Meanwhile, this work opens to several interesting directions for future work. In detail, we will explore the following directions:

1. It is of interest for us to explore other rich text features to enhance the performance in this way.
2. We plan to apply the joint model for combining other text role features and apply it to other community-based question answering sites and forums.

Acknowledgements. This work is supported by NSFC under Grant No. 61532001, and MOE-ChinaMobile under Grant No. MCM20170503.

References

1. Abnar, A., Takaffoli, M., Rabbany, R., Zaïane, O.R.: Ssrm: structural social role mining for dynamic social networks. Soc. Netw. Anal. Min. **5**(1), 56 (2015)
2. Amnueypornsakul, B., Bhat, S., Chinprutthiwong, P.: Predicting attrition along the way: the uiuc model. In: Proceedings of the EMNLP 2014 Workshop on Analysis of Large Scale Social Interaction in MOOCs, pp. 55–59 (2014)
3. Arguello, J., Shaffer, K.: Predicting speech acts in mooc forum posts. In: ICWSM, pp. 2–11 (2015)
4. Crossley, S., et al.: Language to completion: success in an educational data mining massive open online class. In: International Educational Data Mining Society (2015)
5. Fauvel, S., Yu, H.: A survey on artificial intelligence and data mining for moocs. arXiv preprint arXiv:1601.06862 (2016)
6. Gardner, J., Brooks, C.: Dropout model evaluation in moocs. arXiv preprint arXiv:1802.06009 (2018)
7. Gillani, N., Eynon, R.: Communication patterns in massively open online courses. Internet High. Educ. **23**, 18–26 (2014)
8. Hagedoorn, T.R., Spanakis, G.: Massive open online courses temporal profiling for dropout prediction. arXiv preprint arXiv:1710.03323 (2017)
9. Hecking, T., Chounta, I.A., Hoppe, H.U.: Role modelling in mooc discussion forums. J. Learn. Anal. **4**(1), 85–116 (2017)
10. Hou, Y., Zhou, P., Wang, T., Yu, L., Hu, Y., Wu, D.: Context-aware online learning for course recommendation of mooc big data. arXiv preprint arXiv:1610.03147 (2016)

11. Kloft, M., Stiehler, F., Zheng, Z., Pinkwart, N.: Predicting mooc dropout over weeks using machine learning methods. In: Proceedings of the EMNLP 2014 Workshop on Analysis of Large Scale Social Interaction in MOOCs, pp. 60–65 (2014)
12. Malzahn, N., Harrer, A., Zeini, S.: The fourth man: supporting self-organizing group formation in learning communities. In: Proceedings of the 8th iternational conference on Computer supported collaborative learning, pp. 551–554. International Society of the Learning Sciences (2007)
13. Sharkey, M., Sanders, R.: A process for predicting mooc attrition. In: Proceedings of the EMNLP 2014 Workshop on Analysis of Large Scale Social Interaction in MOOCs, pp. 50–54 (2014)
14. Taylor, C., Veeramachaneni, K., O'Reilly, U.M.: Likely to stop? predicting stopout in massive open online courses. arXiv preprint arXiv:1408.3382 (2014)
15. Wang, W., Yu, H., Miao, C.: Deep model for dropout prediction in moocs. In: Proceedings of the 2nd International Conference on Crowd Science and Engineering, pp. 26–32. ACM (2017)
16. Wang, Y.: Mooc leaner motivation and learning pattern discovery. In: EDM, pp. 452–454 (2014)
17. Wen, M., Yang, D., Rose, C.: Sentiment analysis in mooc discussion forums: what does it tell us? In: Educational data mining 2014. Citeseer (2014)
18. Wen, M., Yang, D., Rosé, C.P.: Linguistic reflections of student engagement in massive open online courses. In: ICWSM (2014)
19. Wise, A.F., Cui, Y., Vytasek, J.: Bringing order to chaos in mooc discussion forums with content-related thread identification. In: Proceedings of the Sixth International Conference on Learning Analytics & Knowledge, pp. 188–197. ACM (2016)
20. Lu, X., Wang, S., Huang, J., Chen, W., Yan, Z.: What decides the dropout in MOOCs? In: Bao, Z., Trajcevski, G., Chang, L., Hua, W. (eds.) DASFAA 2017. LNCS, vol. 10179, pp. 316–327. Springer, Cham (2017). https://doi.org/10.1007/978-3-319-55705-2_25
21. Zarra, T., Chiheb, R., Faizi, R., El Afia, A.: Using textual similarity and sentiment analysis in discussions forums to enhance learning. Int. J. Softw. Eng. Its Appl. **10**(1), 191–200 (2016)

MOOC Guider: An End-to-End Dialogue System for MOOC Users

Yuntao Li[✉] and Yan Zhang

Peking University, Beijing, China
{liyuntao,zhyzhy001}@pku.edu.cn

Abstract. With the growth of the amount of MOOC users and course diversity, it becomes a hard work for a new MOOC user to find a suitable course and gather other information. In this paper, we propose a natural language dialogue based MOOC guider, which helps users to find a preferred course and provide more information of courses according to a user's requests. Our method is an end-to-end neural network based method and can be trained efficiently using multi-stage training. Experiments show that our method can understand users' intent well and produce proper response to finish the task.

Keywords: MOOC · Dialogue system · Guidance

1 Introduction

MOOC, being the abbreviation for Massive Open Online Course, has become the leading pattern of online learning. Platforms like Coursera, EdX, Udacity and Xuetang online attracts over millions of learners on every semesters [1]. A MOOC is a course made up of short video lectures combined with computer-graded tests, which is designed to support an unlimited number of participation by accessing video lectures. Influenced by the pattern of online education, some properties are exhibited. First, MOOCs are courses, so that course materials and start and end date for learning are required, with some fixed facilitators. Besides, MOOCs are massive, which means participants attracted by a course can reach more than thousands for one semester and millions enrolled students as a whole [2]. Moreover, MOOC courses are open and online, as a result, course materials should be open to anyone to read in any time and any place, and all the comments and updates from either an instructor or a student can be seen as soon as it is published. Compared with offline learning, MOOCs, in addition to traditional course materials such as lectures, readings and problem sets, often provide forums to support community interactions among students, professors and teaching assistants.

With the increment of both of the amounts of MOOC courses and variety of MOOC users, it becomes harder and harder for a user to select a course which is suitable for the user's requirements. On the one hand, MOOC users have their

© Springer Nature Switzerland AG 2018
L. H. U and H. Xie (Eds.): APWeb-WAIM 2018, LNCS 11268, pp. 270–280, 2018.
https://doi.org/10.1007/978-3-030-01298-4_23

preference on some course properties such as language and duration, according to their background and purpose of learning, on the other hand, finding a lesson that is suitable for a user on either difficulty or topic can be a hard work. Traditional filter based or word search based method can hardly satisfy these requirements. For a participant who is totally new to MOOCs, gathering all the information required to select a course is of great hard. Further more, as a side effect of this phenomenon, low completion rates are often seen on MOOC courses, with a range of 3–10% [6,13].

In this work, we propose a MOOC guider, which guide a user to select a course and provide for information in the form of natural language dialogue. A student can filter their wanted courses in favor of language, duration, university, topic etc., and harvest information about every course via talking with our agent.

2   Task Definition

We define our agent as a MOOC guider, which helps users to select their preferred courses and provide their requested information by talking in a dialogue form with natural texts. In order to better describe functions that our agent can provide, we propose a formal definition of the task. A MOOC guider, which is a natural dialogue system, should be able to accomplish following tasks to form a successful guidance:

- Understanding the user's purpose of each utterance. A user utterance may contain a user's requirement or request on a course, or just be a chit-chat. The model should distinguish from these two types and extract useful information from it.
- Tracking user's intent correctly while communicating with a user. A dialogue is consisting of utterances, and the user can express different requirements on each utterance, as well as change his mind. Our model is required to track the modification of the user's intent during the dialogue procedure.
- Generate proper response for a user to continue a dialogue. To accomplish a dialogue communication, the dialogue agent ought to response a user utterance with a related response or answer an asked question. It is also important to ask for the user for crucial information in order to finish a course recommendation.
- Provide requested information the user asked while answering. The final target of our system is to guide the user with useful information, thus knowing what the user is asking and provide correct answer is a vital criterion for this system. Generating an answer can be done by a simple statistical model, while finding the correct answer requires exact comprehension on user's intent.

A dialogue based service is judged as a successful guidance if every of the above requirements is met. Meanwhile, we also evaluate the performance of our system on these aspects.

3   Methodology

We propose a dialogue generating system called MOOC guider, to help users when selecting a course.

The MOOC guider follows a basic dialogue system architecture, with a number of adaption to become suitable for a dialogue system of MOOC guidance. A dialogue system usually consists of several parts including a natural language understanding (NLU) component, a dialogue manager (DM) component and a natural language generation (NLG) component [15, 23]. The NLU part converts natural language into a sort of form that machine can understand; the DM part record the dialogue state including user's intent, mentioned entities etc.; and the NLG component generates response according to tracked dialogue state and some knowledge predefined by human. Our method follows this architecture, with every components of it is replaced with a deep neural network based model. We named each part of out model as the encoding part, the state tracking part and the dialogue generation part. The architecture overview of our model is shown in Fig. 1.

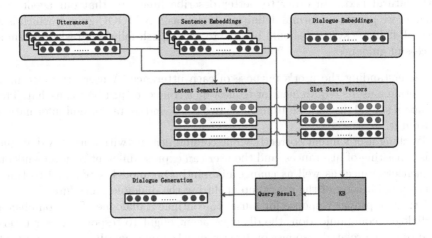

Fig. 1. Architecture overview of MOOC guider

3.1   Encoding

The encoding component is such a part who converts natural language utterances into some fixed-length vectors which is understood easily by machines. The encoder of our model is a hierarchical one made up of a sentence encoder and a dialogue encoder. Vectors generated from encoder is delivered to following state tracking component.

To understand a sentence, what do we think is important is to extract an overview of what the sentence is talking about and some detailed information which is used for exact keyword search. Thus, we use two vectors to represent a sentence following this idea.

For a sentence $x_1, x_2, \cdots, x_m$ with length m, it is first converted into word vectors using word embedding method. A pre-trained word embedding matrix is used.

$$(x_t^1, x_t^2, \cdots, x_t^m) = E(\mathrm{x}_t^1, \mathrm{x}_t^2, \cdots, \mathrm{x}_t^m) \tag{1}$$

To select what is important in this sentence, we employ a simple attention mechanism on those words to compute a weight for each word using a DNN (deep neural network), and summed up to become a vector representing detailed information.

$$\alpha_t^i = DNN(x_t^i)$$
$$s_t^* = \sum_{i=1}^{m} \alpha_t^t x_t^i \tag{2}$$

Besides, an overview vector of a sentence is computed using a bidirectional GRU. The joint of both direction's output vectors of the last time step GRU produces is used to denote the sentence.

$$s_t' = GRU(x_t^1, x_t^2, \cdots, x_t^m) \tag{3}$$

These two vectors are used to represent a sentence.

3.2 State Tracking

The state tracking component tracks dialogue state and user's intent with utterances comes successively. The key point of state tracking is to convert each user input into different spaces which represent state of every slots. Inspired by [10], in which work, a sentence is mapped into several semantic spaces and combined for down-streaming tasks, we map a dialogue into different latent spaces.

Hidden states h_t of sentence encoder are regarded as memories of different time step for a dialogue, and are used for latent semantic vectors generation. Notate a stack of hidden state vectors as H. For any intent slot r_j, a slot specific latent vector according to H is computed with a DNN.

$$H_t = [h_1, h_2, \cdots, h_t]^T$$
$$a_j = softmax(w_{2j}tanh(W_{1j}H_t^T)) \tag{4}$$
$$h_t' = a_j * H_t$$

Where w_{2j} and W_{1j} are parameter matrixes. To stack formulations for each slot up, the above formulation can be simplified as

$$A_j = softmax(W_{2j}tanh(W_{1j}H_t^T))$$
$$H_t' = A_j * H_t \tag{5}$$

H_t' is a matrix, each column of whom is a vector with regard to a slot. Slots are updated according to its current state and latent semantic vector.

$$\alpha_j = tanh(W_1 h_j' + W_2 r_j)$$
$$i_j = W_3(U h_j' + V r_j) \tag{6}$$
$$r_j = r_j + \alpha_j i_j$$

W_1, W_2, W_3, U, V are parameters. Predictions on slot values are computed with each slot vector using a DNN. For a slot with n values, the predictor generates $n+1$ values, representing the probability of each slot value and additionally one for not mentioned.

3.3 Dialogue Generation

In order to summarize a dialogue till the current time step with multiple turns, a dialogue encoder is used. This dialogue encoder is a basic GRU network, who takes sentence vectors of each turn as input, and generate a dialogue vector which is the last output o_T, with hidden states h_t during GRU running as byproducts. This procedure can be formulated as follows:

$$o_t, h_t = GRU(s_1, s_2, \cdots, s_t) \tag{7}$$

A response is produced following a standard sequence to sequence decoding pattern, using an LSTM whose initial state is an encoded dialogue at current time step. Tracked slot states are also converted into an indicator labeling whether a slot is mentioned with a single probability, and passed to dialogue generator. Notate the last output of dialogue encoder as o_T, sentence vector of the last sentence as s_T, probabilities of each slot being mentioned as p, dialogue generator hidden state at generation time step t as $h_{d,t}$, who is initialized to be zeros. Responses are generated according to these vectors.

$$o_{d,t}, h_{d,t} = LSTM(o_T, s_T, p, h_{d,t-1}) \tag{8}$$

Generated responses are template based response, which means that there are placeholders in generated sentences which need to be replaced by its real value acquired from outside knowledge, for example, course language, course duration or recommended courses. We use a knowledge base (KB) to store these kinds of knowledge. Tracked slot states are mapped into KB key-value pairs to form KB query. The KB query returns a list of search results satisfying the user's requirement. Placeholders are replaced by returned values from KB, and a final response is then generated.

4 Training Method

For an intricate end-to-end system consisting of multiple components with complex data interaction such as ours, training directly on the final task target may lead to a situation of loss misconvergence. A main cause for this problem is that, errors from different components may lead to the same mistake and cause similar loss value. When back propagation is conducted on the model, the part with error results and other parts with correct values are updated according to a unified gradient. In our case, either dialogue state tracking error or response generation error can lead to an apparent loss, while back propagation algorithm cannot tell whether it is caused by tracking error or generation error. To prevent this problem from happening, we apply multi-state training on our model.

Multi-stage training is such a kind of training method which split a whole system into some small relatively independent parts and train separately. While these pre-trained models achieves a acceptable results, they are combined and fine-tuned with the final task.

We split our model into three parts just following the description in the chapter methodology. The first part is a word embedding. Notice that few words are used in our generated dataset, thus word embeddings cannot be trained effectively. Here, we train the embedding matrix on a huge corpus using Fast-Text algorithm proposed by Facebook [3], and get those related words appeared. Then, sentence encoder is trained to predict users' intent with regard to each utterances. This part helps the encoder to find a more apparent relationship between words and slot values, since predicting some discrete attribute values with a full dialogue can be hard even for human. And then, the dialogue encoder and state tracker are trained together to generate information for KB querying. This stage avoid interference from dialogue generation error as well as ensures a correct request refined from the dialogue. Last, the dialogue generator is trained together with all of the parts mentioned above, to form a real dialogue to guide users, which is designed to be the original final task of our model.

5   Experiments

5.1   Data Sets

As there is no data set with respect to such a task, we simulate a factitious data set to test the performance of our model, in addition to a public data set named bAbi [4].

The task we designed is to select a course according to user's preference and provide information if the user asks. Three optional course attributes are chosen, which are the duration of course, teaching language and the field the course lie in. And more requestable information slots are selected, which are the adviser of a course and its testing method. Moreover, a filter slot is designed to help the user to choose course with non-optional keywords.

A small knowledge base (KB) is manufactured, which contains information of 16 courses from real MOOC platforms in addition with 16 faked entries to make it more complicated. All the basic information required for dialogue is included in this KB.

Template based agents were used to simulate user and agent's utterances. A user intent is first randomly yield, and two agents generate sentences in turn according to these options. For each turn, one or more attributes may be mentioned, while all the slot values should be mentioned when the dialogue comes to the end. To make the task more complex, some value of slots may change with the dialogue goes.

The bAbi dialogue data set is released by facebook AI research to test a model's performance on dialogue generation. The topic of this data set is restaurant booking, with four slots. A KB with 4200 facts and 600 restaurants is given. This data set is made up of five tasks with gradually increased difficulties.

5.2 Sentence Level Intent Detection

We first evaluate the ability of our model to detect a user's intent on single utterance. To make a prediction of user's intent on a sentence, a classifier is connected after the sentence encoder, which is made up of a two layer DNN. We train the sentence encoder and classifier together, and evaluate on a test data set apart from training data set. Our model achieves 100% of correct prediction rate on the task of extracting slot value on either our factitious courses data set or the bAbi task data set.

5.3 Dialogue Level Intent Tracking

We train a whole dialogue level intent tracker with the pre-trained sentence level intent detector, and evaluate the performance of it. Since this is a harder task compared with sentence level intent detection, some mistake appears. We compare our method with several baseline methods including bag of wards (BOW) based classifier, end-to-end memory networks (E2E MN) [17] as well as rule-based method. Experimental results are shown in Table 1.

Table 1. Dialogue level intent tracking results

Task	Rule-based	BOW	E2E MN	State tracker
Course guidance	100%	57.4%	94.8%	98.2%
bAbi task2	100%	65.3%	97.6%	97.5%

It can be seen from the table that, our model achieves state-of-the art on the task of dialogue state tracking compared with baseline methods. Besides, for almost every cases, the true results are computed correctly, with a low error late less then 4% on course recommendation task. This result ensures a correct KB request and result, which is essential to the final dialogue generation.

5.4 Dialogue Generation

Dialogue generator is trained at last, after all of the other parts being trained well.

We first evaluate our model on the standard of task success rate. A task is successful if and only if all the user's requirements are met, that is, slot values are correctly detected and proper information are returned. Success rate on course recommendation and bAbi dataset is shown in Table 2. As can be seen from the table, rule-based method achieves 100% success rate, since the datasets are human-made ones. Our MOOC guider achieves a high success rate, with a fail rate less then 5%.

Table 2. Success rate on tasks

Task	Rule-based	MOOC guider
Course recommendation	100%	95.4%
bAbi restaurant booking	100%	91.0%

As the correlation between evaluation metrics and the actual generation quality is poor, we directly use human evaluation to judge our system. Evaluation scores include sentence quality, information correctness and overall score. Scores ranging from zero to five are taken and averages. Evaluation results is shown in Table 3.

Table 3. Human evaluation results

Sentence quality	Information correctness	Overall score
4.1	4.7	4.5

6   Related Works

Seldom works were done to guide a MOOC user, especially in dialogue form. Traditional MOOC platforms use hard coded filters to select user's preferred attributes as well as search based filtering.

Neural language processing, being a hot topic in the past few years, is experiencing fast development. Being one of the hottest topic, dialogue system is undergoing great changes. Dialogue systems can be divided into two categories, which are chit-chat dialogue system and dialogue-oriented dialogue system. Chit-chat dialogue systems, also called as non-task oriented dialogue system, talk with user without any purpose aiming at imitating a real human. These models can be regarded as an input-output pair generation, and are often modeled using sequence to sequence methods [8,14,16,18,21]. Compared with chit-chat dialogue systems, task-oriented systems are designed to finish a specific task [5].

Traditional statistical dialogue models often contains NLU components, DM components and NLG components. Among them, partially observable Markov decision process (POMDP) based method can handle complex dialogue state tracking problem by designing dialogue state space, and probabilities to transform from one state to another given an input can be trained using data [22]. However, careful designs are required for every task and can only be done by human, which leads to a huge quantity of human work. End-to-end methods turn up in recent years. Recurrent neural network (RNN) based models are proposed to record sequential data and generate response [7]. In order to optimize the structure of memory, some more complex memory components were used including end-to-end memory network (E2E MN) [17] or neural belief tracker (NBT) [12]. Pipelined end-to-end dialogue systems also appeared. Wen et al. proposed a framework which is made up of an intent network, a belief tracker and a generation network [19]. Reinforcement learning algorithms are introduced into dialogue generation for better generation quality. The work done by Li et al. illustrate an enhancement in both generation quality and information reaction by fine-tuning a standard dialogue generation system using reinforcement learning [9]. [11,20] also shows benefits that reinforcement learning can bring to dialogue systems.

7   Conclusion and Future Works

In this paper, we propose a neural network based MOOC guidance agent to provide information for new MOOC users. By using this agent, users can filter suitable courses according to their preference and search for more information in the form of natural language dialogue. Our method is end-to-end trainable, and achieves state-of-the-art performance, in other words, high accuracy on dialogue state tracking and user intent prediction, and satisfying quality on dialogue generation. Besides, a dialogue simulation method is proposed, which can be used for cold starting of the system.

We will further discover methods to handle out-of-vocabulary problem, as words from texts gathered from crowd source of courses may contains vocabularies not appeared in predefined dictionary. Besides, methods for handling multiple situations with applicability are under considering. We will extent our agent to handle more complex tasks.

Acknowledgment. This work is supported by NSFC under Grant No. 61532001, and MOE-ChinaMobile under Grant No. MCM20170503.

References

1. Adamopoulos, P.: What makes a great mooc? an interdisciplinary analysis of student retention in online courses (2013)
2. Barba, Pd, Kennedy, G.E., Ainley, M.: The role of students' motivation and participation in predicting performance in a mooc. J. Comput. Assist. Learn. **32**(3), 218–231 (2016)
3. Bojanowski, P., Grave, E., Joulin, A., Mikolov, T.: Enriching word vectors with subword information. Trans. Assoc. Comput. Linguist. **5**, 135–146 (2017)
4. Bordes, A., Boureau, Y.L., Weston, J.: Learning end-to-end goal-oriented dialog. arXiv preprint arXiv:1605.07683 (2016)
5. Chen, H., Liu, X., Yin, D., Tang, J.: A survey on dialogue systems: recent advances and new frontiers. arXiv preprint arXiv:1711.01731 (2017)

6. Clow, D.: Moocs and the funnel of participation. In: Proceedings of the Third International Conference on Learning Analytics and Knowledge, pp. 185–189. ACM (2013)

7. Henderson, M., Thomson, B., Young, S.: Word-based dialog state tracking with recurrent neural networks. In: Proceedings of the 15th Annual Meeting of the Special Interest Group on Discourse and Dialogue (SIGDIAL), pp. 292–299 (2014)

8. Li, J., Monroe, W., Ritter, A., Jurafsky, D., Galley, M., Gao, J.: Deep reinforcement learning for dialogue generation. In: Proceedings of the 2016 Conference on Empirical Methods in Natural Language Processing, pp. 1192–1202. Association for Computational Linguistics (2016). https://doi.org/10.18653/v1/D16-1127, http://www.aclweb.org/anthology/D16-1127

9. Li, X., Chen, Y.N., Li, L., Gao, J., Celikyilmaz, A.: End-to-end task-completion neural dialogue systems. In: Proceedings of the Eighth International Joint Conference on Natural Language Processing (Volume 1: Long Papers), pp. 733–743. Asian Federation of Natural Language Processing (2017). http://aclweb.org/anthology/I17-1074

10. Lin, Z., et al.: A structured self-attentive sentence embedding. arXiv preprint arXiv:1703.03130 (2017)

11. Liu, B., Tur, G., Hakkani-Tur, D., Shah, P., Heck, L.: End-to-end optimization of task-oriented dialogue model with deep reinforcement learning. arXiv preprint arXiv:1711.10712 (2017)

12. "Mrkšić, N., Ó Séaghdha, D., Wen, T.H., Thomson, B., Young, S.: Neural belief tracker: data-driven dialogue state tracking. In: Proceedings of the 55th Annual Meeting of the Association for Computational Linguistics (Volume 1: Long Papers), pp. 1777–1788. Association for Computational Linguistics (2017). https://doi.org/10.18653/v1/P17-1163, http://www.aclweb.org/anthology/P17-1163

13. Rieber, L.P.: Participation patterns in a massive open online course (mooc) about statistics. Br. J. Educ. Technol. **48**(6), 1295–1304 (2017)

14. Ritter, A., Cherry, C., Dolan, W.B.: Data-driven response generation in social media. In: Conference on Empirical Methods in Natural Language Processing, pp. 583–593 (2011)

15. Rudnicky, A.I., et al.: Creating natural dialogs in the carnegie mellon communicator system. In: Sixth European Conference on Speech Communication and Technology (1999)

16. Serban, I.V., Sordoni, A., Bengio, Y., Courville, A.C., Pineau, J.: Building end-to-end dialogue systems using generative hierarchical neural network models. AAAI **16**, 3776–3784 (2016)

17. Sukhbaatar, S., Weston, J., Fergus, R., et al.: End-to-end memory networks. In: Advances in neural information processing systems, pp. 2440–2448 (2015)

18. Vinyals, O., Le, Q.V.: A Neural Conversational Model. ICML Deep. Learn. Work. 2015 **37**(13002), 1–6 (2015)

19. Wen, T.H., et al.: A network-based end-to-end trainable task-oriented dialogue system. In: Proceedings of the 15th Conference of the European Chapter of the Association for Computational Linguistics: Volume 1, Long Papers, pp. 438–449. Association for Computational Linguistics (2017). http://aclweb.org/anthology/E17-1042

20. Williams, J.D., Zweig, G.: End-to-end LSTM-based dialog control optimized with supervised and reinforcement learning. arXiv preprint arXiv:1606.01269 (2016)

21. Yao, K., Zweig, G., Peng, B.: Attention with intention for a neural network conversation model. In: NIPS Workshop on Machine Learning for Spoken Language Understanding and Interaction, pp. 1–7 (2015). arXiv:1510.08565

22. Young, S., Gašić, M., Thomson, B., Williams, J.D.: Pomdp-based statistical spoken dialog systems: a review. Proc. IEEE **101**(5), 1160–1179 (2013)
23. Zue, V., et al.: Juplter: a telephone-based conversational interface for weather information. IEEE Trans. Speech Audio Process. **8**(1), 85–96 (2000)

AUnet: An Unsupervised Method for Answer Reliability Evaluation in Community QA Systems

Ruoqing Ren[1,2(✉)], Haimeng Duan[1,2], Wenqiang Liu[1,2], and Jun Liu[1,3]

[1] Shaanxi Province Key Laboratory of Satellite and Terrestrial Network Tech. R&D,
Xi'an Jiaotong University, Xi'an 710049, Shaanxi, China
{renruoqing,dhaimeng,liuwenqiangcs}@stu.xjtu.edu.cn,
iukeen@mail.xjtu.edu.cn
[2] School of Electronic and Information Engineering, Xi'an Jiaotong University,
Xi'an 710049, Shaanxi, China
[3] Guangdong Xian Jiaotong University Academy, Xi'an 528300, China

Abstract. Recently, cQA websites such as Baidu Zhidao and StackExchange have exploded in popularity since everyone can post questions for other users to answer which fully realize the value of exchange. Nevertheless, the answers from different users for a same question may include errors, irrelevant messages or malicious advertisements due to the great different backgrounds of users. Hence, the automatic method for answer reliability evaluation is very important for improving users' experience. However, the weakness of existing supervised methods is the high cost for they need a lot of annotated data. To alleviate such problems, we proposed a novel unsupervised answer evaluation method exploiting Answer-User association Network in this paper. Based on the constructed network, the reliability of answers and users can be obtained simultaneously by an iterative process. The experimental results on real word datasets show that our proposed method outperforms existing approaches.

1 Introduction

Community question answering (cQA) websites, such as the general Baidu Zhidao[1] and Yahoo!Answers[2], and the vertical StackExchange[3] and GuoKe[4], are becoming more and more popular since everyone can ask, answer, edit, and organize questions on the website. Compared to the traditional techniques for information retrieval, cQA has made a headway in solving complex, advice-seeking, reasoning questions based on its user-generated-content.

The fast-growing crowdsourcing Q&A data has a good application and development prospect for understanding complex, implicit and self-organization

---

[1] https://zhidao.baidu.com/.
[2] https://answers.yahoo.com/.
[3] https://stackexchange.com/.
[4] https://www.guokr.com/.

© Springer Nature Switzerland AG 2018
L. H. U and H. Xie (Eds.): APWeb-WAIM 2018, LNCS 11268, pp. 281–292, 2018.
https://doi.org/10.1007/978-3-030-01298-4_24

answers. However, the data quality problem [1–3] still exists due to the great different backgrounds of answerers. The low-quality data makes a portion of data cannot be applied directly. Hence, automatic answer reliability evaluation method is very important for improving user experience and constructing high quality Q&A knowledge base.

However, the existing supervised approaches for user reliability evaluation need large amounts of annotated data which is time consuming and limits the applicability to new domains [4]. Besides, unsupervised methods mainly depend on the answerers reputation and result in low accuracy owing to less factors considered.

Therefore, high accuracy unsupervised methods are needed. In this paper, we proposed a novel unsupervised method for answer reliability evaluation by constructing Answer-User association Network (AUnet). This network can successfully captures a variety of factors that affect the reliability of the answer. The contributions of this paper are as follows:

- We constructed AUnet to capture a variety of factors that affect the reliability of the answer. And then the answer reliability evaluation problem is formalized as computing the reliability of node variables on heterogeneous information network.
- A mutual inference algorithm based on AUnet is proposed to calculate the answer reliability. The reliability of answers and users can be obtained simultaneously by an iterative process without any annotated data.
- Experiments on four real datasets from StackExchange have been conducted to test the effectiveness of our method. The results show our method works well.

2   Related work

Our work relates to the answer reliability evaluation and the network-based trust propagation algorithm.

Researches about evaluating the answer reliability are mainly divided into supervised methods and unsupervised methods. Like Maximum Entropy used in [5], Logistic Regression used in [6] and Rand Forests used in [7], supervised methods mainly evaluate and predict the answer reliability by training the classifier based on the manually annotated features of the answer such as community features, user features, textual features and statistical features. Although supervised methods can achieve excellent results, the cost of labeling data is high. Rather than directly evaluate the answer reliability, unsupervised methods resort to calculates user's authority through mining the relation between users, such as the improved PageRank in [8] and the improved HITS in [9]. Besides, Wu et al. [10] achieved the best results from the current unsupervised methods based on the idea of minimizing the difference among answers. For unsupervised methods, data annotation is not required but the accuracy is relatively low.

The network-based trust propagation algorithm is used to effectively identify the trustworthiness of nodes in the network. At present, the network-based trust propagation algorithm is mainly used for fraud detection, selection of comments with high quality, and the discovery of authoritative users and reliable users [11]. Such as Leman et al. [12] iteratively calculated the reliability of the user by using the trust propagation algorithm on bipartite graph, Li et al. [13] used typed Markov Random Fields to detect the campaign promoters on social media and Ko et al. [2] regarded the marginal probability of each answer inferred by the maximum joint probability distribution on the answer association network as the answer reliability. As far as we know, there is no method constructing the trust network to simultaneously model multiple factors that affect the answer reliability and calculating the answer reliability by the trust propagation algorithm on the network.

3   Approach Overview

3.1   Problem Definition and Data Observation

The problem of evaluating the user reliability is formalized as: Given a set of questions $Q = \{q_1, q_2, ..., q_n\}$, a set of all answers $A = \bigcup_{i=1}^{n} A_i$, where $A_i = \{a_{i1}, a_{i2}, ..., a_{im_i}\}$ is a set of m_i answers of the question $q_i \in Q$ and a set of users $U = \{u_1, u_2, ..., u_k\}$. Our goal is to model multiple factors which affect the answer reliability into a network and output the answer reliability $\tau(a_{ij})$ of each answer a_{ij}.

Definition 1 (Answer Reliability). We let $\tau(a_{ij})$ denote the reliability of the answer a_{ij}, which indicates the extent people trust something [14]. We take the answer reliability $\tau(a_{ij}) \in [0, 1]$, and the answer will be more reliable if the answer reliability is more closer to 1.

Definition 2 (User Reliability). The user reliability $\omega(u_k)$ of a user u_k indicates the probability of the user providing reliable answers, and $\omega(u_k) \in [0, 1]$. The user will be more reliable if the user reliability is more closer to 1.

Through the observation of the data, we found that two direct factors and two indirect factors affect the answer reliability.

Direct Factors

- The number of votes of the answer affect the answer reliability. Answers with more votes tend to be more reliable than those with fewer votes. In order to eliminate the different concerns between questions, the shares of votes is used instead of the number of votes to represent the supporting degree to the answer among all voters in participating for the same question. We let $fvote_{ij} = \dfrac{vote_{ij}}{\sum\limits_{j=1 \to m_i} vote_{ij}}$ denote the share of votes for the answer a_{ij}, where $vote_{ij}$ is the number of votes for the answer a_{ij} and m_i is the number of all answers to question q_i. Our statistics showed that the average share of votes of the best answer is apparently higher than non-best answers.

– The frequency of core words affects the answer reliability. The answer with clearer expression and more information is more likely to be reliable. A sentence is considered consisting of meaningless stop words and informative core words. We use the frequency of core words to represent the amount of the information a sentence conveys: $fcore_{ij} = \frac{\sum_{n=1}^{N_{ij}} I(w_n)}{N_{ij}}$, where N_{ij} is the number of words in the answer a_{ij} and $I(w_n)$ is an indicator function, using 1 or 0 to indicate the word w_n is a core word or not. We found that the frequency of core words of the best answer is apparently higher than non-best answers.

Indirect Factors

– Correlation among answers for the same question affects the answer reliability. The reliability of similar answers is positively correlated and mutually driven. Assume that an answer is reliable, it's similar answers are more likely to be reliable, but the different answers of it are more likely to be unreliable.
– Correlation among answers and corresponding users affects the answer reliability. The answer from a more reliable user is more likely to be reliable. Users who provide reliable answers are more likely to be reliable.

3.2   AUnet Model

Based on the four factors above, we constructed AUnet to model them in a unified framework with the reference to the concept of heterogeneous information networks. The network model is shown in Fig. 1, which is defined as:

$$G = \{V, E, W, P\}$$

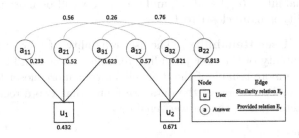

Fig. 1. AUnet model.

– $V = A \cup U$ is a set of all nodes in AUnet, $A = \bigcup_{i=1}^{n} A_i$ is a set of answers to all questions denoted in the blue circle and $U = \{u_1, u_2, ..., u_k\}$ is a set of all users denoted in the black square.
– $E = E_p \cup E_s$ is a set of all edges in AUnet. The similarity relation between answers $E_s \subseteq A \times A$ are denoted as red undirected edges, and the provided relation between users and answers $E_p \subseteq A \times U$ are denoted as black undirected edges.

- $W = \{W_e | e \in E\}$ is a set of the corresponding weights of the edges. $w_s = sim(a_{ij}, a_{ij'})$ is the weight of the similarity relation between answers a_{ij} and $a_{ij'}$, and $w_s \in [0, 1]$. In this paper, we adopted sen2vec [15] and cosine similarity to calculate the semantic similarity w_s between any two answers to the same question. For the weight of the provided relation w_p between the user u_k and the answer a_{ij}, $w_p = prd(a_{ij}, u_k) = 1$ means that all answers provided by the user equally affect the user.
- $P = \{priori(v) | v \in V\}$ is a set of priori reliability of the node $v \in V$ and $priori(v) \in [0, 1]$. The higher the priori reliability is, the more reliable the node is. The priori reliability of the answer a_{ij} is defined on the share of votes $fvote_{ij}$ and the frequency of core words $fcore_{ij}$. $priori(a_{ij}) = \alpha fvote_{ij} + (1 - \alpha) fcore_{ij}$, where α is the influence coefficient between the share of votes and the frequency of core words. The priori reliability of the user $priori(u_k)$ is defined on the reputation, upvotes, downvotes and the homepage views. After the Pearson correlation analysis, we find that the user authority is strongly correlated with the number of the homepage views. Therefore, the normalized user prior reliability is defined as $priori(u_k) = Norm\left(\frac{Reputation}{Views} + Upvote - Downvote\right)$.

4   Mutual Inference Principle

After getting AUnet model, the trust propagation algorithm is used to iteratively update the user reliability and answer reliability based on the mutual inference principle.

4.1   User Reliability Computing

Compared to reliable users, unreliable users have higher error rates. So the reliability of user u_k can be inferred by his/her error rate. Assume that the error rate $\varepsilon(u_k)$ of the user u_k obeys normal distribution, $\varepsilon(u_k) \sim N\left(0, \sigma(u_k)^2\right)$.

Our goal is to make $\varepsilon_{combine} = \frac{\sum_{u_k \in U} \omega(u_k)\varepsilon(u_k)}{\sum_{u_k \in U} \omega(u_k)}$, the variance of the weighted untrustworthiness of all users as small as possible. Since $\varepsilon_{combine}$ also obey normal distribution $\varepsilon_{combine} \sim N\left[0, \frac{\sum_{u_k \in U}(\omega(u_k))^2\sigma^2(u_k)}{\left(\sum_{u_k \in U} \omega(u_k)\right)^2}\right]$. We formulated this goal with the constraint $\sum_{u_k \in U} \omega(u_k) = 1$ into the following optimization problem as:

$$\min_{\{\omega(u_k)\}} \sum_{u_k \in U} (\omega(u_k))^2 \sigma^2(u_k)$$
$$s.t. \sum_{u_k \in U} \omega(u_k) = 1, \omega(u_k) > 0 \tag{1}$$

The optimization problem is a convex function, which can be solved by the Lagrangian multiplier method with a Lagrangian multiplier λ, and the analytical solution is:

$$\omega(u_k) \propto \frac{1}{\sigma^2(u_k)} \tag{2}$$

In Eq. (2), the true variance $\sigma^2(u_k)$ of user u_k can be estimated by the maximum likelihood estimation as:

$$\hat{\sigma}^2(u_k) = \frac{1}{|Q(u_k)|} \sum_{q \in Q(u_k)} \left(x_q^{u_k} - x_q^*\right)^2 \tag{3}$$

Equation (3) means the mean of the squared loss of the errors that user u_k makes. x_q^* is the best answer for the question q which is computed by the weighted average of the answer reliability $x_q^* = \frac{\sum_{u_k \in U_q} \tau\left(a_q^{u_k}\right) \cdot x_q^{u_k}}{\sum_{u_k \in U_q} \tau\left(a_q^{u_k}\right)}$.

According to the statistics, most users give less answers, the method to estimate the users theoretical variance $\sigma^2(u_k)$ by $\hat{\sigma}^2(u_k)$ will be inaccurate when the user provides small number of answers. We solved this long-tail problem by using confidence interval score instead of a single value reference to the work in [16]. Finally, the answer reliability under a certain confidence can be computed as follows:

$$\omega'(u_k) \propto -\frac{1}{\sigma^2(u_k)} = \frac{\chi^2_{1-\frac{\alpha}{2}}(|Q(u_k)|)}{\sum_{q \in Q(u_k)} \left(x_q^{u_k} - x_q^*\right)^2} \tag{4}$$

4.2 Answer Reliability Computing

The answer reliability is affected by the user reliability and other peer answers for the same question [10]. For the reliability, we can get an undirected subgraph for a specific question, consisting of the answers and the corresponding user. Then, we transformed the answer reliability problem to the joint probability distribution of nodes in the undirected probabilistic subgraph. For the undirected subgraph with n random variables, the joint probability distribution can be represented as follows:

$$P(X) = \frac{1}{Z} \prod_{c \in C} \psi_c(X_c) \tag{5}$$

$$Z = \sum_X \prod_{c \in C} \psi_c(X_c) \tag{6}$$

In Eq. (6), $\psi_c(X_c) = exp\{-E(X_c)\}$, and the energy function $E(X_c)$ represents the correlation between variables. Based on the Boltzmann Machines, the probability of the hidden variable y_{ij} of the answer a_{ij} and the probability of the hidden variable y_k of the user u_k are defined as follows:

$$P\left(y_{ij}\right) = \begin{cases} \tau\left(a_{ij}\right), & \text{if } y_{ij} = 1 \\ 1 - \tau\left(a_{ij}\right), & \text{if } y_{ij} = 0 \end{cases} \tag{7}$$

$$P\left(y_{k}\right) = \begin{cases} \omega\left(u_{k}\right), & \text{if } y_{k} = 1 \\ 1 - \omega\left(u_{k}\right), & \text{if } y_{k} = 0 \end{cases} \tag{8}$$

Generally, it's an NP-hard problem to obtain the joint probability distribution on the undirected probabilistic graph [17]. By using the iterated conditional modes ICM [18], we updated the value of the answer node variable in the undirected subgraph step by step based on the idea of gradient ascent as follows:

$$P\left(y_{ij} = \eta\right) = P\left(y_{k} = \eta\right) + \sum_{y_{ij'} \in N(y_{ij})} m_{ij' \to ij}\left(y_{ij} = \eta\right) \tag{9}$$

$$m_{ij' \to ij}\left(y_{ij}\right) = \sum_{y_{ij'}} U\left(y_{ij'}, y_{ij}\right) P\left(y_{ij'}\right) \tag{10}$$

$$U\left(y_{ij'}, y_{ij}\right) = \left[sim\left(a_{ij}, a_{ij'}\right)\right]^{I\left(y_{ij'}, y_{ij}\right)} \cdot \left[1 - sim\left(a_{ij}, a_{ij'}\right)\right]^{1 - I\left(y_{ij'}, y_{ij}\right)} \tag{11}$$

We let $y_{ij'} \in \{0, 1\}$ denote the trustiness transmitted by $a_{ij'}$ to a_{ij}. $U\left(y_{ij'}, y_{ij}\right)$ is the potential function and $sim\left(a_{ij}, a_{ij'}\right)$ denotes the similarity between the answer. When the reliability of similar answers for the same question is consistent, the energy needed by transmission is small and it is easy to happen. In contrast, if the reliability of similar answers for the same question is inconsistent, the energy needed by transmission is big and it is hard to happen.

5   Experiments

5.1   Datasets and Experimental Settings

In order to evaluate the effectiveness of the proposed algorithm in this paper, we conducted experiments on datasets of four domains from the vertical cQA site StackExchange[5], including coffee, movie, music and sports.

The statistics of the four datasets are shown in the first six column in Table 1. To ensure the quality of our dataset, only the question with more than 3 answers are selected.

The dataset of StackExchange only provides the best answer of the question, and doesn't make any judgement on the reliability of other answers. However, answers in cQA often have diversity, so it's not objective to directly treat other answers as negative samples which will cause imbalance between positive and negative examples. Therefore, we randomly selected 50 questions from four domains respectively, totaling 200 questions and 1037 answers, and let two volunteers annotate the answer reliability according to the best answer and relevant

---

[5] https://archive.org/details/stackexchange.

Table 1. Experimental data statistics.

Domain	Questions	Answers	Users	Average answers	Answers per user	Yes	No
Coffee	79	384	221	4.86	1.74	171	70
Movie	960	5135	3104	5.35	1.65	182	87
Music	2255	12689	3481	5.63	3.65	192	91
Sports	171	842	523	4.92	1.61	169	75
Total	3465	19050	7329	5.5	2.6	714	323

information. Each volunteer annotated 125 questions and all answers are annotated as "Yes"(reliable) and "No"(not reliable). After verifying the consistency of the labeling results, the final statistics for all areas is shown in the last two columns in Table 1.

All the experiments were conducted over a server equipped with core i7-4790 CPU on 16 GB RAM, four cores and 64-bit Windows 10 operating system.

5.2 Baseline and Metrics

Four methods Vote, LR, TDM and LQ are selected as the comparison in this paper.

- **Vote**, the basic voting method, directly ranks answers according to the number of votes of the answer.
- **LR**, proposed by Shah et al. [6], trains the Logistic regression model based on non-textual information of answers to evaluate and predict the answer reliability in cQA. The output of LR is a trust value of the answer between 0 and 1.
- **TDM**, a method proposed in [19] based on the iterative idea of TruthDiscovery, estimates the trustworthiness of the answer. TDM smoothes the long-tail user with the priori reliability of the user, and it uses basic iterative methods to update the user reliability and the answer reliability.
- **LQ**, an unsupervised answer reliability evaluation method, is proposed in [10], which detects the low quality answer using the relation between peer answers and label answers through minimizing the variance of the question. We represented the answer by 121 relevant features categorized in 5 types including the statistical characteristics and textual features of the answer, user features and similar features between peer answers.

For the evaluation of answer reliability, we focused on whether the model can effectively filter and return reliable answers, that is, whether the top few answers in the answer list presented to the user are more reliable. Therefore, we evaluated the performance of five models using the two indicators MRR and MAP which are commonly used in information retrieval and question-answering.

MRR (Mean Reciprocal Rank) measures the average of the reciprocal of the position of the best answer in the answer list, which is defined as follows:

$$MRR = \frac{1}{|Q|} \sum_{q \in Q} \frac{1}{bp_q} \tag{12}$$

where $|Q|$ is the total number of questions, and bp_q is the position of the most reliable answer in the answer list. MRR can evaluate whether the algorithm can effectively filter out the best answer.

MAP (Mean Average of Precision) measures the average accuracy of the ranking of answers for each question. That is to say, not only the position of the most reliable answer, but also the position of other reliable answers in the final ranking result are measured. MAP is defined as follows:

$$MAP = \frac{1}{|Q|} \sum_{q \in Q} \left(\frac{1}{TN_q} \sum_{i=1}^{TN} \frac{i}{p_i} \right) \tag{13}$$

where TN_q is the number of reliable answers labeled as positive samples of the question q, and p_i is the position of the ith reliable answer in the final ranking result.

5.3   Performance and Results Analysis

The main parameters of our AUnet method are the window size of sen2vec and α in calculating the answer reliability. The DM model of sen2vec is adopted to represent the answer as a 300-dimensional vector, and the window size is 5. After experiments, the best value of α in coffee domain is 0.6, in movie domain is 0.7, in music domain is 0.6 and in sports domain is 0.5.

We firstly verified the convergence of the algorithm. Figure 2 shows the change in the cumulative value of the answer reliability with iterations in each iteration. When the reliability change of each answer between two iterations is less than 0.001, the algorithm is considered to have reached a steady state.

It can be seen in Fig. 2 that the data of four domains all reach a steady state after 15 iterations in the experiment. Among them, the convergence speed of data of the Music domain is obviously faster than other domain. This is because the number of per capita answers of the Music domain is relatively large, and the number of answers under each question is also large.

The MRR and MAP of five models in four domains are shown in Table 2.

From the experimental results of MRR, we can see that using the voting method alone can filter out about 70% of the best answers. After adding user information and statistical information, the trained LR method can filter out about 80% of the best answers. To improve the ability to filter the best answers in the case of few votes to a certain extent, TDM smooths the long-tail users and LQ introduces similarity relation between peer answers. AUnet achieves the best screening ability in all four areas, and it can effectively return the best answer of more than 86% of problems.

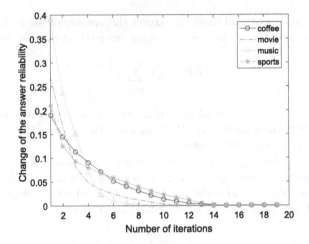

Fig. 2. The number of iterations in four domains.

Table 2. MRR(%) and MAP of five models in four domains

Indicator	Domain	Vote	LR	TDM	LQ	AUnet
MRR	Coffee	70.11	78.42	82.32	82.03	**86.67**
	Movie	71.83	80.03	83.24	83.17	**86.21**
	Music	70.83	79.53	83.43	84.25	**87.66**
	Sports	69.95	78.91	81.5	82.89	**86.33**
MAP	Coffee	0.679	0.774	0.835	0.862	**0.902**
	Movie	0.729	0.83	0.834	0.869	**0.91**
	Music	0.679	0.799	0.843	0.851	**0.908**
	Sports	0.681	0.789	0.833	0.853	**0.893**

The experimental results of MAP count the average sorting accuracy of all questions in each domain, and it measures the ability and accuracy of the algorithm for returning reliable answers.

On the whole, the average ranking performance of Vote which only considers the number of votes is the worst. This is because the number of votes can be affected by factors such as release time and malicious voting, and the reliability of the answer cannot be effectively evaluated without considering the influence of other factors. Because in addition to the community features, the statistical features of the answer and user features are also considered, the performance of LR is slightly improved on the basis of Vote. However, because a large part of answers are long-tail users with less number of votes, the prediction result for the answer with sparse features is poor in LR. This can cause some reliable answers to be sorted backwards, so except Movie, the MAP value of LR in other three domains are below 80%. TDM uses the iterative method to evaluate the reliability of the answer, and the value of MAP is about 83%, which is stable and unaffected by community information. LQ introduces the similarity features and the textual features of the answer on the basis of LR, which can effectively filter the low-quality answer, so the MAP value has greatly improved compared to the other three methods. AUnet models the relation between the user and the answer simultaneously, and utilizes the priori reliability based on the community information and the statistical information, achieving the highest average sorting accuracy in all four domains. For 90% questions, the top three answers returned by AUnet are all reliable. In addition, AUnet achieves the largest performance improvement in the Music domain. That is to say, when the number of answers to the question and the number of answers per capita are large, evaluating answer reliability by AUnet is significantly better than characteristic methods.

6   Conclusion

To alleviate the high cost of labeling data in supervised methods and the low performance in unsupervised methods, we proposed an unsupervised method based on AUnet to evaluate the answer reliability in this paper. On the basis of the probabilistic graphical model and the mutual inference algorithm, our AUnet method can calculate the answer reliability and user reliability simultaneously without supervision and automatically rank the answer in cQA. Results of experiments on four domains in StackExchange verified the convergence and effectiveness of our algorithm and showed our method is superior to other methods in the screening ability of the best answer and the ability to discriminate between reliable and unreliable answers. The potential direction for future research if focusing on evaluating the answer reliability under the multi-source conflict.

Acknowledgments. This work was supported by National Key Research and Development Program of China (2018YFB1004500), Science and Technology Planning Project of Guangdong Province, China (2017A010101029), National Natural Science Foundation of China (61532015, 61532004, 61672419, and 61672418), Innovative Research Group of the National Natural Science Foundation of China(61721002), Innovation Research Team of Ministry of Education (IRT_17R86), Project of China Knowledge Centre for Engineering Science and Technology, and Teaching Reform Project of XJTU (No. 17ZX044).

References

1. Yao, Y., Tong, H., Xie, T., Akoglu, L., Xu, F., Lu, J.: Detecting high-quality posts in community question answering sites. Inf. Sci. **302**(C), 70–82 (2015)
2. Ko, J., Nyberg, E., Luo, S.: A probabilistic graphical model for joint answer ranking in question answering. In: International ACM SIGIR Conference on Research and Development in Information Retrieval, pp. 343–350 (2007)
3. Nie, L., Wei, X., Zhang, D., Wang, X., Gao, Z., Yang, Y.: Datadriven answer selection in community qa systems. IEEE Trans. Knowl. Data Eng. **29**(6), 1186–1198 (2017)

4. Tymoshenko, K., Bonadiman, D., Moschitti, A.: Learning to rank non-factoid answers: comment selection in web forums. In: ACM International on Conference on Information and Knowledge Management, pp. 2049–2052 (2016)
5. Jeon, J., Croft, W.B., Lee, J.H., Park, S.: A framework to predict the quality of answers with non-textual features. In: International ACM SIGIR Conference on Research and Development in Information Retrieval, pp. 228–235 (2006)
6. Shah, C., Pomerantz, J.: Evaluating and predicting answer quality in community qa. In: International ACM SIGIR Conference on Research and Development in Information Retrieval, pp. 411–418 (2010)
7. Dalip, D.H., Cristo, M., Calado, P.: Exploiting user feedback to learn to rank answers in q & a forums: a case study with stack overflow. In: International ACM SIGIR Conference on Research and Development in Information Retrieval, pp. 543–552 (2013)
8. Zhang, J., Ackerman, M.S., Adamic, L.: Expertise networks in online communities: structure and algorithms. In: International Conference on World Wide Web, pp. 221–230 (2007)
9. Jurczyk, P., Agichtein, E.: Discovering authorities in question answer communities by using link analysis, pp. 919–922 (2007)
10. Wu, H., Tian, Z., Wu, W., Chen, E.: An unsupervised approach for low-quality answer detection in community question-answering. In: Candan, S., Chen, L., Pedersen, T.B., Chang, L., Hua, W. (eds.) DASFAA 2017. LNCS, vol. 10178, pp. 85–101. Springer, Cham (2017). https://doi.org/10.1007/978-3-319-55699-4_6
11. Akoglu, L., Tong, H., Koutra, D.: Graph based anomaly detection and description: a survey. Data Min. Knowl. Discov. **29**(3), 626–688 (2014)
12. Akoglu, L., Chandy, R., Faloutsos, C.: Opinion fraud detection in online reviews by network effects, pp. 2–11 (2013)
13. Li, H., Mukherjee, A., Liu, B., Kornfield, R., Emery, S.: Detecting campaign promoters on twitter using markov random fields. In: IEEE International Conference on Data Mining, pp. 290–299 (2014)
14. Fogg, B.J., Tseng, H.: The elements of computer credibility. In: Proceeding of the CHI '99 Conference on Human Factors in Computing Systems: the CHI Is the Limit, Pittsburgh, Pa, USA, pp. 80–87 (May 1999)
15. Le, Q.V., Mikolov, T.: Distributed representations of sentences and documents, vol. 4, pp. II-1188 (2014)
16. Li, Q., et al.: A confidence-aware approach for truth discovery on long-tail data. Proc. Vldb Endow. **8**(4), 425–436 (2014)
17. Liu, W., Liu, J., Duan, H., Hu, W., Wei, B.: Exploiting source-object networks to resolve object conflicts in linked data. In: Blomqvist, E., Maynard, D., Gangemi, A., Hoekstra, R., Hitzler, P., Hartig, O. (eds.) ESWC 2017. LNCS, vol. 10249, pp. 53–67. Springer, Cham (2017). https://doi.org/10.1007/978-3-319-58068-5_4
18. Kittler, J., Hater, M., Duin, R.P.W.: Combining classifiers. In: International Conference on Pattern Recognition, vol. 2, pp. 897–901 (1998)
19. Li, Y., et al.: Extracting medical knowledge from crowdsourced question answering website. IEEE Trans. Big Data **PP**(99), 1–1 (2016)

Model and Practice of Crowd-Based Education

Xinjun Mao[✉], Yao Lu, Liangze Yin, Tao Wang, and Gang Yin

Key Laboratory of Software Engineering for Complex Systems,
College of Computer, National University of Defense Technology,
Changsha, China
xjmao@nudt.edu.cn

Abstract. Based on connectivism pedagogy crowd-based education provides a practical method to extensively exploit wisdoms of core learners in education organization and external crowds on Internet. However, when applying such a method in education field, several design questions about *why, what, when, who, where and how* to adopt such method should be clarified and elaborated. In this paper, we introduce our successful applications of the crowd-based method in software engineering course for undergraduates. We design an organization structure consisting of "small-core crowd" and "large-external crowd" for our course projects in which both learners and crowds on Internet work together to contribute their wisdoms. Two kinds of wisdoms of crowds are exploited in our practices. One is the high-quality open source software (OSS) developed by crowds on the Internet, the other is the diverse software development issues, knowledges, experiences, expertise, etc., that are discussed and interacted by crowds in OSS communities and course communities. We design several course practice activities to exploit crowds' wisdoms, including reading high-quality OSS, searching and reusing OSS in course project, joining and getting helps from OSS communities. The results show that the crowd-based method applied in our software engineering course can significantly improve learner's engineering capabilities of developing high-quality and large-scale software.

Keywords: Crowd-based education · Course teaching · Open source software Community

1 Introduction

Information technology such as (mobile) Internet and social medias (e.g., Facebook, Twitter, WeChat), create great technical opportunities to innovate education ideas that may disrupt traditional education pedagogy, and provide new solutions to address the above challenges (Mark 2013). They provide various Internet-based tools for learners to enrich their learning skills (e.g., collaborative learning) and accommodate new education methods, and thereby result in the emergence of connectivism and the changes of pedagogy moving from andragogy/constructivism towards heutagogy / connectivism (Reese 2015). The evolution of Internet technologies can be used as metaphors of how education should be evolving. The convincing examples include the

© Springer Nature Switzerland AG 2018
L. H. U and H. Xie (Eds.): APWeb-WAIM 2018, LNCS 11268, pp. 293–305, 2018.
https://doi.org/10.1007/978-3-030-01298-4_25

use of mobile learning (Helen 2017), MOOCs and recent SPOCs and their worldwide applications, which support a great number of learners to study on Internet, and motivate some new teaching methods such as flipped classroom and blended learning (Fox 2013; Weld 2012).

Current emerging education pedagogy like connectivism emphasizes learning is the process of building networks of information, contacts, and resources that are applied to real problems (Terry 2011). Coined as the learning theory for the digital age (Reese 2015), connectivism focuses on building and maintaining networked connections that are current and flexible enough to be applied to existing and emergent education problems. It is founded on individual ideas and opinions, valuing diversity in the perspectives of others, building relationships, interdisciplinary connections and current information (Siemens 2004). The education based on connectivism and modern information technologies should be conceived as delocalized, decentralized, online, and collaborative ways (Levy 2015), in which essential teaching and learning interactions occur in virtual cyber spaces in form of various online social medias (e.g., WebChat, Facebook), learning platforms (e.g., Coursera, edX) on the Internet, and social network structure (Shelley 2017). Recently there are several efforts to apply connectivism theory in modern educations like MOOC (Marc 2013) by various Internet technologies like cloud computing (Kultawanich 2015; Dunaway 2011) and on-line education platforms (Reese 2015).

Crowdsourcing is defined as the act of an organization taking a function once performed by employees and outsourcing it to an undefined network of people (Zhao 2014). The essence of crowdsourcing is to harness the dispersed and collective wisdoms (e.g., competences, expertise and skills) of distributed crowds on the Internet (Xinjun 2015; Brabham 2008) to accomplish certain tasks in a more effective and efficient manner, which is essentially consistent with connectivism. Its success mainly relies on the Internet technology and platform, which enables crowds on the Internet to connect, interact, access and share with each other (Howe 2006). Over the years crowdsourcing has attracted great attentions from both industrial practitioners and academic researchers, and has been successfully applied in several domains, e.g., identifying chemical structure, designing mining infrastructure, estimating mining resources, medical drug development, logo design, software development, etc. (Li 2015; Al-Jumeily 2015; Sohibani 2015). One of the well-known domains for crowd-sourcing are the crowdsourcing-based software engineering and the massive high-quality open source software (OSS) developed by crowdsourcers on the Internet. Crowdsourcing presents an effective way to build connections with crowds on Internet and utilize their wisdoms in term of collaborations to solve specific problem (Levy 2015). The success of crowdsourcing manifests its potential impacts and applications on education field (Foulger 2014; Monika 2012; Graziosi 2015). To exploit crowd forces on the Internet for tackling education issues has become an important trend in modern education researches and practices (Sylaiou 2013). Especially it gives us inspirations to develop novel and practical education methods to accomplish connectivism pedagogy.

2   Crowd-Based Education as Operating Model of Connectivism

Essentially teaching and learning are collaborative processes in which various people with diverse knowledges, skills, expertise and even artifacts are involved and cooperated with each other (Huang-Yao 2016). In addition to learners (e.g., student) and teachers, the participants in such process should be diverse to satisfy various education requirements and tackle different education issues. They play various roles in the education process, e.g., some of them play the roles of provider and contributor, others play the roles of consumer and beneficiary.

The goal of crowd-based education is to extensively utilize the wisdoms of crowds on Internet to support the learning and teaching, so as to address issues of traditional education methods, e.g., closure of organization, limitation of knowledge sources, etc. By means of utilizing as many crowds as possible, crowd-based education not only brings the necessary diversity (ranging from knowledges, answers, experiences, and creativity) into the education process, but also provide more learning opportunities and resources for learners. Essentially, crowd-based education acts as the operating model and practical method of connectivism, in which information and resources are in form of various wisdoms of crowds, they are encapsulated as autonomous crowds, connection networks are built in form of virtual crowds community on the Internet, and the learning based on crowds is achieved in term of the collaborations between learners and crowds. Different from traditional education methods, crowd-based education has several distinct features. (1) Learners are not only the education consumer but also the education contributor, which means in learning process learners should positively provide their valuable learning resources for sharing with other learners. (2) The contributors of learning may come from other organizations or societies, especially the crowds on Internet, which means the openness of learning organization in crowd-based education; (3) The learning resources provided by various contributors may be diverse ranging from knowledges, artefacts, skills, expertise and experiences, which means the variety of wisdoms to be utilized in the crowd-based education process.

The main features of crowd-based education are the introduction of crowds and the utilization of their wisdoms in the learning process. Therefore, the success of crowd-based education largely depends on the following factors including the chosen education problems, the crowds to be selected, their wisdoms to be utilized, the innovation of education methods, and the corresponding platforms supporting the crowd-based education. Typically, when applying the crowd-based method in education, several aspects of education should be elaborately designed and the corresponding issues should be addressed.

– *Why*, the goals of applying the crowd-based method in education, e.g., what kinds of education problems are expected to be solved in term of crowd-based method, and what the education results (e.g., education quality, scale, efficiency, effectiveness, etc.) are expected to obtain as the adoption of the crowd-based method.
– *Who*, the crowds that participate in the crowd-based education process. In addition to the traditional education participators (e.g., teachers and students), who are

expected to participate in the education process, provide their wisdoms and contribute to the education activities, what the roles they should play and the responsibility they should take.

- *What*, the kinds of crowds' wisdoms to be utilized in the education. In order to solve specific education problems and achieve education goals, the educator should consider what educational resources are outsourced to the crowds on Internet, what kinds of crowd wisdoms are beneficial for the learners, and what kind of forms of these wisdoms should present to students, e.g., domain knowledge, expertise, skills and products.
- *When*, the occasion to apply the crowd-based education method. Education is a complex process composed of multiple phases and activities. Therefore, we should clarify the occasions and identify the opportunity of applying the crowd-based method. For example, in specific learning phases (e.g., course practices), when some learning events occur, or specific learning conditions are satisfied.
- *How*, the means and solution to apply crowd-based education method. There are several practical issues that should be addressed when carrying out the crowd-based method in education, e.g., how to organize the crowds and learners together, how to design mechanisms to encourage crowds to contribute their wisdoms and participate in the education process, how to manage and utilize crowd's wisdoms, etc.
- *Where*, the platform to support the crowd-based education. Typically crowd-based education is performed on Internet as infrastructure that aggregates great number of crowds and wisdoms. Internet-based education platforms should be provided as the cyber education place to organize crowds, manage wisdoms, and support their learning activities.

3   Model of Applying Crowd-based Method in Software Engineering Course Project

In order to address the education issues of the software engineering course project, we propose an organization structure that consists of small inner crowd and large external crowd. The former is small in number and consists of the learners and lectures/TAs, the latter are large in number and consists of the crowds on Internet. Students are encouraged to act as inner crowds to actively share their experiences and contribute their wisdoms in their course projects. The individuals in OSS communities on the Internet are regarded the external crowds whose knowledges and intellectual outcomes (e.g., OSS) can be utilized as supporting resources in the course project.

- *"small-core crowd"*: It consists of learners (e.g., students) and instructor /TAs in the course education. Comparing with the crowds on the Internet, the organization members in the "small-core crowd" are limited in scale. However, they are actually the main body of the course education. Different from traditional course teaching and practices, the individuals in the "small-core crowd" are required to contribute their wisdoms to the course practice. The wisdoms may take various forms such as issues to be encountered, answers to some development problems, specific knowledge and development skills, software development experiences, domain expertise, etc. Therefore, they are not only the consumer and demander, but also the providers and contributors of the entire process of the course project.

- *"large-external crowd"*: It consists of large number of crowds on the Internet. Normally, they are large in scale but actually are the external entities of the course teaching and learning organization. The individuals in the "large-external crowd" are typically either the developers of OSS projects, or the members of OSS communities. They are experts in software development with abundant software development skills, experiences and can provide high-quality software products (like OSS) as value learning materials for course students. In the software engineering course project, large-external crowds can provide diverse wisdoms to assist the learners in an active or passive way. Therefore, they play the roles of providers and contributors of the course education.

In software engineering literature, there are millions of skilled crowdsourcers on the Internet. They create millions of OSS for various applications, and are normally active to provide a great amount of valuable wisdoms about how to develop software in OSS communities such as Github, and Stack Overflow, etc. It is essential to utilize their wisdoms as important sources to support the software engineering course practices.

4 Learning Activities to Utilize Crowds' Wisdoms

Course project of software engineering is complex enough with several engineering aspects and tasks to be completed, ranging from project management to software development, e.g., conceiving software requirements, designing and modeling software, programming source codes, testing software system, etc. In the whole process, students may meet various development problems and have to seek solutions to tackle these problems. We design the following three course practice tasks that can exploit crowds' wisdoms to improve students' software engineering capabilities.

4.1 Reading High-Quality OSS Codes Developed by Crowds

Though students have been taught various technologies related with quality assurance like design pattern, software testing, code specification and convention, etc. in previous courses, it is still a great challenge for students to construct high-quality medium-size software systems. One effective way is to provide students with high-quality software source codes as references so that they can thoroughly read the software in order to obtain necessary skills and experiences of software construction.

In the course project, we choose some high-quality OSS projects developed by crowds from OSS communities, and require students to read them. Typically, the selected OSSs should have the following features: (1) high-quality in both design and code, e.g., code convention, modularization, well-organization, maintainability, etc., (2) the scale is large enough, e.g., at least 10000+ lines of code; (3) the functions provided by the software and the application domain are known to the students.

When reading large scale and high-quality OSS projects, several learning activities are required to be performed by the learners: (1) reading the program codes of OSS and understand their semantics. (2) commenting the program codes based on their understanding. (3) writing blogs to describe the quality feature of OSS and the relevant development skills in OSS. (4) Reverse-designing the OSS and presenting design models based on their codes reading. (4) maintaining OSS by adding new functions and fixing found issues by adopting the learned development skills.

4.2 Developing Software by Utilizing Crowds' Creative Ideas and OSS

To develop large-scale and high-quality software system is the main task of software engineering course project. We desire the software projects of the course are creative in requirements and functions. They provide novel methods by adopting and integrating emerging computing technologies (mobile computing, service computing, etc.) to solve some interesting problems (service robot application, augment virtual reality, etc.). Moreover, these creative software requirements are feasible in technology. However, given the limited knowledge about software engineering and computing, most of students are not able to initialize the suitable software requirements. They may be either too conservative in idea creation when conceiving software requirements or too aggressive to present advance features in the requirements that are infeasible to implement within the scenario of the course . One effective way to tackle the problem is to utilize the wisdoms of crowds. We require every student to discuss their course projects, especially software requirements and technical feasibility, in both the "small-core crowd" community and "large-external crowd" community. The former is actually the course community established in the Trustie platform, the latter is the OSS communities (e.g., robot operating system community). Such discussions in communities are helpful for learners to get constructive suggestions and feedbacks from the skilled and experienced crowds. Typically, the process may last 3–4 weeks or even more.

Another important issue in the course project comes from the technical challenge to design and implement the conceived software requirements. For example, in the course project sample of home reception robot, robot should recognize visitors in term of recognizing facial images and voices sensed by the robot. Obviously, it is a hard challenge for undergraduates to design professional recognition algorithms and develop intelligent software to detect and identify facial features and voice information. One effective way to solve the problem is to exploit crowds' forces and reuse their OSS libraries. In our course practice, we require students to master OSS software development approach. Students are encouraged to find appropriate OSSs on the Internet, evaluate their applicability for the course project, reuse and maintain these OSSs to satisfy the technical and functional requirements.

4.3 Joining Course Community and OSS Community

In the process of course project, various problems may occur ranging from technical solutions, domain knowledge, usages of development tools. The variety and complexity of these problems means that it is a great challenge for teachers to provide approaches to solving the problems and for students to obtain satisfactory answers. We have to take advantage of crowd wisdoms to enable students to find possible solutions by themselves. Typically, these crowd wisdoms are accumulated in term of discussions and communications in OSS communities. In our crowd-based course project, we encourage students to join the course community and relevant OSS communities and actively interact and collaborate with various community crowds to learn a variety of wisdoms such as programming skills, design patterns, debugging expertise, domain knowledge.

5 Evaluation of Learning Results

We have applied the crowd-based method in software engineering courses for 2 years with 22 and 25 undergraduates of computer science respectively with the support of on-line platform Trustie. The learning results are positive. Students have made impressive progress on QoS (Quality of Software), software project development, and modern software development technologies.

5.1 On-Line Education Platform Trustie for Supporting Crowd-Based Education

In order to support the crowd-based education method and their practices, we have developed an on-line platform called Trustie. It is deployed on Internet and provides three core functions related with the crowd-based education. (1) aggregation of wisdoms, it supports to aggregate various wisdoms generated in the learning process, including knowledges, software artefacts, found issues, OSS, etc. (2) organization of crowds and learners, it organizes the learning individuals (e.g., learners and crowds) in term of community mechanism in order to promote their connections and interactions, e.g., learners' connections cross classes and grades, connection between the learners in course community and the crowds in OSS communities. Trustie supports diverse interactions among individual in course community and OSS community, e.g., submitting an issue, answering a question, providing an OSS, presenting a proposal, etc., and all of interactions are aggregated as valuable wisdoms. (3) evaluation of learning, Trustie records all of the individuals' learning activities and the corresponding wisdoms as evaluating evidences, especially their interactions, and evaluate the scores of learners. For example, we can evaluate the contributions of learners based on their participations and the provided wisdoms.

In our practice of applying the crowd-based method in software engineering course project, learners should register in the platform and all of learning activities related with course project are required to be performed in the platform, including publishing tasks of course project, submitting assignments, interacting with each other, querying wisdoms and resources, etc. Teachers can use the platform to trace the learning process and activities of learning individuals and evaluate their scores. Students however can use the platform to gain various off-line wisdoms (e.g., OSS codes) and on-line wisdoms (e.g., valuable and suggestive answers to some proposed questions).

5.2   Reading high-quality OSS codes developed by crowds on Internet

We have chosen two high-quality OSSs as wisdom artifacts of crowds in OSS communities for reading. One is the "XiaoMi Note" App developed by XiaoMi Company, a famous mobile Internet company in China. The software is developed in Java and operated in Android-based mobile devices (like intelligent cell) to provide note functions for users. It is a classical Android based software system with about 10000+ lines of code (LOC) and little comments. The second is the "OSChina" App. It is actually a software platform for developers to search, use and share OSS codes and techniques. This software has 3400 stars and 3100 forks on Github, and is continuously updated. The "OSChina" App is much larger than "XiaoMi Note". It has about 80000+ lines of code. It is worthful and manageable for undergraduates due to the high-quality of its design and codes. The purposes of reading OSS are to learn good programming skills and experiences provided by crowds on Internet and apply them to develop high-quality software.

All students in our course are required to read one of the selected OSSs and complete required learning activities such as commenting, maintaining, etc., in two months. However, it is actually impossible for student to intensively read and maintain all of the 10000+ or 80000+ lines of codes within just two months. Hence, we require each student to roughly read the whole OSS and make detailed analysis of about 2000+ LOC of the OSS. At this phase, two students are paired up to complete the code reading task together. The aim of rough reading is to understand the whole design framework of the software. The detailed reading however requires students to deeply understand the codes and make semantics comments, including critical variables, statements, algorithms, all functions and classes of OSS. We also require students to rewrite, modify, and test those codes. The two students in pair collaborate with each other to complete the code reading task and share their experiences in code reading. We also require each pair to add 2–3 new functions into OSS. Table 1 depicts the results of the course practice in the code reading phase. Averagely, each student roughly reads 8000+ LOC, finishes the detailed reading about 2000+ LOC, and comments for about 800 LOC, reuses 1200 LOC and writes extra 500 LOC to enrich the functions of OSS. Each pair contributes 142 submissions of code comments on average. The added codes in OSS are verified to be high-quality in program convention and design. According to our investigation, most of students believe they know about what the high-quality software is, how to develop high-quality software, and how to maintain existing software in term of code reading.

Table 1. The efforts and results in code reading.

Learning activities	Max.	Min.	Average
Rough Code Reading (LOC)	10000+	5000+	8000+
Detailed Code Reading (LOC)	3000+	1200+	2000+
Code Commenting (LOC)	1500	400	800
OSS Reusing (LOC)	2000	0	1200
OSS Coding (LOC)	1000	200	500
Submitting comments	153	131	142

5.3 Developing Software by Utilizing Crowds' Creative Ideas and OSS

In the second phase of the course project, we require students to develop a software system based on the software engineering knowledges and the learning development skills and experiences in code reading. 4–5 students are organized as a project team and all the students are organized as a course community. Each project team proposes their software requirements based on the discussions in the course community and corresponding OSS communities. As a result, two teams decided to develop Android apps (one was 3D navigation and the other was campus information sharing APP), two groups focused on developing robot-based software systems (one was for smart home and the other was for library service), one group decided to develop a battle-simulation application. All of these software systems are creative in requirements, large enough in scale, and need to integrate various software engineering technologies, languages and platforms. One purpose of course project at this phase is to extensively utilize crowds' forces to support the development of large enough and high-quality software system.

Table 2 depicts the course project statistics for 5 teams' software projects. All the software projects have employed OSSs to support the implementation of some core functions, and to improve software quality. Various OSSs have been used, ranging from software development framework to artificial intelligence software packages. Most of the course projects are large enough in scale and have more than 10000 LOC. Many of them reuse OSS with 4000+ to 80000+ LOC.

Table 2. The efforts and results of software development in course projects

Project Name	OSS Used in course project	LOC of software	LOC of OSS	LOC of new codes
AR-based 3D navigation	Software framework, API of MapABC	22000+	17000+	5000
Campus information sharing	Software framework	9000+	4000+	5000
Smart home	Naoqi development framework, Face++ image recognition	12500+	5000+	7500
Smart library	Naoqi development framework, Naomark recognition	13500+	5000+	8500
Battle-simulation application	Software framework	87000+	80000+	7000

5.4 Joining Course Community and OSS Community

All of the learning activities are performed in the on-line Trustie platform in which students in course are organized as community to interact with each other about their wisdoms of software development. They can submit their encountered problems, understanding of development technologies, found OSS, development skills, creations on software requirements, etc. Students in course community may play either provider or consumer of wisdoms in the whole process. All of the interactions in course community are recorded by Trustie to trace the development of student individuals and to evaluate their course scores.

We require students in course to participate in Internet OSS communities in order to perform the following learning activities: (1) publishing issues to report encountered bugs in software and discuss their desired requirements; (2) asking various technical questions about code reading and their software development project; (3) submitting pull requests to core developers in order to add new features and fix bugs. We connected the user accounts in CodeCloud with our students, and analyzed their learning activities. As a result, our students published 17 issues and 1 pull request to the project community. The pull request is to add an enhancement on the user interface. Although it was not accepted by the core team, it received high recognition and acknowledgements.

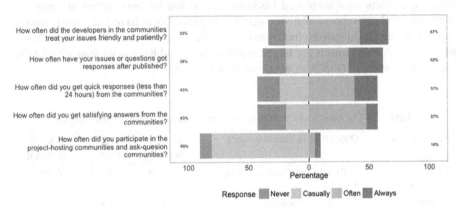

Fig. 1 Investigation on Students' participations in OSS community

Further, we conducted a post-course online survey in order to understand students' learning behaviors in the course project, their attitudes towards the crowd-based methods and their encountered issues in the development process. Figure 1 shows that 90% of the students have participated in the OSS communities, while the participating frequency is not high enough. Besides, most of them get positive responses from the crowds in the communities: over 60% of the issues get responses and patient answers; around 60% of the issues get quick responses (less than 24 h) and satisfactory answers as well.

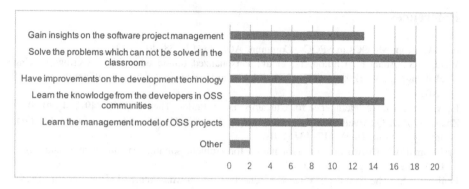

Fig. 2 Students' perception of the crowd-based method

In terms of the students' perceptions on our methods, over 70% of the students believe that they benefit from this novel education method. As shown in Fig. 2, the students mainly benefit from solving problem which cannot be well solved in the traditional classroom settings. Furthermore, the survey results show that 90% of them will continue to use our methods in the future software engineering courses.

6 Conclusions

Obviously, the crowd-based method provides a novel approach to extensively utilizing collective wisdoms of open crowds to solve specific education problems. It can facilitate the participation of crowds into education, build connectivity between crowds and learners, promote their collaboration in term of community mechanism, and enhance the exploitation of crowds' wisdoms in learning. Such method is inspired by the crowdsourcing and its success in software development and highly depends on Internet technologies and corresponding on-line education platforms like Trustie. Various wisdoms of crowds can be exploited depending on the selected education problems, ranging from knowledges, skills, expertise to artefacts. Crowd-based education presented in this paper provides a practical method and operating model to accomplish connectivism pedagogy. Our practices of applying crowd-based education method in software engineering course show its positive results and convince its effectiveness. We believe such method is heuristic, and can be extended or adjusted to satisfy specific education requirements and problems.

Acknowledgments. We greatly thank the project supports of National Key R&D Program of China (2018YFB1004202) and National Science Foundation of China (61532004), the under-graduates that participate in the practices, the Trustie development and technical support team.

References

Fox, A.: From MOOCs to SPOCs. Commun. ACM **56**(12), 38–40 (2013)

Weld, D.S., Adar, E., Chilton, L., et al.: Personalized online education - a crowdsourcing challenge In: Workshops at the 26th AAAI Conference on Artificial Intelligence, pp. 1–31 (2012)

Levy, P.: Collective intelligence for educators. Educ. Philos. Theory **47**(8), 749–754 (2015)

Zhao, Y., Zhu, Q.: Evaluation on crowdsourcing research: Current status and future direction. Inf. Syst. Front. **16**(3), 417–434 (2014)

Brabham, D.C.: Crowdsourcing as a model for problem solving: an introduction and cases. Convergence **14**(1), 75–90 (2008)

Foulger, T.S.: The 21st-century teacher educator and crowdsourcing. J. Digit. Learn. Teach. Educ. **30**(3), 110 (2014)

Li, W., Tsai, W.-T., Wu, W.: Crowdsourcing for large-scale software development. In: Li, W., Huhns, Michael N., Tsai, W.-T., Wu, W. (eds.) Crowdsourcing. PI, pp. 3–23. Springer, Heidelberg (2015). https://doi.org/10.1007/978-3-662-47011-4_1

Graziosi, S., Ferrise, F.: Crowdsourcing and organizational forms: emerging trends and research implications. Appl. Cogn. Psychol. **19**(1), 137–138 (2015)

Al-Jumeily, D., Hussain, A., Alghamdi, M., et al.: Educational crowdsourcing to support the learning of computer programming. Res. Pract. Technol. Enhanc. Learn. **10**(1), 1–15 (2015)

Sylaiou, S., Tampaki, S.: Crowdsourcing in education: challenges and perspectives, In: International Conference on Reimagining Schooling (2013)

Howe, J.: The Rise of Crowdsourcing; Jenkins, H.: Convergence Culture Where Old & New Media Collide **14**(14), 15 (2006)

Al Sohibani, M., Al Osaimi, N., Al Ehaidib, R., Al Muhanna, S., Dahanayake, A.: Factors that influence the quality of crowdsourcing. In: Bassiliades, N., Ivanovic, M., Kon-Popovska, M., Manolopoulos, Y., Palpanas, T., Trajcevski, G., Vakali, A. (eds.) New Trends in Database and Information Systems II. AISC, vol. 312, pp. 287–300. Springer, Cham (2015). https://doi.org/10.1007/978-3-319-10518-5_22

Anderson, T., Dron, J.: Three generations of distance education pedagogy. Int. Rev. Res. Open Distrib. Learn. **12**(3), 1–8 (2011)

Kultawanich, K., Koraneekij, P., Na-Songkhla, J.: A proposed model of connectivism learning using cloud-based virtual classroom to enhance information literacy and information literacy self-efficacy for undergraduate students. Procedia - Soc. Behav. Sci. **191**, 87–92 (2015)

Dunaway, M.K.: Connectivism: learning theory and pedagogical practice for networked information landscapes. Ref. Serv. Rev. **39**(4), 675–685 (2011)

Reese, S.A.: Online learning environments in higher education: Connectivism vs. dissociation. Educ. Inf. Technol. **20**(3), 579–588 (2015)

Siemens, G.: Connectivism: a learning theory for the digital age. Int. J. Instr. Technol. Distance Learn. (2004). Retrieved from http://www.elearnspace.org/Articles/connectivism.htm

Salter, M.B.: Crowdsourcing: student-driven learning using web 2.0 technologies in an introduction to globalization. J. Polit. Sci. Educ. **9**(3), 362–365 (2013)

Crompton, H., Burke, D., Gregory, K.H., Gräbe, C.: The use of mobile learning in science: a systematic review. J. Sci. Educ. Technol. **25**(2), 149–160 (2017)

Clarà, M., Barberà, E.: Learning online: massive open online courses (MOOCs), connectivism, and cultural psychology. Distance Educ. **34**(1), 129–136 (2013)

Mao, X., Hou, F., Wu, W.: Multi-agent system approach for modeling and supporting software crowdsourcing. In: Li, W., Huhns, M.N., Tsai, W.-T., Wu, W. (eds.) Crowdsourcing. PI, pp. 73–89. Springer, Heidelberg (2015). https://doi.org/10.1007/978-3-662-47011-4_5

Skaržauskaitė, M.: The application of crowd sourcing in educational activities. Soc. Technol. **15** (1), 43–53 (2012)

Hong, H.-Y., Chai, C.S., Tsai, C.-C.: College students constructing collective knowledge of natural science history in a collaborative knowledge building community. J. Sci. Educ. Technol. **24**(5), 549–561 (2016)

Exploring Business Models and Dynamic Pricing Frameworks for SPOC Services

Zhengyang Song[✉], Yongzheng Jia, and Wei Xu

Institute of Interdisciplinary Information Sciences,
Tsinghua University, Beijing, China
songzy16@mails.tsinghua.edu.cn

Abstract. MOOCs provide irreplaceable opportunities of making high-quality courses accessible to everybody. However, MOOCs are often criticized for lacking sustainable business models, and academic research for business strategies for MOOCs is also a blind spot currently, especially for the B2B markets. As the primary B2B business model, *SPOC services* can help the institutional users to improve their in-classroom teaching outcomes, as well as bring considerable revenue to the MOOC platforms. In this work, we formulate the economic model and pricing strategy for SPOC services in a theoretical way and further present the future work of applying our model in real markets.

1 Introduction

MOOCs bring an unprecedented revolution to the worldwide higher education of producing high-quality online courses and make them accessible to everybody. At the same time, universities can also adopt flipped classroom learning to improve the teaching quality of the on-campus education by using *SPOCs* (i.e. *Small Private Online Courses*) through various blended teaching and learning methodologies. MOOC is an ecosystem involving efforts from many parties. The *MOOC platforms* are the core of the ecosystem. Every MOOC platform is a marketplace where *MOOC producers* (usually universities) deliver their MOOCs and corresponding education services to the *users*. The users in the MOOC ecosystem consist of both *Internet users* and *institutional users*. From an industrial perspective, the profitability and financial stability of the MOOC platforms become a critical problem associated with the sustainable development of the entire MOOC ecosystem.

As we know, the B2C (i.e., business-to-customer) services are the basic business models for MOOC platforms of making money from the Internet users with a *freemium* strategy: Where the basic materials of MOOCs are open and free to all users, and the MOOC platforms also offer fee-based *online value-added services* to the users including the *Verified Certificates*, *Specializations*, *Online Micro Masters*, *Advanced Placement* (i.e. AP) courses and so forth. However, for some MOOC platforms, both the completion rate (i.e., the percentage of the users to pass the basic requirement of a course) and the paying rate (i.e., the

© Springer Nature Switzerland AG 2018
L. H. U and H. Xie (Eds.): APWeb-WAIM 2018, LNCS 11268, pp. 306–317, 2018.
https://doi.org/10.1007/978-3-030-01298-4_26

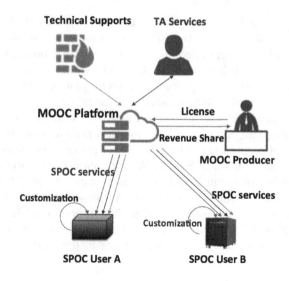

Fig. 1. Business model and market structure for the SPOC services.

percentage of users to pay for value-added services such as verified certificates) is not promising [1]. These MOOC platforms can hardly sustain by only providing value-added services to the Internet users. In China, some MOOC platforms (e.g., xuetangX, icourse163) make profits by providing B2B (i.e., business-to-business) education services to institutional users (e.g., universities, professional training institutions, etc) by sub-licensing MOOC contents and on-campus SPOC platforms. In this work, we focus on the B2B services, which attract less attention from both the industries and academics, but also play an important role in the MOOC ecosystem, and sometimes bring more revenue to some early-stage MOOC platforms than B2C services. In the B2B *course sub-licensing* market, the MOOC platform is the seller in the market, and buyers are *institutional users* with the demand of using the MOOC materials from the platform and deploy them as SPOCs for purposes of blended learning.

In practice, the sub-licensing services always exist in the pattern of *SPOC services* by allowing the *institutional users* to import MOOC materials from the platform and use them as SPOCs with blended teaching and learning approaches. To guarantee the quality of service, the *SPOC services* are dynamic and highly customized, and a user's demand is a *bundle* of education services including MOOC contents, teaching assistant services, SaaS services, technical supports and so forth. We illustrate the business model and market structure for the SPOC services in Fig. 1. On one hand, each institutional user pays for a bundle of customized education services and the MOOC platform makes revenue from the services. On the other hand, as the copyrights of the licensed MOOCs do not belong to the platform, the MOOC platform should get sub-licensing approvals from the MOOC producers and share revenue with them.

To the best of our knowledge, this is the first work to study the business model and pricing strategy for the SPOC services with economic models. As the users' demands are dynamic, we design interactive business process and dynamic pricing framework to analyze the SPOC services, and further present the profit maximization strategy for the MOOC platform by formulating an integer programming with resource capacity constraints.

In this work, we propose a theoretical model to analyze the business process and pricing strategy for SPOC services, and also present ideas of applying our model in the MOOC industry. The rest of the paper is organized as follows: We first review related work in Sect. 2. We formulate the theoretical model for the dynamic pricing framework and business process of the SPOC services in Sect. 3. To solve the optimization problem, We present the theoretical analysis to maximize the MOOC platform's total profits through combinatorial auction mechanisms in Sect. 4. We then give ideas of applying our model in practice in Sect. 5. Finally, we present the directions of our future work and conclude the paper in Sect. 6.

2   Related Work

SPOC (i.e., Small Private Online Course) refers to another version of a MOOC (i.e., Massive Open Online Course) which localizes the instances of a MOOC on campus through business-to-business contexts. The concept of the SPOC is initiated in the University of California Berkeley by Armando Fox [2]. Soon after that, Chinese pioneers start to deploy the SPOC services and apply blended learning methodologies on campus from 2013 [3]. To the best of our knowledge, this work is the first to study the business model and dynamic pricing framework with theoretical analysis for SPOC services. In [1], we analyze the flat-rate pricing strategies for B2C markets with both theoretical models and data-driven analysis. [4] presents the ideas of involving adaptive learning into the business model design of MOOCs. There are also discussions on future trends of business development of MOOCs from the industry. For instance, [5] shows the latest experience of finding niche and business model for MOOC in 2016.

To analyze the business models for the B2B services, there have been lots of research done in the area of resource allocation using auction models. One early work dates back to [6], where they use the combinatorial auction to deal with the problem of airport time slot allocation by utilizing the concept of shadow price. Recent work such as [7], designs a heuristic greedy deterministic auction as well as a randomized linear optimization based auction for allocating wireless spectrum in secondary network. The authors in [8] present a discriminative second price auction technique to motivate users in peer-to-peer video-on-demand streaming applications. [9] presents a reverse auction framework to motivate the smartphone users to join mobile crowdsourcing applications. The authors in [10] show how we can deal with resource provisioning in cloud computing with an online combinatorial auctions framework. These studies bring us inspirations of applying various mechanisms to optimize the MOOC B2B market.

We further study the theoretical methodologies for auction mechanism design. [11] shows the strong links between combinatorial auction and Lagrangian-based decomposition. The authors in [12] utilize mixed integer programming to manage general combinatorial auction problems efficiently. [13] presents a new family of preference elicitation algorithms to prevent bidders to bid on all combinations. In another way, [14] shows how to use boosting to automatically modifying existing mechanisms to increase expected revenue. [15] exhibits three approximation algorithms for the allocation problem in combinatorial auctions with different ratios under different assumptions. [16] dives deeper into the question of whether polynomial-time truthful mechanisms for combinatorial auctions are provably weaker in terms of approximation ratio than non-truthful ones. We use some of the above-mentioned techniques to improve the performance of our combinatorial auction model.

3   Modeling the SPOC Services

In this section, we analyze the business model of the SPOC services by using an auction-based pricing framework. In the auction-based market, each user attaches a bid to her *bundle* that signifies her willingness to pay for the SPOC services, and then the MOOC platform decides whether to accept this bundle. If yes, the MOOC platform makes a contract with the user under a dynamic pricing mechanism; if no, the business negotiation between the user and the platform continues.

Then we formulate the B2B market with one MOOC platform as the seller within a certain period (e.g. a semester, or a fiscal year). There is a total of C courses with sub-licensing approvals which can be used as *SPOCs*. We use $[X] = \{1, 2, \ldots, X\}$ to denote the set of X elements throughout the paper, and therefore $[C] = \{1, 2, \cdots, C\}$ is the set of sub-licensing courses. For each course $c \in [C]$, the MOOC platform provides SaaS services, teaching assistant services, technical support, and other education services. Due to the resource capacity constraint (e.g., the limited number of TAs, or limited computational resources of the SPOC platform) for each course, course c can support at most q_c students from all the SPOCs. There are N users in the market, and the business process is a series of negotiations between the users and the platform. We assume that each user-platform negotiation completes within K steps, otherwise the negotiation fails. In the k-th step ($k \in [K]$) of the negotiation, the user $n \in [N]$ submits a bid with a bundle $B_{n,k}$ of SPOC services and her valuation $v_{n,k}$ (i.e., willingness to pay) to the bundle. Each bundle $B_{n,k}$ contains a vector of C integers indicating the enrollments for the C courses, and we denote the number of enrollments for the SPOC c in bundle $B_{n,k}$ as $s_{n,k,c}$. If the negotiation terminates in K' rounds and $K' < K$, then let $B_{n,k} = \{0, 0, \cdots, 0\}$ and $x_{n,k} = 0$ for all $K' < k \leq K$. In the following analysis, we first consider the offline setting, where we know the information for all the bids of each user in advance, and we take all the data of $B_{n,k}$ as the input.

We further consider two types of cost to deploy and operate the SPOC services: The *capital cost* is the expenditures to support the basic features of SPOCs (e.g. the cost to build the customized SPOC platform), and we denote the capital cost for bundle $B_{n,k}$ as $d_{n,k}$, which is not associate with each SPOC; The *operational cost* is the cost to operate and support each SPOC including the TA's labor cost, video traffics cost and so forth, and we denote the *operational cost* for the course c in bundle $B_{n,k}$ as $\omega_{n,k,c}$.

The platform need a *decision algorithm* $\mathcal{A}$ to decide whether to accept a bid and a pricing mechanism $\mathcal{P}$ to maximize the total profit from all the users. We use $\mathcal{R}$ to denote the current resource capacity which is a vector of C integers indicating the remaining capacities for the C courses. The decision algorithm $\mathcal{A}$ is a function of $B_{n,k}$, $v_{n,k}$ and $\mathcal{R}$. We use the binary variable $x_{n,k}$ to denote whether bundle $B_{n,k}$ is accepted by the platform, so we have:

$$x_{n,k} = \mathcal{A}(B_{n,k}, v_{n,k}, \mathcal{R}) = \begin{cases} 1 \text{ Accept} \\ \\ 0 \text{ Reject} \end{cases} \quad \forall k \in [K], n \in [N]$$

Let $p_{n,k} = \mathcal{P}(B_{n,k}, v_{n,k}, \mathcal{R})$ denote the price for bundle $B_{n,k}$, then we formulate the profit maximization strategy for the platform by using an integer programming as follows:

$$\text{maximize:} \sum_{n \in [N], k \in [K]} (p_{n,k} - d_{n,k} - \sum_{c \in [C]} \omega_{n,k,c}) \cdot x_{n,k} \qquad (1)$$

s.t.

$$\sum_{k \in [K]} x_{n,k} \leq 1, \quad \forall n \in [N]; \qquad (2a)$$

$$\sum_{k \in [K]} \sum_{n \in [N]} s_{n,k,c} \cdot x_{n,k} \leq q_c, \quad \forall c \in [C]; \qquad (2b)$$

$$x_{n,k} \in \{0, 1\}, \quad \forall n \in [N], \forall k \in [K]. \qquad (2c)$$

The objective function (1) is the total profits gained by the MOOC platform. Constraint (2a) means that each buyer wins at most one bundle of the SPOC services and constraint (2b) is the resource capacity constraint, showing that the total enrollments of each course are smaller than its capacity.

4   Combinatorial Auction Mechanisms

In this section, we present combinatorial auction mechanisms to solve the offline problem to optimize the MOOC platform's total profits, where we know all the information for the bids of each user (i.e, $B_{n,k}, \forall n, \forall k$) in advance.

To maximize the platform's total profits from the *bundled education services*, the combinatorial auction mechanisms are promising techniques to fit our settings. In a typical one round combinatorial auction, there are n bidders, each of them will bid for k bundles of items. Then the outcome (i.e., total profits in our settings) of this auction (x, p) will be decided by a specific mechanism, where $x_{i,k}$ is the allocation for bidder i and bundle k and $p_{i,k}$ is the price that bidder i should pay for bundle k. In our setting, the courses and services represent the items to be sold in the general combinatorial auction setting, and each institutional user bids for several bundles of services (i.e., at most k bundles). Constraints come from the limits of teaching assistants, computing resources and so on. In this work, We apply three combinatorial mechanisms: the *VCG Mechanism*, the *Virtual Valuation Mechanism*, and the *Shadow Price Mechanism*.

4.1 VCG Mechanism

We first apply the famous VCG mechanism [17] in our setting. VCG mechanism employs an allocation rule to maximize the social warfare, i.e. the sum of all the valuations of users who win the bidding bundles. The formulation of allocation rule is as follows:

$$\max \quad \sum_{n\in[N]}\sum_{k\in[K]} v_{n,k}x_{n,k}$$

$$\text{s.t.} \quad Constraints\ (2a)\ \text{-}\ (2c)$$

Then the payment p_i for each bidder i by VCG mechanism is:

$$p_i = \sum_{j\neq i}\sum_{k\in[K]} v_{j,k}\tilde{x}_{j,k} - \sum_{j\neq i}\sum_{k\in[K]} v_{j,k}x_{j,k}$$

where

$$\tilde{x}_{j,k} = \arg\max_{x_{j,k}} \sum_{j\neq i}\sum_{k\in[K]} v_{j,k}x_{j,k}$$

The intuition of the above mechanism is that we set the price of the SPOC services to one particular bidder as the decrease of all the other bidders' gain due to the participation of this bidder. Note that VCG auctions are generally computational intractable, existing work such as [18] incorporate the use of VCG-style pricing with exploited greedy allocation schemes.

4.2 Virtual Valuation Mechanism

VCG Mechanism may not be optimal in respect of the seller's revenue. Inspired by Myerson mechanism [19] for the optimal single item auction, and the authors of VVCA (virtual valuation combinatorial auction) [14] further introduce two kinds of virtual valuation forms to boost seller's revenue in combinatorial auction. Instead of maximizing the sum of all the real valuations of users who won the bidding bundle, the allocation is decided by maximizing the sum of all the virtual valuations. Then the price is decided by calculating the decrease of other bidders' virtual gain due to the participation of this bidder.

VVCA mechanisms is parameterized by a bunch of preset parameters μs and λs, corresponding to the bidder weighting technique and allocation boosting technique respectively: the former assign priorities to bidders with higher valuations, while the latter assign priorities to a specific bundle for a bidder. The allocation is computed by solving:

$$\max \quad \sum_{n\in[N]}\sum_{k\in[K]} (\mu_n v_{n,k} x_{n,k} + \lambda_{n,k} x_{n,k})$$

$$\text{s.t.} \quad \textit{Constraints (2a) - (2c)}$$

where μ are positive real numbers. The payment rule is

$$p_i = \frac{1}{\mu_i}\left(\sum_{j\neq i}\sum_{k\in[K]} (\mu_j v_{j,k}\tilde{x}_{j,k} + \lambda_{j,k}\tilde{x}_{j,k} - \mu_j v_{j,k} x_{j,k} - \lambda_{j,k} x_{j,k}) - \sum_{k\in[K]} \lambda_{i,k} x_{i,k}\right)$$

where

$$\tilde{x}_{j,k} = \arg\max_{x_{j,k}}\left(\sum_{j\neq i}\sum_{k\in[K]} \mu_j v_{j,k} x_{j,k} + \lambda_{j,k} x_{j,k}\right)$$

The intuition is that we substitute the valuations in VCG to virtual valuations here. Then we determine the parameters λ and μ by using numerical methods such as hill-climbing to maximize the seller's expected revenue.

4.3   Shadow Price Mechanism

The main idea of the Shadow Price Mechanism [6] is to determine a shadow price for each available resource using the optimal Lagrangian multiplier of the primal integer program, and determine the bid rejection prices (D_R) and bid acceptance prices (D_A) using two pseudo-dual programs.

$$\min_{x^*_{n,k}=0} \sum y_{n,k}$$

$$\text{s.t.} \sum_c w_c s_{n,k,c} \leq v_{n,k} \qquad\qquad \forall x^*_{n,k}=1$$

$$y_{n,k} \geq v_{n,k} - \sum_c w_c s_{n,k,c} \qquad\qquad \forall x^*_{n,k}=0$$

$$y_{n,k} \geq 0 \qquad\qquad\qquad\qquad \forall x^*_{n,k}=0$$

$$w_c \geq 0$$

Here $x^*_{n,k}$ is the optimal solution to the primal integer program, and the set of lower bound prices is w^*_c, and we denote the exceeding price of a rejected bundle comparing to the market price as $y_{n,k}$.

$$\min_{x^*_{n,k}=1} \sum y'_{n,k}$$

$$\text{s.t.} \sum_c u_c s_{n,k,c} \geq v_{n,k} \qquad\qquad \forall x^*_{n,k} = 0$$

$$y'_{n,k} \geq \sum_c u_c s_{n,k,c} - v_{n,k} \qquad\qquad \forall x^*_{n,k} = 1$$

$$y'_{n,k} \geq 0 \qquad\qquad\qquad\qquad \forall x^*_{n,k} = 1$$

$$u_c \geq 0$$

Here the set of upper bound prices is u^*_c, and we also denote the exceeding price of a rejected bundle comparing to the market price as $y'_{n,k}$.

Then the allocation rule has the following properties: (i) if a bid is greater than the sum of its component values in the set $\{u^*\}$, then it is accepted. (ii) if a bid is less than the sum of its component values in the set $\{w^*\}$, then it is rejected. (iii) all bids between are determined by considering the primal integer program regardless of the shadow prices.

The pricing strategy is to set the price for any bidder whose bundle k is accepted in the solution of the primal problem to the sum of the shadow prices for the resources in the package.

4.4   Examples

Now we give an example to show why VVCA mechanism can generate higher revenue than VCG mechanism. Consider the following scenario, there are three bidders $\{A, B, C\}$ and two items $\{P, Q\}$, the valuation of bidders to bundles are listed below:

$$v_{A,\{P\}} = 5, \ v_{B,\{Q\}} = 1, \ v_{C,\{P,Q\}} = 16$$

Then with VCG mechanism, the allocations are decided by solving the following integer programming:

$$\max \quad 5 \cdot x_{A,\{P\}} + x_{B,\{Q\}} + 16 \cdot x_{C,\{P,Q\}} \tag{3}$$

s.t.

$$x_{A,\{P\}} + x_{C,\{P,Q\}} \leq 1 \tag{4a}$$

$$x_{B,\{Q\}} + x_{C,\{P,Q\}} \leq 1 \tag{4b}$$

$$x_{A,\{P\}}, \ x_{B,\{Q\}}, \ x_{C,\{P,Q\}} \in \{0,1\} \tag{4c}$$

A simple enumeration of the feasible solutions gives the optimal solution

$$x_{A,\{P\}} = x_{B,\{Q\}} = 0, \ x_{C,\{P,Q\}} = 1$$

So the final allocation would be give item $\{P, Q\}$ to C. Now we compute what price would C pay. Without the presence of C, we have the integer programming:

$$\max \quad 5 \cdot x_{A,\{P\}} + x_{B,\{Q\}} \tag{5}$$

s.t.

$$x_{A,\{P\}} \leq 1 \tag{6a}$$

$$x_{B,\{Q\}} \leq 1 \tag{6b}$$

$$x_{A,\{P\}},\ x_{B,\{Q\}} \in \{0,1\} \tag{6c}$$

The optimal solution would be

$$\tilde{x}_{A,\{P\}} = \tilde{x}_{B,\{Q\}} = 1$$

So the price C needs to pay is the decrease of social ware-fare due to the presence of himself, which is

$$
\begin{aligned}
P_c &= \left(\tilde{x}_{A,\{P\}} \cdot v_{A,\{P\}} + \tilde{x}_{B,\{Q\}} \cdot v_{B,\{Q\}} \right) \\
&\quad - \left(x_{A,\{P\}} \cdot v_{A,\{P\}} + x_{B,\{Q\}} \cdot v_{B,\{Q\}} \right) \\
&= (1 \cdot 5 + 1 \cdot 1) - (0 \cdot 5 + 0 \cdot 1) \\
&= 6
\end{aligned}
$$

Thus the revenue of VCG mechanism would be 6. Then we show how VVCA would boost the revenue of above VCG mechanism. The main idea is to use virtual valuations to bring down the difference of valuations between strong bidders and weak bidders, thus creating an artificial competition to extract more revenue from strong bidders. We assign the following λ, μ:

$$\mu_C = 0.5,\ \lambda_{B,\{Q\}} = 1$$

Now the integer programming would become:

$$\max\quad 5 \cdot x_{A,\{P\}} + x_{B,\{Q\}} + x_{B,\{Q\}} + 0.5 \cdot 16 \cdot x_{C,\{P,Q\}}$$
$$\text{s.t.}\quad \textit{Constraints (4a) - (4c)}$$

Without the presence of C, we have:

$$\max\quad 5 \cdot x_{A,\{P\}} + x_{B,\{Q\}} + x_{B,\{Q\}}$$
$$\text{s.t.}\quad \textit{Constraints (6a) - (6c)}$$

The optimal solution would still be

$$x_{A,\{P\}} = x_{B,\{Q\}} = 0,\ x_{C,\{P,Q\}} = 1$$

$$\tilde{x}_{A,\{P\}} = \tilde{x}_{B,\{Q\}} = 1$$

The difference lies in the price that C would pay:

$$
\begin{aligned}
p'_C &= \frac{1}{\mu_C} \left(\tilde{x}_{A,\{P\}} \cdot v_{A,\{P\}} + \tilde{x}_{B,\{Q\}} \cdot v_{B,\{Q\}} + \lambda_{B,\{Q\}} \tilde{x}_{B,\{Q\}} \cdot v_{B,\{Q\}} \right) \\
&\quad - \frac{1}{\mu_C} \left(x_{A,\{P\}} \cdot v_{A,\{P\}} + x_{B,\{Q\}} \cdot v_{B,\{Q\}} \lambda_{B,\{Q\}} x_{B,\{Q\}} \cdot v_{B,\{Q\}} \right) \\
&= \frac{1}{0.5}(1 \cdot 5 + 1 \cdot 1 + 1 \cdot 1 \cdot 1) - \frac{1}{0.5}(0 \cdot 5 + 0 \cdot 1 + 1 \cdot 0 \cdot 1) \\
&= 14
\end{aligned}
$$

Thus the revenue of VVCA mechanism would be 14, which is much higher than the revenue of VCG mechanism, i.e., 6.

5 Business Process in the MOOC Industry

In the previous discussion, we only consider the offline setting for the combinatorial auction in only one round. In practice, the business process of the user-platform negotiation is *online*, indicating that the platform must give quick (or even instant) response to the user when she submits a bid. Moreover, the response message to the user is interactive, including not only the result of whether the bid is accepted or not, but also the reasons of why the bid is rejected, or even suggestions on the combinations of $B_{n,k}$ and $v_{n,k}$. For instance, a bid may be rejected due to the low valuation, or unmet resource capacities for some SPOCs. When the user receives the message, she will adjust her bid by reducing the demand or increasing the valuation to the bundle in the next step of negotiation. We further present Algorithm 1 to demonstrate the business process of the user-platform negotiation in a systematic way.

To solve the online setting of the problem, we apply the iterative combinatorial auctions in [20], and further use the *iBundle* mechanism [21] for detailed analysis.

iBundle maintains *ask prices* on bundles, which is the lowest price at which a bundle may be sold, and also the *provisional allocation*, which is the possible allocation for the current bids. The process goes through multiple rounds, in each round a bidder can submit bids on bundles, where at most one bid will be chosen. A bid is called competitive if it is at or above the current ask price. A bidder is called competitive if at least one of his bids is competitive. Then a winner-determination algorithm computes the provisional allocation to maximize the seller's revenue. iBundle terminates when each competitive bidder receives a bundle in the provisional allocation. Otherwise, prices are increased by a preset parameter ϵ above the bid price on all bundles that receive a bid from some losing bidder in the current round and the allocation and new prices are provided as feedback to bidders. On termination, the provisional allocation becomes the final allocation, and the bidders pay their final bid prices.

Algorithm 1: Negotiation between user n and the platform

---

1 Initialization: Set $t = 1$ and $flag = 0$. Suppose the current status of resource capacity is $\mathcal{R}$.

2 **while** $t \leq T$ **do**

3   (a) User n submits his bids $(B_{n,k}, v_{n,k})$ to the platform.

4   (b) The platform calculates $x_{n,k}$ and $p_{n,k}$, and sends the response message to the user.

5   (c) **If** *accepted*, **then** the negotiation succeeds, update $\mathcal{R}$, set $flag = 1$, and **break. Else** (i.e. *rejected*) the negotiation continues with $t = t + 1$.

6 **end**

7 **If** $flag = 0$, **then** the negotiation fails.

---

6   Conclusion Remarks

In this work, we focus on analyzing the business model and pricing strategy for the SPOC services, and maximize the MOOC platform's total profits by using combinatorial auctions. We present formulations and solutions for both the one-round offline scenario, and also the iterative multi-round scenario. We present several mechanisms for the allocation rules and pricing strategies.

Working in the MOOC industry for the past four years, we have gained valuable marketing experience of selling MOOC and SPOC services. For the B2C services, we can directly get sales data from the online purchasing records and use data-driven approaches to better analyze the real market. For the SPOC services, we successfully deploy the services to 125 real institutional users, including 90 universities, 20 corporations, 7 high schools and 8 government organizations. Even though there are no automatic ways to collect sales data for SPOC services, we can use the *Customer Relationship Management* (i.e. *CRM*) system to keep track of the marketing data and the business process. It is also practical to conduct surveys to the key users to better understand the buyers' behavior. We will apply these methodologies in the MOOC industry to improve the marketing performance of the models in our future work.

References

1. Jia, Y., Song, Z., Bai, X., Xu, W.: Towards economic models for mooc pricing strategy design. In: Bao, Z., Trajcevski, G., Chang, L., Hua, W. (eds.) DASFAA 2017. LNCS, vol. 10179, pp. 387–398. Springer, Cham (2017). https://doi.org/10.1007/978-3-319-55705-2_31
2. Fox, A.: From MOOCs to SPOCs. Communications of the ACM (2013)
3. Xu, W., Jia, Y., Fox, A., Patterson, D.: From MOOC to SPOC: lessons from MOOC at Tsinghua and UC Berkeley. In: Modern Distance Education Research (2014)
4. Daniel, J., Cano, V., Cervera, G.: The future of MOOCs: adaptive learning or business model? Int. J. Educ. Technol. High. Educ. (2015)
5. Morrison, D.: Need-to-Know MOOC News: MOOCs Find Their Niche and Business Model in 2016. https://onlinelearninginsights.wordpress.com/2016/02/03/need-to-know-mooc-news-the-mooc-business-model-gets-its-teeth-in-2016/ (2016)
6. Rassenti, S.J., Smith, V.L., Bulfin, R.L.: A combinatorial auction mechanism for airport time slot allocation. Bell J. Econ., 402–417 (1982)
7. Zhu, Y., Li, B., Li, Z.: Truthful spectrum auction design for secondary networks. In: INFOCOM, 2012 Proceedings IEEE, pp. 873–881. IEEE (2012)
8. Wu, C., Li, Z., Qiu, X., Lau, F.: Auction-based p2p vod streaming: Incentives and optimal scheduling. ACM Trans. Multimed. Comput. Commun. Appl. (TOMM) 8(1S), 14 (2012)
9. Feng, Z., Zhu, Y., Zhang, Q., Ni, L.M., Vasilakos, A.V.: Trac: truthful auction for location-aware collaborative sensing in mobile crowdsourcing. In: IEEE INFOCOM 2014-IEEE Conference on Computer Communications, pp. 1231–1239. IEEE (2014)
10. Shi, W., Zhang, L., Wu, C., Li, Z., Lau, F.: An online auction framework for dynamic resource provisioning in cloud computing. ACM SIGMETRICS Perform. Eval. Rev. 42(1), 71–83 (2014)

11. Kutanoglu, E., Wu, S.D.: On combinatorial auction and lagrangean relaxation for distributed resource scheduling. IIE Trans. **31**(9), 813–826 (1999)
12. Andersson, A., Tenhunen, M., Ygge, F.: Integer programming for combinatorial auction winner determination. In: Proceedings. Fourth International Conference on MultiAgent Systems, 2000, pp. 39–46. IEEE (2000)
13. Conen, W., Sandholm, T.: *Differential*-revelation VCG mechanisms for combinatorial auctions. In: Padget, J., Shehory, O., Parkes, D., Sadeh, N., Walsh, W.E. (eds.) AMEC 2002. LNCS (LNAI), vol. 2531, pp. 34–51. Springer, Heidelberg (2002). https://doi.org/10.1007/3-540-36378-5_3
14. Likhodedov, A., Sandholm, T.: Methods for boosting revenue in combinatorial auctions. AAA **I**, 232–237 (2004)
15. Dobzinski, S., Nisan, N., Schapira, M.: Approximation algorithms for combinatorial auctions with complement-free bidders. In: Proceedings of the thirty-seventh annual ACM symposium on Theory of computing, pp. 610–618. ACM (2005)
16. Dobzinski, S., Vondrák, J.: The computational complexity of truthfulness in combinatorial auctions. In: Proceedings of the 13th ACM Conference on Electronic Commerce, pp. 405–422. ACM (2012)
17. Jia, J., Zhang, Q., Zhang, Q., Liu, M.: Revenue generation for truthful spectrum auction in dynamic spectrum access. In: Proceedings of the Tenth ACM International Symposium on Mobile Ad Hoc Networking and Computing, pp. 3–12. ACM (2009)
18. Nisan, N., Ronen, A.: Algorithmic mechanism design. Games Econ. Behav. **35**(1–2), 166–196 (2001)
19. Myerson, R.B.: Optimal auction design. Math. Oper. Res. **6**(1), 58–73 (1981)
20. Parkes, D.C.: Iterative Combinatorial Auctions. MIT press (2006)
21. Parkes, D.C., Ungar, L.H.: Iterative combinatorial auctions: theory and practice. AAAI/IAAI **7481** (2000)

APWeb-WAIM Data Science Workshop

Speed-Up Algorithms
for Happiness-Maximizing Representative Databases

Xianhong Qiu[1], Jiping Zheng[1,2(✉)], Qi Dong[1], and Xingnan Huang[1]

[1] College of Computer Science and Technology, Nanjing University of Aeronautics
and Astronautics, Nanjing, China
{qiuxianhong,jzh,dongqi,huangxingnan} @nuaa.edu.cn
[2] Collaborative Innovation Center of Novel Software Technology
and Industrialization, Nanjing, China

Abstract. Helping user identify the ideal results of a manageable size k from a database, such that each user's ideal results will take a big picture of the whole database. This problem has been studied extensively in recent years under various models, resulting in a large number of interesting consequences. In this paper, we introduce the concept of minimum happiness ratio maximization and show that our objective function exhibits the property of *monotonictity*. Based on this property, two efficient polynomial-time approximation algorithms called Lazy NWF-Greedy and Lazy Stochastic-Greedy are developed. Both of them are extended to exploit lazy evaluations, yielding significant speedups as to basic RDP-Greedy algorithm. Extensive experiments on both synthetic and real datasets show that our Lazy NWF-Greedy achieves the same minimum happiness ratio as the best-known RDP-Greedy algorithm but can greatly reduce the number of function evaluations and our Lazy Stochastic-Greedy sacrifices a little happiness ratio but significantly decreases the number of function evaluations.

Keywords: Minimum happiness ratio · Representative skyline
Lazy evaluation

1 Introduction

When users query the entire database trying to find an item that they are interested in, they may not be willing to search through all of the results storing a multitude of items. Given this, multiple operators have been proposed to reduce output size of query results while still effectively representing the entire database. Among these, top-k [1–4] and skyline [5–9] are two well-studied operators that can effectively reduce output size of query results. Top-k operator takes as input a dataset, a utility function and a user-specified value k, then outputs k data points with the highest utility scores. In other words, top-k returns a customized

L. H. U and H. Xie (Eds.): APWeb-WAIM 2018, LNCS 11268, pp. 321–335, 2018.
https://doi.org/10.1007/978-3-030-01298-4_27

set of data points to users. Unlike top-k operator, skyline returns a set of interesting data points without the need to appoint a utility function. The concept of domination is crucial in skyline operator. In fact, skyline returns data points that are not dominated by any other point in the database. Specifically, a point p dominates another point q if p is as good or better in all dimensions, and strictly better in at least one dimension. However, both operators suffer from some drawbacks. Top-k operator asks for users to specify their utility functions and the size of the result set, but users may not provide their utility functions precisely. Skyline operator finds all points that are not dominated by other points in the database, so the exact number of results is uncontrollable, and cannot be foreseen before the whole database is accessed. In addition, the output size of skyline operator will increase rapidly with the dimensionality.

Fortunately, Nanongkai et $al.$ [10] first introduced k-regret query and developed the best-known algorithm based on the framework of Ramer–Douglas–Peucker algorithm. We called the Greedy algorithm proposed in [10] RDP-Greedy. Given an integer k and a database D, RDP-Greedy algorithm returns a set S of k skyline points of D that minimizes the maximum regret ratio of S. However, as k-regret query exploits linear utility function space to simulate all possible utility functions that users may have, it has a side effect of performing too many function evaluations through Linear Programming (LP). In general, the number of function evaluations of RDP-Greedy is nk, where n is the size of whole database. Besides, for RDP-Greedy, LP for function evaluations is the overwhelming majority of the running time. Thus, reducing the number of function evaluations by LP should speed up the algorithm to a great extent.

Table 1(a) shows the classic skyline example of best hotels for a sample set of hotels where each hotel has two attributes, namely $Distance$ and $Price$. Suppose that user's utility functions are the class of $U = \{u_{(0.4,0.6)}, u_{(0.5,0.5)}, u_{(0.6,0.4)}\}$ where $u_{(x,y)} = x \cdot Distance + y \cdot Price$. The utilities of each hotel are shown in Table 1(b). Given $k = 3$, the k-regret query will report a solution $S = \{p_8, p_1, p_5\}$ under the aforementioned class of utility functions U. The implementation of k-regret query is as follows, it picks the point that maximizes the first coordinate and then iteratively adds the $worst$ point, $i.e.$, the point that is still outside the current solution and contributes the most to the maximum regret ratio of current solution. Unfortunately, even in this simple example, k-regret query performs 13 times of function evaluations and each function evaluation results in $O(k^2d)$ running time by LP [11]. However, the maximum regret ratio obtained by adding a point to a larger set isn't greater than adding the same point to a smaller set. The process of RDP-Greedy algorithm doesn't take this property into consideration thus resulting in an unnecessary function evaluation to p_7 which occurs to the selection of the 3rd point. For further explanation, please refer to Example 2 in Sect. 5.1 for details.

As mentioned above, the deficiency of k-regret query is performing too many function evaluations. Motivated by this, in this paper, the concept of happiness ratio is first introduced. The utility provided by the best hotel in the result is user's happiness and the happiness ratio is by dividing happiness by the utility of her ideal hotel. The minimum happiness ratio is a measurement which

Table 1. Hotel example

(a) Hotel database

Hotel	Distance	Price
p_1	125	1000
p_2	250	800
p_3	312.5	750
p_4	375	650
p_5	562.5	625
p_6	625	450
p_7	750	250
p_8	1000	75

(b) Hotel utilities

Hotel	$u(0.4, 0.6)$	$u(0.5, 0.5)$	$u(0.6, 0.4)$
p_1	650	562.5	475
p_2	580	525	470
p_3	575	531.25	487.5
p_4	540	512.5	485
p_5	600	593.75	587.5
p_6	520	537.5	555
p_7	450	500	550
p_8	445	537.5	630

measures how happy the user will be after displaying k hotels instead of the whole database. Our purpose is to select k hotels that maximize the minimum happiness ratio of a user. Moreover, we demonstrate that our objective function for maximizing the minimum happiness ratio exhibits the property of *monotonicity*. Based on this property, two efficient algorithms by extending to exploit lazy evaluations, yielding significant speedups, called Lazy NWF-Greedy and Lazy Stochastic-Greedy are proposed. The former is an improvement of RDP-Greedy and the latter essentially follows the lazier idea of [12]. In our extensive experiments, Lazy NWF-Greedy achieves the same minimum happiness ratio as RDP-Greedy but can greatly reduce the number of function evaluations and Lazy Stochastic-Greedy sacrifices a little happiness ratio but significantly decreases the number of function evaluations.

The main contributions of this paper are listed as follows:

1. We propose the concept of minimum happiness ratio maximization[1] and show that our objective function for maximizing the minimum happiness ratio is a monotone non-decreasing function.
2. Based on the monotonicity of our objective function, we introduce two efficient greedy algorithms called Lazy NWF-Greedy and Lazy Stochastic-Greedy respectively. The former achieves the same minimum happiness ratio as RDP-Greedy but runs much faster, the latter sacrifices a little happiness ratio but offers a tradeoff between minimum happiness ratio and the number of function evaluations.
3. Extensive experiments on both synthetic and real datasets are conducted to evaluate our methods and the experimental results confirm that both of our proposed algorithms are superior to RDP-Greedy algorithm proposed by [10].

---

[1] Our minimum happiness ratio maximization is consistent with the k-regret proposed in [10]. However, k-regret denotes different things by Nanongkai et al. [10] and Chester et al. [13]. In the former, k-regret is the representative set of k objects, whereas in the latter, k-regret is used to denote the regret between the scores of top 1 and top k. To avoid confusion, we refer k-regret in [13] to $kRMS$.

The remainder of this paper is organized as follows. In Sect. 2, previous work related to this paper is described. The formal definitions of our problem are given in Sect. 3. In Sect. 4, important properties which are applied in our algorithms are discussed. Followed by two accelerated greedy algorithms in Sect. 5. The performance of our algorithms on synthetic and real datasets is presented in Sect. 6. Finally, Sect. 7 concludes this paper and points out our future work.

2   Related Work

Top-k [1–4] and skyline [5–9] operators have received considerable attentions during last two decades as they play an important role in multi-criteria decision making. However, top-k operator requires users to specify their utility functions and it may be too hard for users to specify their utility functions precisely while skyline operator has a potential large output problem which may make users feel overwhelmed. Motivated by the deficiencies of these two operators, a lot of alternatives have been proposed in recent years.

From top-k perspective, Mindolin et al. [8] asked users to specify a small number of possible weights each indicating the importance of a dimension. Lee et al. [4] asked users to specify some pair-wise comparisons between two dimensions to decide whether a dimension was more important than the other for a comparison. However, these studies ask users to specify their utility functions, which may be a heavy burden on users. In [7], skyline points were ranked according to the skyline frequency, which was a measure of how often points appear as skyline points in each particular subspace. The frequency ranking skyline query thus returned the k skyline points with the highest skyline frequency. However, this sort of quality measure is highly subjective and hard to verify.

Other works are in view of skyline operator. Researchers attempt to reduce the output size of the skyline operator. The k representative skyline proposed by Lin et al. [14] represented the whole skyline with only k skyline points which dominated the most non-skyline points in the database. Tao et al. [15] demonstrated the approach provided by [14] was not *stable*. Instead, they proposed distance-based representative skyline borrowing the idea of solving the k-center problem and provided a solution that the maximum distance from any skyline point to its nearest representative skyline was minimized. This method captures the contour of the full skyline well, but is not *scale-invariant*. Magnani et al. [16] introduced an approach, which was trying to make the diversity of the k representative skyline points returned as large as possible. A recent approach based on the diversity measure was proposed by Søholm et al. [17], which returned the k skyline points, such that the coverage was maximized. Also, [18] proposed a new criterion to choose k skyline points as the k representative skyline for data stream environments, termed the k largest dominance skyline. Unfortunately, all these methods are not stable, scale-invariant or with deficiencies of top-k or skyline operators.

To alleviate the burden of top-k for specifying accurate utility functions and skyline operator for outputting too many results, regret-based k representative query was first proposed by Nanongkai *et al.* [10] to minimize user's maximum regret ratio. The stable, scale-invariant approach returned an approximation of the contour of the full skyline without asking users to input their utility functions. However, the approach proposed by them suffers from a heavy burden on the function evaluations by Linear Programming to seek the point with maximum regret ratio in linear utility space. Several works extended Nanongkai *et al.* [10] to some extent. Peng *et al.* [19] proposed the concept of *happy points* considered as candidate points for k-regret query and showed that *happy points* were better used as candidate points compared with skyline points due to their small size resulting in more efficient algorithms. However, the overall time complexity of finding all *happy points* is $O(d^2 n^2)$ which is undesirable when n is very large. To reduce the bounds of regret ratio, [20] combined user's interactions into the process of selection. [13] introduced the relaxation to k-regret minimizing sets, [21] extended linear utility functions to non-linear utility functions for k-regret queries and [22] proposed the metric of *average regret ratio* to measure user's satisfaction. Recently, [23] developed a compact set to efficiently compute the k-regret minimizing set and [24] studied the $kRMS$, which returned r tuples from the database which minimized the k-Regratio. [25] and [26] developed efficient algorithms for $kRMS$. Rank-regret representative was proposed as a way of choosing a small subset of the database guaranteed to contain at least one good choice for every user in [27]. Xie *et al.* [28] proposed an elegant algorithm which has a restriction-free bound on the maximum regret ratio. Our research is from another respective: lazy evaluations by reducing LP calls thus improving the efficiency.

3   Problem Definition

Let D be a set of n d-dimensional points over positive real values. Each point in D can be regarded as a tuple in the database. For each point $p \in D$, the value on the i-th dimension is represented as $p[i]$. We assume that users prefer to the smaller values. Before we define our problem, some definitions of utility function, happiness ratio and minimum happiness ratio are given.

Definition 1 (Utility Function). *A utility function u is a mapping $u: \mathbb{R}_+^d \to \mathbb{R}_+$. The utility of a user with utility function u is $u(p)$ for any point p and shows how satisfied the user is with the point.*

Definition 2 (Happiness Ratio). *Given a dataset D, a set of S with points in D and a utility function u. The happiness ratio of S, represented as $H_D(S, u)$, is defined to be*

$$H_D(S, u) = \frac{\max_{p \in S} u(p)}{\max_{p \in D} u(p)}$$

The happiness ratio is in the range (0, 1]. According to the definition of happiness ratio, the larger the value of happiness ratio is, the happier the user feels.

Since we do not ask users for utility functions, we know nothing about users' preferences for the attributes. For each user, she may have arbitrary utility function. In this paper, we assume that user's utility functions are a class of functions, denoted by U. Usually, four kinds of utility functions [21] are considered, which are Linear, Convex, Concave and CES. In this paper, only the case of function class U consisting of all liner utility functions is considered since it is widely used in modeling user's preferences.

Definition 3 (Linear Utility Function [10]). Assume that existing some non-negative reals v_1, v_2, $\cdots$, v_d which denote the user's preferences for the i-th dimension, then a linear utility function can be represented as $u(p) = \sum_{i=1}^{d} v_i \cdot p[i]$ for any d-dimensional point with a liner utility function u. We can also say that a linear utility function can be expressed as a weight vector, *i.e.* $v = (v_1, v_2, ..., v_d)$, so the utility of any point p can be expressed as the dot product of v and p, namely, $u(p) = v \cdot p$.

The formal definition of minimum happiness ratio of a set S is as follows.

Definition 4 (Minimum Happiness Ratio). *Given a dataset D, a set of S with points in D and a class of utility functions U. The minimum happiness ratio of S, represented as $H_D(S, U)$, is defined to be*

$$H_D(S, U) = \inf_{u \in U} H_D(S, u) = \inf_{u \in U} \frac{\max_{p \in S} u(p)}{\max_{p \in D} u(p)}$$

Since U is allowed to be an infinite class of functions and the minimum value may not exist, so we use $\inf(S, U)$ to represent the minimum happiness ratio.

Example 1. In the following, we present an example for the illustration. Consider the hotel database containing 8 hotels as shown in Table 1(a) in the previous section. The utility function class U is $\{u_{(0.4,0.6)}, u_{(0.5,0.5)}, u_{(0.6,0.4)}\}$ where $u_{(x,y)} = x \cdot Distance + y \cdot Price$. The utilities of hotels for the utility functions in U are shown in Table 1(b). Consider p_1 in Table 1(b). Its utility for the utility function $u_{(0.4,0.6)}$ is $u_{(0.4,0.6)} = 125 \times 0.4 + 1000 \times 0.6 = 650$. The utilities of the remaining points are computed in a similar way. Consider a selection set $S = \{p_3, p_4\}$. Then, the maximum utility of S for the utility function $u_{(0.4,0.6)}$ is 575 which is achieved by p_3 while the maximum utility of whole database is 650 which is achieved by p_1. Then the happiness ratio of S for the utility function $u_{(0.4,0.6)}$ is $H_D(S, u_{(0.4,0.6)}) = 575/650 = 0.8846$. Similarly, we can get $H_D(S, u_{(0.5,0.5)}) = 531.25/593.75 = 0.8947$ and $H_D(S, u_{(0.6,0.4)}) = 487.5/630 = 0.7738$. Hence, the minimum happiness ratio of S is 0.7738.

Problem Definition: Given a dataset D, a positive integer k, our problem of minimum happiness ratio maximization is trying to find a subset S of D containing at most k points such that the minimum happiness ratio is maximized while simultaneously keeping the number of Linear Programming as small as possible.

Our aim is to maximize user's minimum happiness ratio while reducing the times of Linear Programming to return k representatives efficiently compared with RDP-Greedy in [10].

4   Properties of The Objective Function

The minimum happiness ratio function $H_D(S, U)$ has an intuitive and important property which can be exploited to improve the efficiency while solving our problem. First, when $S = \emptyset$, $H_D(S, U) = 0$, this means that we do not obtain any happiness if we do not select any point. Secondly, $H_D(S, U)$ is a non-decreasing function, i.e., $H_D(S_1, U) \leq H_D(S_2, U)$ for all $S_1 \subseteq S_2 \subseteq D$. Hence, adding some points into a subset can increase the happiness ratio or at least keep it unchanged (not decreasing the minimum happiness ratio).

Definition 5 (Monotonicity). *A set function* $f : 2^D \rightarrow \mathbb{R}_+$ *is monotone if for every* $S_1 \subseteq S_2 \subseteq D$, *it holds that* $f(S_1) \leq f(S_2)$. *In addition,* $f(S)$ *is non-negative if* $f(S) \geq 0$ *for any set* S.

A set function satisfies the above properties, then we say it is a monotone non-decreasing function.

Lemma 1. *Our minimum happiness ratio maximization function is a monotone non-decreasing function, namely, it satisfies the property of monotonicity.*

The proof of monotonicity that our minimum happiness ratio maximization function meets is given below.

Proof. Suppose that there exist two arbitrary non-empty subsets S_1, S_2 where $S_1 \subseteq S_2 \subseteq D$. According to Definition 4, we have

$$H_D(S_1, U) = \inf_{u \in U} \frac{\max_{p_1 \in S_1} u(p_1)}{\max_{p_1 \in D} u(p_1)}$$

$$H_D(S_2, U) = \inf_{u \in U} \frac{\max_{p_2 \in S_2} u(p_2)}{\max_{p_2 \in D} u(p_2)}$$

Since $S_1 \subseteq S_2$, it's obvious that the minimum happiness ratio of S_1 is not greater than that of S_2, i.e., $H_D(S_1, U) \leq H_D(S_2, U)$. $\square$

The calculation of the maximum regret ratio based on the method introduced in [10] has to run LP at most nk times. In practice, an LP solver (such as variations of the Simplex method) will result in $O(k^2 d)$ running time per LP call and hence $O(nk^3 d)$ for the RDP-Greedy. This, however, can be improved using the property of monotonicity of our objective function. Detailed description is presented in the following section.

5   Proposed Algorithms

We are ready to present our algorithms, Lazy NWF-Greedy and Lazy Stochastic-Greedy, both of them are boosted in performance compared with RDP-Greedy.

5.1 Lazy NWF-Greedy Algorithm

The RDP-Greedy introduced in [10] picks the point that maximizes the first coordinate and adds the point that currently contributes the most to the maximum regret ratio in the subsequent iterations. Unfortunately, evaluating the minimum happiness ratio or the maximum regret ratio is time-consuming, in this section, we introduce Lazy NWF-Greedy which is an improvement of the RDP-Greedy, providing the same greedy solution since they actually share a same selection strategy. But Lazy NWF-Greedy is with fewer function evaluations of the minimum happiness ratio compared with RDP-Greedy as some lazy strategies are exploited.

Algorithm 1. Lazy-Evaluation($D, S_{i-1}, i, \{\rho(p_j)\}$)

Input: A set of n d-dimensional points $D = \{p_1, p_2, \cdots, p_n\}$, a current solution S_{i-1}, an integer i(the current number of points to find), and a list stored the minimum happiness ratio of each point $p_j \in D \backslash S_{i-1}$.

Output: A point p^*, where $\rho(p^*)$ is minimized.

1 let $h = 1$ and $p^* = NULL$;
2 $\rho(p_{j_0}) = \min_{p_j \in D \backslash S_{i-1}} \{\rho(p_j)\}$;
3 **if** p_{j_0} has already been selected at step i **then**
4 $\quad$ $h = \rho(p_{j_0})$;
5 $\quad$ **if** $h < 1$ **then**
6 $\quad\quad$ $p^* = p_{j_0}$;
7 $\quad$ return p^*;
8 **else**
9 $\quad$ calculate the value of $H_{S_{i-1} \cup \{p_{j_0}\}}(S_{i-1}, U)$ using Linear Programming;
10 $\quad$ $h = H_{S_{i-1} \cup \{p_{j_0}\}}(S_{i-1}, U)$;
11 $\quad$ $\rho(p_{j_0}) = h$;
12 $\quad$ **if** $h > \min_{p_j \in D \backslash S_{i-1}, p_j \neq p_{j_0}} \{\rho(p_j)\}$ **then**
13 $\quad\quad$ Lazy-Evaluation($D, S_{i-1}, i, \{\rho(p_j)\}$);
14 $\quad$ **else**
15 $\quad\quad$ **if** $h < 1$ **then**
16 $\quad\quad\quad$ $p^* = p_{j_0}$;
17 $\quad\quad$ return p^*;

Speeding up Lazy NWF-Greedy with lazy evaluations. Since we have no idea of user's utility functions, the number of utility functions is infinite in the linear utility space. In order to estimate the maximum regret ratio or the minimum happiness ratio of a subset when we add a point into, a large number of function evaluations by Linear Programming need to be performed when we run RDP-Greedy algorithm. Fortunately, monotonicity of our minimum happiness ratio maximization function can be exploited algorithmically to implement an accelerated variant of RDP-Greedy to reduce the number of function evaluations. In each iteration i, RDP-Greedy must identify the point p whose $H_{S_{i-1} \cup \{p\}}(S_{i-1}, U)$ is minimized(equivalent to maximizing the maximum regret ratio), where S_{i-1} is the set of points selected in the previous iterations. The key insight from the monotonicity of H, the minimum happiness ratio obtained by any fixed point $p \in D$ is monotonically non-decreasing during the iterations of adding points, i.e., $H_D(S_i \cup \{p\}, U) \leq H_D(S_j \cup \{p\}, U)$, whenever $i \leq j$. Instead of recomputing for each point $p \in D$, we can use lazy evaluations to maintain a list of lower bounds $\{\rho(p)\}$ on the minimum happiness ratio sorted in ascending order. Then in each iteration, the accelerated algorithm needs to extract the minimal point $p \in \arg\min_{p':S_{i-1} \cup \{p\}}\{\rho(p')\}$ from the ordered list and then updates the bound $\rho(p) \leftarrow H_{S_{i-1} \cup \{p\}}(S_{i-1}, U)$. After this update, if $\rho(p) \leq \rho(p')$, then $H_{S_{i-1} \cup \{p\}}(S_{i-1}, U) \leq H_{S_{i-1} \cup \{p'\}}(S_{i-1}, U)$ for all $p \neq p'$, and therefore we have identified the point that contributes the least to the minimum happiness ratio, without having to compute $H_{S_{i-1} \cup \{p'\}}(S_{i-1}, U)$ for a potentially large number of point p'. We set $S_i \leftarrow S_{i-1} \cup \{p\}$ and repeat until there is no further feasible point which can be added. This idea of using lazy evaluations is useful to our algorithm and can lead to orders of magnitude performance speedups. The pseudocodes of lazy evaluation and Lazy NWF-Greedy are shown in Algorithms 1 and 2 respectively. Example 2 is given to show how to combine the procedure of calculating minimum happiness ratio with lazy evaluations to boost the performance of our Lazy NWF-Greedy.

Algorithm 2. Lazy NWF-Greedy(D, k)

Input: A set of n d-dimensional points $D = \{p_1, p_2, \cdots, p_n\}$ and an integer k, which is the desired output size.
Output: A result set S, $|S| = k$.

1 Initially, let $S_1 = \{p_1^*\}$, where $p_1^* = \arg\max_{p \in D} p[1]$;
2 **for** $(i = 2; i \leq k; i++)$ **do**
3   let $p^* = NULL$;
4   **if** $i = 2$ **then**
5   **for** each $p_j \in D \backslash S_{i-1}$ **do**
6   calculate the value of $H_{S_{i-1} \cup \{p_j\}}(S_{i-1}, U)$ using Linear Programming;
7              $\rho(p_j) = H_{S_{i-1} \cup \{p_j\}}(S_{i-1}, U)$;

8      $p^* =$ Lazy-Evaluation$(D, S_{i-1}, i, \{\rho(p_j)\})$;
9   **if** $p^* = NULL$ **then**
10   return S_{i-1};

11   **else**
12          $S_i = S_{i-1} \cup \{p^*\}$;

13 return S_k;

Example 2. Consider the example in Table 1, we want to select 3 points among 8 points and first pick p_8 into the current solution S_1, namely $S_1 = \{p_8\}$. In next iteration, we have a list on the minimum happiness ratio sorted in ascending order as shown in Table 2. The second point Lazy NWF-Greedy selects is p_1 as $H_{S_1 \cup \{p_1\}}(S_1, U)$ is the minimum, then $H_{S_1 \cup \{p_1\}}(S_1, U)$ will remove from the list and $S_2 = \{p_8, p_1\}$. After this operation, the algorithm begins to select the third point. As $H_{S_1 \cup \{p_5\}}(S_1, U)$ is the minimum in the current list, so Lazy NWF-Greedy computes the minimum happiness ratio p_5 will obtain if it is added to S_2 and gets that $H_{S_2 \cup \{p_5\}}(S_2, U)$ is 0.947 which is not the minimum in the list as

$H_{S_1 \cup \{p_2\}}(S_1, U)$ to $H_{S_1 \cup \{p_6\}}(S_1, U)$ are all less than $H_{S_2 \cup \{p_5\}}(S_2, U)$, hence the minimum happiness ratio obtained by adding them to S_2 should be calculated. In our example, they are all equal to 1. At this moment, this algorithm can directly get the smallest minimum happiness ratio is 0.947 which is achieved by p_5 with no need to compute $H_{S_2 \cup \{p_7\}}(S_2, U)$ since $H_{S_2 \cup \{p_7\}}(S_2, U) \geq H_{S_1 \cup \{p_7\}}(S_1, U) = 0.989$. In Example 2, Lazy NWF-Greedy reduces the function evaluations for 1 time. Obviously, in the settings of large datasets, our Lazy NWF-Greedy can reduce overall times of function evaluations dramatically compared with RDP-Greedy.

Table 2. $H_{S_i \cup \{p\}}(S_i, U)$ for iteration i

Points	p_1	p_5	p_2	p_3	p_4	p_6	p_7
$H_{S_1\{p\}}(S_1, U)$	0.685	0.742	0.767	0.774	0.824	0.856	0.989
$H_{S_2\{p\}}(S_2, U)$	×	0.947	1	1	1	1	×

5.2 Lazy Stochastic-Greedy Algorithm

As described in Sect. 5.1, Lazy NWF-Greedy identifies an ideal point from $D \backslash S_{i-1}$ which is almost the whole dataset. When k points need to be picked out, the algorithm has to go through the whole dataset for k times. In this section, we show how the performance of Lazy NWF-Greedy can be boosted by a random sampling phase thus leading to a randomized greedy algorithm called Lazy Stochastic-Greedy. Lazy Stochastic-Greedy essentially follows the framework of STOCHASTIC-GREEDY algorithm proposed in [12]. The algorithm offers a tradeoff between minimum happiness ratio and the number of function evaluations. It means that this algorithm sacrifices a little happiness ratio while reducing the number of LPs to improve its efficiency.

We present our Lazy Stochastic-Greedy algorithm in Algorithm 3. Initially, similar to our Lazy NWF-Greedy algorithm, the algorithm starts with a set containing the point that maximizes the first coordinate, then adds a point to our solution whose minimum happiness ratio is minimized. Note that the difference between Lazy Stochastic-Greedy and Lazy NWF-Greedy is that Lazy NWF-Greedy finds a point from $p \in D \backslash S_{i-1}$ directly, but Stochastic-Greedy samples a subset R of size $(n/k)log(1/\epsilon)(\epsilon > 0$ is an arbitrarily small constant) from $D \backslash S_{i-1}$ randomly and then finds the point in R whose minimum happiness ratio is minimized. We can combine the the random sampling procedure with lazy evaluations to speed up the implementation of the algorithm as the randomly sampled sets can overlap and we can exploit the previously evaluated minimum happiness ratio. Hence in line 9 of Algorithm 3 we can apply lazy evaluations as described in Sect. 5.1.

Lazy Stochastic-Greedy has two important features. The first feature is that it is almost identical to Lazy NWF-Greedy in terms of minimum happiness ratio.

Algorithm 3. Lazy Stochastic-Greedy(D, k)

Input: A set of n d-dimensional points $D = \{p_1, p_2, \cdots, p_n\}$ and an integer k, which is the desired output size.

Output: A result set S, $|S| = k$.

1 Initially, let's $S_1 = \{p_1^*\}$, where $p_1^* = \arg\max_{p \in D} p[1]$;
2 **for** $(i = 2;\ i \leq k;\ i++)$ **do**
3   let $p^* = NULL$;
4   obtain a random subset R by sampling s random points from $D \backslash S_{i-1}$;
5   **for** each $p_j \in R$ **do**
6   **if** p_j *has not been sampled* **then**
7   calculate the value of $H_{S_{i-1} \cup \{p_j\}}(S_{i-1}, U)$ using Linear Programming;
8              $\rho(p_j) = H_{S_{i-1} \cup \{p_j\}}(S_{i-1}, U)$;

9      $p^* =$ Lazy-Evaluation($R, S_{i-1}, i, \{\rho(p_j)\}$);
10   **if** $p^* = NULL$ **then**
11   return S_{i-1};

12   **else**
13         $S_i = S_{i-1} \cup \{p^*\}$;

14 return S_k;

The second feature is that it is more efficient than Lazy NWF-Greedy, let alone the best-known algorithm, RDP-Greedy.

Theorem 1. *Suppose that the size of random sampling set R is $s = (n/k)log(1/\epsilon)$, then* Lazy Stochastic-Greedy *provides a greedy solution to our minimum happiness ratio maximization problem with at most $O(nlog(1/\epsilon))$ function evaluations of the minimum happiness ratio H.*

Since there are $k - 1$ iterations in total and at each iteration we have $(n/k)log(1/\epsilon)$ points, the total number of function evaluations cannot be more than $(k - 1) \times (n/k)log(1/\epsilon) \leq k \times (n/k)log(1/\epsilon) = nlog(1/\epsilon)$.

6 Experimental Results

In this section we show the performance of the proposed algorithms via experiments. All algorithms were implemented in C++ and the experiments were all conducted on a 64-bit 3.3 GHz Intel Core machine which was running Ubuntu 14.04 LTS operating system.

We ran our experiments on both synthetic and real datasets. The synthetic datasets were created using the dataset generator of [5]. Unless otherwise stated, our synthetic dataset consists of a 6-dimensional anti-correlated dataset of 10,000 points. The real-world dataset we have used are a 6-dimensional *Household* of 127,391 points, an 8-dimensional *NBA* of 17,265 points and a 9-dimensional *Color* of 68,040 points. Moreover, like studies in the literature [10,19–21], we computed the skyline first and our queries on these datasets returned anywhere from 10 to 60 points and evaluated the minimum happiness ratio using Linear Programming implemented in the GUN Linear Programming Kit[2].

---

[2] https://www.gnu.org/software/glpk/.

In our experiments, we consider both Lazy NWF-Greedy and Lazy Stochastic-Greedy algorithms introduced in this paper. To verify the superiority of our proposed algorithms, we compared them with RDP-Greedy [10]. Due to the number of function evaluations by Linear Programming accounting for the majority of the total time of RDP-Greedy, we measure the computational cost in terms of the number of function evaluations by Linear Programming performed instead of the running time of CPU. In addition, for Lazy Stochastic-Greedy, different values of $\epsilon(\epsilon = 0.01, 0.1, 0.3)$ have been chosen. In the following experimental result figures, we abbreviate RDP-Greedy as RDP, Lazy NWF-Greedy as LNWF, Lazy Stochastic-Greedy as LS1 when $\epsilon = 0.1$, Lazy Stochastic-Greedy as LS2 when $\epsilon = 0.01$, Lazy Stochastic-Greedy as LS3 when $\epsilon = 0.3$.

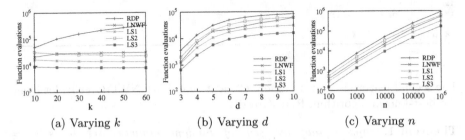

Fig. 1. Function evaluations for anti-correlated data

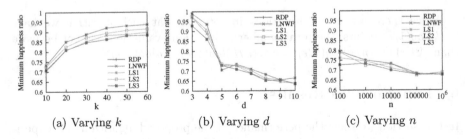

Fig. 2. Minimum happiness ratio for anti-correlated data

Results on Synthetic Datasets: The effects on function evaluations and minimum happiness ratio on anti-correlated datasets for different k are presented in log scale in Figs. 1(a) and 2(a) respectively. Lazy NWF-Greedy and Lazy Stochastic-Greedy have negligible number of function evaluations which do not increase with k and keep a relative small stable number of evaluations even for very large inputs. However, RDP-Greedy performs linear function evaluations as k increases, that's because it performs all possible evaluations of user's utilities, namely, for a total of nk times. The minimum happiness ratio of Lazy NWF-Greedy is the same as that of RDP-Greedy as they share the same greedy

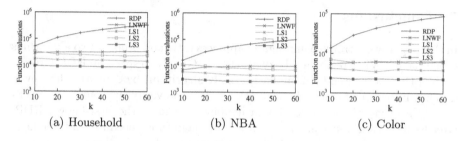

Fig. 3. Function evaluations on real datasets

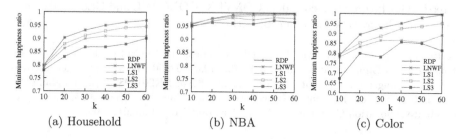

Fig. 4. Minimum happiness ratio on real datasets

skeleton and increases when k increases, appearing to much above the theoretical bound proposed by [10]. For Lazy Stochastic-Greedy, different values of ϵ result in a performance close to that of Lazy NWF-Greedy. It means that Lazy Stochastic-Greedy provides very compelling tradeoffs between the number of function evaluations and minimum happiness ratio compared with RDP-Greedy and our Lazy NWF-Greedy.

We varied the number of dimensions on anti-correlated data when the number of points was fixed to $n = 10,000$ and $k = 10$. As illustrated in Fig. 1(b) and in Fig. 2(b), the number of function evaluations increases with the increase of dimensions, but our proposed algorithms still have less function evaluations than RDP-Greedy since they are extended with lazy evaluations. Due to the curse of dimensionality, the minimum happiness ratio resulted in by all algorithms decreases with the increase of d. In Figs. 1(c) and 2(c), the effects of varying n are presented. The effects are similar to those of varying dimensions.

Results on Real Datasets: Figs. 3(a), (b) and (c) show that the number of function evaluations of RDP-Greedy increases dramatically with k, but our proposed algorithms keep much less function evaluations and maintain a stable level. The minimum happiness ratio of all algorithms are shown in Figs. 4(a), (b) and (c) respectively. We observe similar trends. Besides, similar to the experiments on synthetic datasets, Lazy Stochastic-Greedy achieves near-maximal minimum happiness ratio with substantially less function evaluations compared with the other algorithms.

7   Conclusions

In this paper, we introduce minimum happiness ratio and propose two algorithms called Lazy NWF-Greedy and Lazy Stochastic-Greedy to speed up RDP-Greedy, both of them are extended from basic greedy algorithms by exploiting lazy evaluations. Experiments on real and synthetic datasets verify that our Lazy NWF-Greedy achieves the same minimum happiness ratio as the best-known RDP-Greedy algorithm, but can lead to orders of magnitude speedups and our Lazy Stochastic-Greedy sacrifices a little happiness ratio but significantly decreases the number of function evaluations compared with RDP-Greedy or Lazy NWF-Greedy. Our future work considers CONVEX, CONCAVE and CES utility functions with the framework of our algorithms.

Acknowledgment. This work is partially supported by the National Natural Science Foundation of China under grant Nos. U1733112,61702260, the Natural Science Foundation of Jiangsu Province of China under grant No. BK20140826, the Fundamental Research Funds for the Central Universities under grant No. NS2015095, Funding of Graduate Innovation Center in NUAA under grant No. KFJJ20171605.

References

1. Ilyas, I.F., Beskales, G., Soliman, M.A.: A survey of top-k query processing techniques in relational database systems. ACM Comput. Surv. **40**(4), 11:1–11:58 (2008)
2. Lian, X., Chen, L.: Top-k dominating queries in uncertain databases. In: EDBT, pp. 660–671 (2009)
3. Soliman, M.A., Ilyas, I.F., Chang, K.C.-C.: Top-k query processing in uncertain databases. In: ICDE, pp. 896–905 (2007)
4. Lee, J., You, G.W., Hwang, S.W.: Personalized top-k skyline queries in high-dimensional space. Inf. Syst. **34**(1), 45–61 (2009)
5. Börzsöny, S., Kossmann, D., Stocker, K.: The skyline operator. In: ICDE, pp. 421–430 (2001)
6. Chan, C.-Y., Jagadish, H.V., Tan, K.-L., Tung, A.K.H., Zhang, Z.: Finding k-dominant skylines in high dimensional space. In: SIGMOD, pp. 503–514. ACM (2006)
7. Chan, C.-Y., Jagadish, H.V., Tan, K.-L., Tung, A.K.H., Zhang, Z.: On high dimensional skylines. In: Ioannidis, Y., et al. (eds.) EDBT 2006. LNCS, vol. 3896, pp. 478–495. Springer, Heidelberg (2006). https://doi.org/10.1007/11687238_30
8. Mindolin, D., Chomicki, J.: Discovering relative importance of skyline attributes. In: VLDB, pp. 610–621 (2009)
9. Papadias, D., Tao, Y., Greg, F., Seeger, B.: Progressive skyline computation in database systems. TODS **30**(1), 41–82 (2005)
10. Nanongkai, D., Sarma, A.D., Lall, A., Lipton, R.J., Xu, J.: Regret-minimizing representative databases. In: VLDB, pp. 1114–1124 (2010)
11. Bertsimas, D., Tsitsiklis, J.: Introduction to Linear Optimization. Athena Scientific, Belmont (1997)
12. Mirzasoleiman, B., Badanidiyuru, A., Karbasi, A., Vondrak, J., Krause, A.: Lazier than lazy greedy. In AAAI, pp. 1812–1818 (2015)

13. Chester, S., Thomo, A., Venkatesh, S., Whitesides, S.: Computing k-regret minimizing sets. In: VLDB, pp. 389–400 (2014)
14. Lin, X., Yuan, Y., Zhang, Q., Zhang, Y.: Selecting stars: the k most representative skyline operator. In: ICDE, pp. 86–95 (2007)
15. Tao, Y., Ding, L., Lin, X., Pei, J.: Distance-based representative skyline. In: ICDE, pp. 892–903, 2009
16. Magnani, M., Assent, I., Mortensen, M.L.: Taking the big picture: representative skylines based on significance and diversity. VLDB J. **23**(5), 795–815 (2014). October
17. Søholm, M., Chester, S., Assent, I.: Maximum coverage representative skyline. In: EDBT, pp. 702–703 (2016)
18. Bai, M., et al.: Discovering the k representative skyline over a sliding window. TKDE **28**(8), 2041–2056 (2016)
19. Peng, P., Wong, R.C.W.: Geometry approach for k-regret query. In: ICDE, pp. 772–783 (2014)
20. Nanongkai, D., Lall, A., Sarma, A.D., Makino, K.: Interactive regret minimization. In: SIGMOD, pp. 109–120 (2012)
21. Faulkner, T.K., Brackenbury, W., Lall, A.: k-regret queries with nonlinear utilities. In: VLDB, pp. 2098–2109 (2015)
22. Zeighami, S., Wong, R.C-W.: Minimizing average regret ratio in database. In: SIGMOD, pp. 2265–2266 (2016)
23. Asudeh, A., Nazi, A., Zhang, N., Das, G.: Efficient computation of regret-ratio minimizing set: a compact maxima representative. In: SIGMOD, pp. 821–834 (2017)
24. Cao, W. et al.: k-regret minimizing set: efficient algorithms and hardness. In: ICDT, pp. 11:1–19 (2017)
25. Agarwal, P.K., Kumar, N., Sintos, S., Suri, S.: Efficient algorithms for k-regret minimizing sets. In: International Symposium on Experimental Algorithms, pp. 7:1–7:23 (2017)
26. Kumar, N., Sintos, S.: Faster approximation algorithm for the k-regret minimizing set and related problems. In: Proceedings of the 20th Workshop on Algorithm Engineering and Experiments, pp. 62–74 (2018)
27. Asudeh, A., Nazi, A., Zhang, N., Das, G., Jagadish, H.V.: RRR: rank-regret representative. CoRR (2018)
28. Xie, M., Wong, R.C.-W., Li, J., Long, C., Lall, A.: Efficient k-regret query algorithm with restriction-free bound for any dimensionality. In: SIGMOD, pp. 959–974 (2018)

Multi-location Influence Maximization in Location-Based Social Networks

Zhen Zhang[1]([✉]), Xiangguo Zhao[1], Guoren Wang[1], and Xin Bi[2]

[1] College of Computer Science and Engineering, Northeastern University,
Shenyang, Liaoning, China
zhangzhenneu@gmail.com
[2] Sino-Dutch Biomedical and Information Engineering,
Northeastern University, Shenyang, Liaoning, China

Abstract. With the development of location-based social networks
(LBSNs), location property has been gradually integrated into the influence maximization problem, the key point of which is to bring the users
in social networks (online phase) to the product locations for consuming in the real world (offline phase). However, the existing studies considered that a company dependent on the viral marketing only has a
product location in the real world and could not suit the situation that
there is more than one product location. In this paper, first, we propose
a new propagation model, called multiple factors propagation (MFP)
model which can work in the situation that there are multiple product
locations in the real world. Meanwhile, the definition of multi-location
influence maximization (MLIM) problem is presented. Then, we design
a hybrid index structure to improve the search efficiency of offline phase,
called hybrid inverted R-tree (HIR-tree). Furthermore, we propose the
enhanced greedy algorithm for solving MLIM problem. Finally, we conduct a set of experiments to demonstrate the effectiveness and efficiency
of enhanced greedy algorithm.

Keywords: Influence maximization · Location-based social networks
Viral marketing · Propagation model

1 Introduction

The traditional influence maximization problem in social networks aims to select
a set of users as seeds to achieve the greatest propagation [1,2]. However, it
only tends to focus on the online social networks and ignores the product acceptance in the real world. With the development of GPS technology, location-based
social network (LBSN) services have become very popular, such as Foursquare,
Gowalla. In LBSNs, the location property sets up the bridge between the online
social networks and the real world [3]. In fact, LBSN services can be used to
promote the influence maximization problem, which brings the users in social
networks (online phase) to the product locations for consuming in the real world

© Springer Nature Switzerland AG 2018
L. H. U and H. Xie (Eds.): APWeb-WAIM 2018, LNCS 11268, pp. 336–351, 2018.
https://doi.org/10.1007/978-3-030-01298-4_28

(offline phase). In other words, a set of users are selected as seeds to attract more and more users to visit the product locations in the real world [4–6]. The existing studies considered that a company dependent on the viral marketing only has one product location in the real world and the distance is the only factor that determines the user's possibility of consumption. In fact, there may be more than one product location in the real world. For example, there are many chain KFC restaurants around the world. Now, a LBSN service (e.g., Foursquare or Gowalla) helps KFC find k users to advertise. The purpose is to maximum the number of users who will visit KFC restaurants in the real world instead of being limited to a certain KFC restaurant. However, the existing studies could not suit this situation. In order to solve this problem, for each user, how to find his potential consuming location from all KFC restaurants is important. The rules should consider multiple factors instead of just depending on the distance.

Therefore in this paper, we study the multi-location influence maximization (MLIM) problem in LBSNs with a new propagation model, called multiple factors propagation (MFP) model. Our target is to maximize the number of users who will visit product locations in the real world instead of being limited to a certain product location. MFP model contains online and offline phases. In the online phase, we still use IC model to influence spread and calculate the user's online influenced probability [7]. In the offline phase, MFP model considers more factors to determine whether the user will visit product locations: (1) the distance; (2) the user's interest; (3) the friends' evaluation. By using these factors, the product location where the user has a maximum visiting probability from all product locations is defined as the user's potential consuming location. And this user's maximum visiting probability on his potential consuming location will be approximated as the user's offline consuming probability.

There are two main challenges for MLIM problem with MFP model. The first is to find each user's potential consuming location from all product locations quickly. The second is to design an algorithm with the high-performance to select top-k seed users. For the first challenge, we propose a hybrid index structure called hybrid inverted R-tree (HIR-tree) to explore the spatial, social, and textual pruning techniques in LBSNs. For the second challenge, when finishing the query of each user's potential consuming location, the enhanced greedy algorithm is proposed to find top-k seed users, which only utilizes 1-hop or 2-hop friend relationships to estimate a lower bound of influence spread. The lower bound can help avoid massive calculation and improve the efficiency of algorithm.

Our main contributions of the paper are summarized as:

- We propose MFP model which can work in the situation that there are multiple product locations in the real world. And under MFP model, we define MLIM problem in LBSNs, which is NP-hard.
- To solve MLIM problem, the essential part is to find each user's potential consuming location. We propose HIR-tree index to speed up the query.
- When finishing the query, we propose the enhanced greedy algorithm to find top-k seed users, which estimates a lower bound of influence spread to improve the efficiency of algorithm.

– We conduct a set of experiments on three real LBSN datasets. The experimental results show that the enhanced greedy algorithm has a good performance for MLIM problem.

The remainder of this paper is organized as follows. In Sect. 2, the detail of MFP model is introduced and MLIM problem is defined. In Sect. 3, the index structure of HIR-tree and enhanced greedy algorithm are given. In Sect. 4, the empirical studies are conducted. In Sect. 5, the related works are introduced. Finally, in Sect. 6, our work is summarized.

2   MFP Model and Problem Formulation

In this section, we first give MFP model, including online and offline phases. Then, we give the definition of MLIM problem under MFP model.

2.1   MFP Model

Influence Spread in Social Networks (online). For the influence spread in social networks, we still use the propagation rules of IC model. For each user v, we select a set of initial seeds from seed set S to influence it. $p(u, v)$ is defined as the online probability that v is influenced by seed u. Thus, $p(S, v)$ is defined as the final online probability that v is influenced by all the seeds in current seed set S. For the calculation of $p(u, v)$ and $p(S, v)$, we can refer to other literatures which solve the traditional influence maximization problem using IC model on social networks, such as the literature [7–9].

User's Consuming Behavior in the Real World (offline). Once the user v is influenced in social networks, we consider three factors to determine whether v will visit product locations in the real world.

For the first factor, existing studies have showed that the probability that a user may move from a given location to another location has relationship with the distance between two locations [10]. And the mobile probability will decrease with the increasing of distance. In LBSNs, the user's every historical check-in is a reality of user's mobile behavior. Therefore, for each user v, according to the check-in time, we sort his check-in sequence set $C_v = \{c_{v1}, c_{v2}, \cdots, c_{vi}\}$. The corresponding check-in location sequence set can be defined as $L_v = \{l_{v1}, l_{v2}, \cdots, l_{vi}\}$. Give a product location set L with a set of product locations. Assume that v's every historical check-in location l_{vi} is a starting location where he may move to the product location $l \in L$ [6]. We define the distance score of offline as follows:

$$ds_l(v) = (1 + \alpha^{\sum_{l_{vi} \in L_v} ||l_{vi}l||/N_v})/DsmaxDist \tag{1}$$

where $\sum_{l_{vi} \in L_v} ||l_{vi}l||$ is the sum of Euclidean distance from every historical check-in location l_{vi} to the product location l (Other metric distance functions can also be used). N_v is the number of the user v's check-in locations. Constant

1 insures that the distance score is never equation 0. $\alpha \in [0, 1)$ is a damping factor. $DsmaxDist$ is the user's maximal calculated value of distance factor.

For the second factor, a user's interest can determine his check-in behavior [11]. For each user v, we extract the textual keyword set W_v from his historical check-in records to represent v's interest preferences. And for each product location $l \in L$, there is a textual keyword set W_l to describe l's attributes. The overlapping number of textual keywords between the user v and the location l can determine the probability that the user v visits the location l. For example, a product location l is described "spicy" and "restaurant". The user v's textual keyword set contains one "spicy" and two "restaurant". The overlapping number of textual keywords between the user v and the product location l is 3. The interest relevance score of offline can be defined as follows:

$$irs_l(v) = \sum_{w \in W_l} |N_v^w|/IrmaxDist \tag{2}$$

where $|N_v^w|$ is the number of textual keyword w in W_v. $\sum_{w \in W_l} |N_v^w|$ is the overlapping number of textual keywords between v and l. $IrmaxDist$ is the user's maximal calculated value of interest factor.

For the third factor, a user's social relationship can determine his check-in behavior [12]. In a social network, for users u and v, $||uv||_s$ can be used to define the social distance which is length of the shortest path between u and v. For example in Fig. 1, the social distance between u_1 and u_2 is 2. For each product location $l \in L$, there is a user set $l \cdot F$ where these users have positive attitude towards l. In this paper, if a user has had check-in records on the product location l, we consider that this user has positive attitude towards l. The social relationship score of offline can be defined as follows:

$$srs_l(v) = (1 + \sum_{u \in l \cdot F} \alpha^{||uv||_s})/SrmaxDist \tag{3}$$

where $||uv||_s$ is the social distance from the positive user u towards l to the user v. Constant 1 insures that the social score is never equation 0. $\alpha \in [0, 1)$ is a damping factor. $SrmaxDist$ is the user's maximal calculated value of relationship factor.

By the above mentioned, assume that the user v is influenced in social networks and the probability that he will visit each product location $l \in L$ in the real world can be calculated, which is a linear combination of these three scores:

$$pscore_l(v) = \beta ds_l(v) + \gamma irs_l(v) + (1 - \beta - \gamma)srs_l(v) \tag{4}$$

where β, γ are parameters for general weighting functions, and β, $\gamma \in [0, 1]$.

In order to improve the efficiency, we only find each user v's potential consuming location $\tilde{l}$ with maximum probability $pscore_{\tilde{l}}(v)$ from all product locations in L. $pscore_{\tilde{l}}(v)$ will be approximated as v's offline consuming probability.

2.2 Problem Formulation

We model a location-based social network $< G, C >$ which contains a social network $G < V, E >$ and a check-in set C. In the social network $G < V, E >$, V is the set of users, E is the set of user's connection edges. $C = \{(u, l, t)\}$ is the user's check-in set, where (v, l, t) represents that a user v has a check-in activity at location l at time t. Formally, given a query $Q = (L, k)$ with the product location set L and the number k, the purpose is to find k seed users forming the seed set S to make market promotion and maximize the number of users who will visit product locations in L in the real world. With MFP model, first, for every user $v \in V$ in social networks, the online probability of a set of seeds in set S to v can be denoted as $p(S, v)$. Once v is influenced in social networks, the offline probability that v will visit product locations in the real world can be approximated as $pscore_{\bar{i}}(v)$. The final influence spread is denoted by $\sigma(S, V)$. In this paper, we use the function $\varphi(S, V)$ to approximate the value of $\sigma(S, V)$.

$$\varphi(S, V) = \sum_{v \in V} p(S, v) \cdot pscore_{\bar{i}}(v) \tag{5}$$

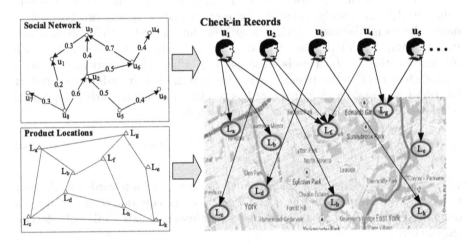

Fig. 1. A query example in a location-based social network

Definition 1 (*Multi-Location Influence Maximization*). Given a LBSN and a query $Q = (L, k)$, we need to find a set S with k-vertexes, so that for any other set K with k-vertexes, $\varphi(S, V) \geq \varphi(K, V)$.

Example 1. In Fig. 1, we give an example of MLIM problem in a LBSN $< G, C >$. The users' social network $G(V, E)$ is shown in the top-left of Fig. 1, where there are 9 users. The users' check-in records on 9 product locations are shown in the right of Fig. 1 (apostrophe indicates that we only show the part users' check-ins). There is a $Q = (L, 2)$ with the product location set L, in which the number and latitude/longitude coordinates of product locations are also given, shown in the bottom-left of Fig. 1. An undirected edge represents the physical distance between two product locations. Our purpose is to find 2 users as seeds and make the largest value of $\varphi(s, V)$.

Problem Hardness. We show the hardness of the problem by considering a simple case with $pscore_i(v) = 1$. In this case, when a user v is influenced in social networks, it must be sure to visit product locations in the real world. Thus, MLIM becomes a traditional influence maximization (IM) problem which is NP-Hard [7]. Therefore, MLIM is also an NP-Hard problem.

3   Solutions

In this section, we first give HIR-tree index to find each user's potential consuming location $\tilde{l}$. Then, when finishing the query, we propose the enhanced greedy algorithm to select top-k seed users.

3.1   HIR-Tree Index

HIR-tree extended from R-tree [13] provides the simultaneous computation of three factors in the offline phase, which can prune the search space and speed up the query of each user's potential consuming location $\tilde{l}$. Each leaf node covers some entries of form $\langle pt, F, \Lambda, \Psi \rangle$, where pt represents the location, F represents the positive users who have had check-ins on the pt, Λ is the minimum bounding rectangle (MBR) of locations, Ψ includes a pointer to point an inverted file which notes the textual keywords of locations. Non-leaf node covers some entries of form $\langle pt, F, \Lambda, \Psi \rangle$, where pt is a reference in the child nodes, F represents the positive users who are the union of the positive users of entries of the child nodes, Λ is the MBR of all rectangles in entries of the child nodes, Ψ includes a pointer to point an inverted file of the entries stored in the child nodes.

Example 2. Figure 2 (a) illustrates the spatial partition of HIR-tree for the promoted location set L in Fig. 1. Figure 2 (b) illustrates HIR-tree structure based on the spatial partition in Fig. 2 (a). Each entry has a set of positive users who have had check-in records on these object locations subordinating to the right of Fig. 1. For example, the positive users of object locations L_e, L_f, and L_g are $\{u_5...\}$, $\{u_1, u_2, u_4...\}$, and $\{u_4, u_5...\}$ respectively (apostrophe indicates that we only give the part positive users shown in the right of Fig. 1). The parent R_2 is the union of sets of positive users for L_e, L_f, and L_g, i.e., $\{u_1, u_2, u_4, u_5...\}$. Figure 2 (c) gives the inverted files for each node. For example, the location L_k is described as "Fair price" and "Cinema". Figure 2 (d) shows each user's inverted textual keywords file extracted from the user's historical check-in records to reflect users' hobbies and interests.

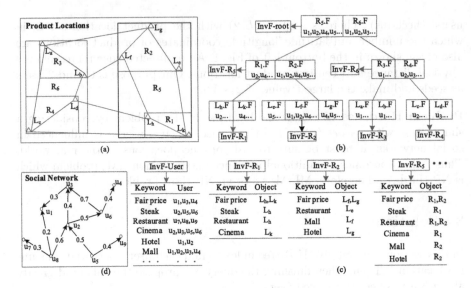

Fig. 2. An example of HIR-tree index structure

HIR-tree possesses the significant characteristic from R-tree. The distance score, interest relevance score, and social relationship score in a non-leaf entry is an upper bound to any object contained in the subtree at the entry.

LEMMA 1. Given an object location l and a non-leaf entry e with its rectangle $e \cdot \Lambda$, we have $\forall l \in e \cdot \Lambda \quad ds_l(v) \leq ds_e(v), irs_l(v) \leq irs_e(v), srs_l(v) \leq srs_e(v)$.

PROOF. Since l is in the child node of e, l is enclosed in the rectangle of e. In addition to the object location l, e may contain other object locations. It can conclude that $\alpha^{\sum_{l_{vi} \in L_v} ||l_{vi}l||/N_v} \leq \alpha^{\sum_{l_{vi} \in L_v} ||l_{vi}e||/N_v}$. Therefore, $ds_l(v) \leq ds_e(v)$. The construction of textual keywords of a non-leaf entry is an upper bound on textual keywords of any object in the entry of subtree. It can conclude that $\sum_{w \in W_l} |N_v^w| \leq \sum_{w \in W_e} |N_v^w|$. Therefore, $irs_l(v) \leq irs_e(v)$. The construction of the sets of positive users for a non-leaf entry is an upper bound on the sets of positive users for any object in the entry of subtree. It can conclude that $\sum_{u \in l \cdot F} \alpha^{||uv||_s} \leq \sum_{u \in e \cdot F} \alpha^{||uv||_s}$. Therefore, $srs_l(v) \leq srs_e(v)$.

LEMMA 2. Given an object location l and a non-leaf entry e with its rectangle $e \cdot \Lambda$, the following is true: $\forall l \in e \cdot \Lambda \quad pscore_l(v) \leq pscore_e(v)$.

PROOF. Since the object location l is a child node of e, according to Lemma 1, we have $ds_l(v) \leq ds_e(v), irs_l(v) \leq irs_e(v), srs_l(v) \leq srs_e(v)$. Hence, we have

$$
\begin{aligned}
pscore_l(v) &= \beta ds_l(v) + \gamma irs_l(v) + (1 - \beta - \gamma) srs_l(v) \\
&\leq \beta ds_e(v) + \gamma irs_e(v) + (1 - \beta - \gamma) srs_e(v) = pscore_e(v)
\end{aligned}
\tag{6}
$$

Query Processing. Algorithm 1 shows the query processing of each user v's potential consuming location $\tilde{l}$ with HIR-tree (Lines 1–12). It takes the user v, product location set L, and HIR-tree as input. It takes the user v's potential consuming location $\tilde{l}$ as output. First, we build a priority queue to keep recording the entries of nodes or object locations which haven't been visited (Lines 1–2). When the queue is not empty, we start to dequeue the elements in the queue (Lines 3–4). The best-first traversal algorithm[14] is used to search for $\tilde{l}$. The probability score $pscore(.)$ of entries is the most crucial factor according to Eq. 4. If e is a non-leaf node, the algorithm picks the entry in the node e with the largest probability score $pscore(.)$ to visit next (Lines 5–8). Otherwise, if e is a leaf node and refers to an object location l, This object location l is the potential consuming location $\tilde{l}$. And the algorithm will terminate (Lines 10–12).

Algorithm 1. HIR-tree Based Query Processing Algorithm

Input: User v, L, HIR-tree index;
Output: $\tilde{l} \in L$;

1 Queue← NewPriorityQueue();
2 Query.Enqueue(index.RootNode,0);
3 **while** $Queue \neq \phi$ **do**
4 | Entry $e \leftarrow$ Queue.Dequeue();
5 | **if** e *is a non-leaf node* **then**
6 | | **for** *each entry é in the node e* **do**
7 | | | Compute its $pscore_é(v)$ using BFS;
8 | | | Queue.Enqueue($é$, $pscore_é(v)$);
9 | **else**
10 | | **if** e *is a leaf node and refers to an object l* **then**
11 | | | $\tilde{l} \leftarrow l$;
12 | | | return $\tilde{l}$;

3.2 Enhanced Greedy Algorithm

Enhanced greedy algorithm uses a greedy idea to retrieve top-k initial seeds. The basic solution is to select a vertex into the seed set S every time with the maximum influence spread at the present stage. The algorithm will be terminated after k vertexes are selected into the seed set S. We know that the credibility of information gradually declines over the spread of 2-hop links in social networks[15]. Based on these observations, we first use 1-hop or 2-hop friend relationships to estimate a lower bound of influence spread. In the online phase, we use $p(u, v)$ to estimate the user u's influence for the user v, where the user v is only the user u's 1-hop or 2-hop neighbor. In the offline phase, we use 1-hop or 2-hop friend relationships with v to estimate the social relationship score of offline $srs_l(v)$ which is defined $\overline{srs_l(v)}$.

$$\overline{srs_l(v)} = (1 + \sum_{u \in e \cdot F \wedge ||uv||_s \leq 2} \alpha^{||uv||_s})/SrmaxDist \qquad (7)$$

$pscore_l(v)$ is defined as $\overline{pscore_l(v)}$ by using $\overline{srs_l(v)}$. Thus, for each user u in LBSNs, we can give his spread capacity.

Definition 2 (*Spread Capacity*). Given a LBSN $< G, C >$, for each user u, his spread capacity in the whole social network is denoted as $\Gamma(u)$:

$$\Gamma(u) = \sum_{v \in \Upsilon} \overline{p(u,v)} \cdot \overline{pscore_{\bar{i}}(v)} \qquad (8)$$

where Υ is a user set that has 1-hop or 2-hop friend relationships with u in the whole social network.

Algorithm 2. Enhanced Greedy Algorithm

Input: A LBSN graph $< G, C >$, $Q(L, k)$;
Output: S:k-vertex set;
1 Initialize: $S = \phi$;
2 **for** *each vertex* $u \in V$ **do**
3   Calculate $\varphi(u, V) = \sum_{v \in V} p(u, v) \cdot pscore_{\bar{i}}(v)$;

4 Build the max-heap for each vertex u in V with $\varphi(u, V)$;
5 **for** $i = 1$ *to* k **do**
6   **if** *(i == 1)* **then**
7        $s = Heap \cdot pop()$;
8        $S = S \cup s$;

9   **else**
10       $\alpha = Heap \cdot pop()$;
11   **if** α *and any vertex in S do not have spread overlapping* **then**
12          $S = S \cup \alpha$;

13   **else**
14   **if** α *with* $\varphi(\alpha, V)$ **then**
15   Calculate $\widehat{\Gamma(\alpha)} = \Gamma(s \cup \alpha) - \Gamma(s)$;
16   Add $\widehat{\Gamma(\alpha)}$ into the max-heap;
17   Update the heap;

18   **if** α *with* $\widehat{\Gamma(\alpha)}$ **then**
19             $S = S \cup \alpha$;
20   Update the max-heap;

21 Return S;

Therefore, we can say $\Gamma(u) \leq \varphi(u,V)$ and $\Gamma(u)$ is a lower bound of u's influence spread $\varphi(u,V)$.

Definition 3 (*Spread Overlapping*). So far, we have selected a set of seeds in seed set S with the influence spread $\varphi(S,V)$. When a new vertex u wants to add into set S, $\varphi(S \cup u, V) < \varphi(S,V) + \varphi(u,V)$. We can say that u has spread overlapping with the seeds in S.

Definition 4 (*Incremental Influence*). If a new vertex u has spread overlapping with the current seed set S, $\widehat{\varphi(u,V)} = \varphi(S \cup u, V) - \varphi(S,V)$, called u' incremental influence. In the same way, $\widehat{\Gamma(u)} = \Gamma(S \cup u) - \Gamma(S)$, called estimated incremental influence.

Next, we devise the enhanced greedy algorithm and the pseudo-code is shown in Algorithm 2. It takes a LBSN, the given query $Q(L,k)$ as input. It takes k-vertex seed set S as output. First, we put the seed set S into empty (Line 1). Then, for each vertex $v \in V$, we need to find its potential consuming location $\tilde{l}$ with HIR-tree index by Algorithm 1. Thus, for each vertex $u \in V$, we can calculate the initial influence spread $\varphi(u,V)$ (Lines 2–3). Next, the max-heap is built for each vertex u with the value $\varphi(u,V)$ (Line 4). If the vertex is the top vertex and the seed set S is still empty, pop this top vertex assigned as s. Then, put s into the seed set S (Lines 6–8). Otherwise, pop the top vertex assigned as α. Then, we need to judge whether α has spread overlapping with the vertexes in set S. If they do not have spread overlapping, α is the next seed (according to the result of influence spread in Lines 2–3). Add α into the set S (Lines 11–12). Otherwise, if α is only with initial $\varphi(\alpha,V)$, we need to estimate its $\widehat{\Gamma(\alpha)}$ and put this value into the max-heap (Lines 14–17). If α has estimated its $\widehat{\Gamma(\alpha)}$, we pop it and put α into the seed set S. Update max-heap (Lines 18–20). The algorithm will stop until the set S has existed k elements.

Example 3. Next, we give a complete example of Algorithm 2. In Fig. 1, there are 9 users and 9 locations in a LBSN. Give a query $Q = (L,2)$. Our purpose is to find 2 users as seeds to make market promotion and maximize the number of users who visit product locations in L in the real world. First, we use HIR-tree index to search for each user's potential consuming location $\tilde{l}$ and calculate 9 users's initial influence spread. To simplify the example, we assume that each user v's offline probability score $pscore_i(v)$ is 0.50. Thus, 9 users are $u_2, u_5, u_6, u_8, u_3, u_1, u_4, u_7$ with initial influence spread of 0.715, 0.675, 0.655, 0.55, 0.15, 0, 0, 0, respectively. In the first iteration, u_2 is the first seed. Then the next user is u_5. Since it has spread overlapping with u_2, we will calculate its $\widehat{\Gamma(u_5)} = 0.20$. Then we add $< u_5, 0.20 >$ into the max-heap. u_6 will become the top vertex. Because u_6 also has spread overlapping with u_2, we will also calculate its $\widehat{\Gamma(u_6)} = 0.268$. Then we add $< u_6, 0.268 >$ into the max-heap. Next, the top vertex is u_8. Since it has spread overlapping with u_2, we will calculate its $\widehat{\Gamma(u_8)} = 0.1585$. Then we add $< u_8, 0.1585 >$ into the max-heap. The top vertex is u_6. Because u_6 has been calculated $\widehat{\Gamma(u_6)} = 0.268$, u_6 is the next seed. The final results are u_2, u_6.

LEMMA 3. $\varphi(S, V)$ is monotonic and submodular. Enhanced greedy algorithm has $1 - 1/e$ approximation ratio.

PROOF. First, since $\varphi(S, V)$ is an approximate value of spread influence by S, $\varphi(\phi, V) = 0$ and incremental influence of φ is always greater than 0. If $S_1 \subseteq S_2$, it is sure that $\varphi(S_1, V) \leq \varphi(S_2, V)$. Therefore, the value function $\varphi(S, V)$ is monotone. Second, the traditional IM problem with $\varphi(S, V)$ under IC model is submodular. MLIM problem with MFP model only adds the offline probability, that is, for any $S_1 \subseteq S_2$, a user u, $\varphi(S_1 \cup u, V) - \varphi(S_1, V) \geq \varphi(S_2 \cup u, V) - \varphi(S_2, V)$. Therefore, this function is submodular. Since the object function $\varphi(.)$ is monotonic and submodular, let S be the result of enhanced greedy algorithm and S^* be an optimal set that maximizes the value of $\varphi(.)$. Although the enhanced greedy algorithm uses 1-hop or 2-hop friend relationships to estimate a lower bound of influence spread, it does not impact the property of diminishing returns in the traditional IM problem. According to [7], $\varphi(S, V) \geq (1 - 1/e)\varphi(S^*, V)$. Therefore, enhanced greedy algorithm has $1 - 1/e$ approximation ratio.

Time and Space Complexity. First, the time complexity of finding potential consuming location $\tilde{l}$ of all users in V with HIR-tree index is $O(|V| \log |M|^{|N|})$. $|N|$ is the number of nodes in HIR-tree and $|M|$ is the maximum number of child nodes. Second, when finishing the query, we need to cost $O(|V|)$'s time complexity to calculate the initial influence spread of each vertex $u \in V$. Third, the max-heap construction complexity is $O(|V|)$, and the max-heap adjustment complexity is $O(k \log |V|)$. Therefore, the time complexity of enhanced greedy algorithm is $O(|V| \log |M|^{|N|} + 2|V| + k \log |V|)$. First, the space complexity of HIR-tree is $O(|N|)$. Second, the space complexity of max-heap is $O(|V|)$. Therefore, The space complexity of enhanced greedy algorithm is $O(|N| + |V|)$.

4   Experiments

In this section, we conduct a set of experiments to evaluate our proposed enhanced greedy algorithm on three real-world LBSN datasets. In the following, we will describe experiment settings and analyze experiment results.

4.1   Experiments Settings

Dataset Description: We use three real LBSN datasets Foursquare, Gowalla, and Brightkite. The three datasets consist of users' check-in records on the different locations. Each location has some brief textual keyword descriptions. The positive users of each location are the users who have had check-in records on this location. Table 1 summarizes the characteristics of these three datasets.

***Algorithm and Parameters*:** Enhanced greedy algorithm is abbreviated as E-G. E-G algorithm will be compared with the state-of-the-art algorithms IPH [6] and RIS-DA [4]. In this paper, IPH and RIS-DA algorithms will regard the product location l that has the average shortest Euclidean distance from the user's every historical check-in location as the user's potential consuming location $\tilde{l}$. The parameter α of MFP model is set 0.3 (by experiment). The parameter β and γ are set 1/3 (by experiment). The number of initial seeds k is set 10, 20, 30, 40, 50 with 10 as the default value. The number of product locations $|L|$ is set 10, 20, 30, 40, 50 with 10 as the default value.

***Workload*:** The index structure of HIR-tree is disk resident. And the page size is 8KB. Enhanced greedy algorithm utilizes the index structure of HIR-tree, where leaf nodes contain at most 500 vertexes and the height is 7. Our experiment

Table 1. The users' check-in characteristics of three datasets

Characteristic	User personality		
	Foursquare	Gowalla	Brighkite
Users $|V|$	200 K	50 K	20 K
Edges $|E|$	950 K	214 K	116 K
Locations $|L|$	1.3 M	773 K	43 K
Check-ins $|C|$	6.4 M	4.5 M	2 M

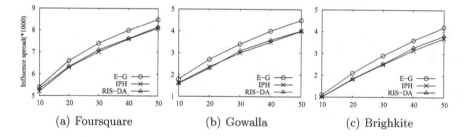

Fig. 3. The effectiveness by varying the number of seeds k

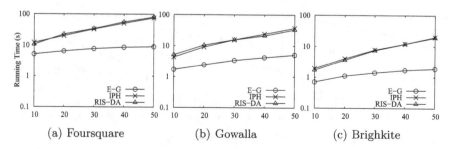

Fig. 4. The efficiency by varying the number of seeds k

carried out 2.67 GHz Intel Xeon CPU E5640 with 32 GB of RAM and 64 bit Windows 7 operating system. The experiment program is coded in C++.

4.2 Experiment Results and Analysis

First, we compare the effectiveness and efficiency of E-G, IPH, and RIS-DA algorithms by varying the number of seeds k. In Fig. 3(a), (b), (c), we show the influence spread of three algorithms in three datasets. In Fig. 4(a), (b), (c), we also show the running time of three algorithms in three datasets. We can conclude two results. The first result is that whatever dataset is, both the influence spread and the running time of three algorithms will increase with the increasing of the number of seeds k. The second result is that the influence spread of E-G is more than other two algorithms, and the whole running time of E-G algorithm is less than other two algorithms. For the first result, the larger the seed set, it is bound that the more users can be influenced. And with the increasing of the influenced target users, the running time of three algorithms must be extended. For the second result, E-G algorithm uses MFP model to make the influence spread. MFP model considers more factors to determine whether users accept the influence spread in the online phase and visit product locations in the offline phase, such as user's interest and friends' evaluation. However, IPH, and RIS-DA algorithms only use the distance factor to determine the user's possibility of consumption. Therefore, the influence spread of E-G algorithm is more than the influence spread of IPH and RIS-DA algorithms. And E-G algorithm uses

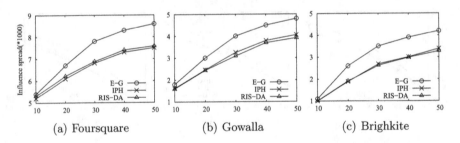

Fig. 5. The effectiveness by varying the number of product locations $|L|$

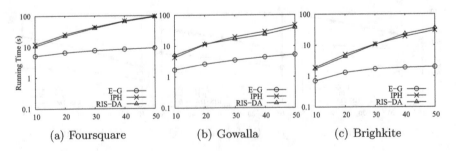

Fig. 6. The efficiency by varying the number of product locations $|L|$

HIR-tree index to speed up the query of each user's potential consuming location $\tilde{l}$ and only utilizes 1-hop or 2-hop friend relationships to decrease computational cost efficiently. Therefore, the whole running time of E-G algorithm is less than the whole running time of other two algorithms.

Second, we compare the effectiveness and efficiency of E-G, IPH, and RIS-DA algorithms by varying the number of product locations $|L|$. In Fig. 5 (a), (b), (c), we show the influence spread of three algorithms in three datasets. In Fig. 6 (a), (b), (c), we also show the running time of three algorithms in three datasets. We can conclude two results. The first result is that whatever dataset is, both the influence spread and the running time of three algorithms will increase with the increasing of the number of product locations $|L|$. The second result is that the influence spread of E-G algorithm is more than other two algorithms, and the whole running time of E-G algorithm is less than other two algorithms. For the first result, the more product locations, it is bound that the more users can be influenced and visit product locations. Therefore, the number of influenced users will increase. Thus, the running time of three algorithms must be extended. For the second result, the reason is the same as the last comparison experiment.

5   Related Works

Influence Maximization in Social Networks. Influence maximization in social networks is a very popular topic, which was proposed in the literatures [7, 9,16]. In the literature [7], Kempe et al. proposed independent cascade (IC) and linear threshold (LT) propagation models and used a greedy algorithm to solve the influence maximization problem with $1 - 1/e$ approximation ratio. In the literature [9], Leskovec et al. proposed the cost-effective lazy forward (CELF) selection to improve the greedy method. In the literature [16], Goyal et al. exploited submodularity to improve CELF greedy method. Some studies have presented competitive extensions under IC and LT models in viral marketing. In the literature [17], Yaron defined a variation of influence maximization for seeds with a budget limit in viral marketing.

Influence Maximization in LBSNs. LBSN services give the new opportunities and challenges for the traditional influence maximization problem [4,6,8]. In the literature [8], Li et al. first considered the user's location information. But, it made some unrealistic assumption, e.g., each user had a known fixed location. And in the literature [6], Zhou et al. improved the above-mentioned problem and proposed a new idea that the promotion of products considered both the online and offline phases. The key point was to bring the online consumers to the offline shops. In the literature [4], Wang et al. defined distance-aware influence maximization which also confirmed the significance of distance for the influence maximization in LBSNs. However, the existing studies have some the shortcomings. They considered that a company dependent on the viral marketing only has a product location in the real world and ignored a fact that it may have more than one product location. And in the real world, there are multiple factors to determine whether the user will visit product locations.

6 Conclusions

In this paper, we first propose a new propagation model, called MFP model. Next, we define MLIM problem under MFP model. In order to solve MLIM problem, we first propose the index structure of HIR-tree based on R-tree and inverted index to improve the search efficiency of offline phase. Furthermore, we propose the enhanced greedy algorithm to select top-k seed users. Finally, based on the real LBSN datasets, a series of experiments show that the efficiency and effectiveness of enhanced greedy algorithm are better than other methods.

Acknowledgement. This research is partially supported by the National Natural Science Foundation of China under Grant Nos. 61672145, 61572121, 61602323, 61702086, and U1401256.

References

1. Li, H., Bhowmick, S.S., Sun, A.: Cinema: conformity-aware greedy algorithm for influence maximization in online social networks. In: Proceedings of the 16th International Conference on Extending Database Technology, EDBT 2013, pp. 323–334. ACM, New York, NY, USA (2013)
2. Zhou, C., Zhang, P., Zang, W., Guo, L.: On the upper bounds of spread for greedy algorithms in social network influence maximization. IEEE Trans. Knowl. Data Eng. **27**(10), 2770–2783 (2015)
3. Bao, J., Zheng, Y., Wilkie, D., Mokbel, M.F.: Recommendations in location-based social networks: a survey. GeoInformatica **19**, 525–565 (2014)
4. Wang, X., Zhang, Y., Zhang, W., Lin, X.: Efficient distance-aware influence maximization in geo-social networks. IEEE Trans. Knowl. Data Eng. **29**(3), 599–612 (2017)
5. Zhu, W.-Y., Peng, W.-C., Chen, L.-J., Zheng, K., Zhou, X.: Modeling user mobility for location promotion in location-based social networks. In: Proceedings of the 21th ACM SIGKDD International Conference on Knowledge Discovery and Data Mining, Sydney, NSW, Australia, 10–13 August 2015, pp. 1573–1582 (2015)
6. Zhou, T., Cao, J., Liu, B., Xu, S., Zhu, Z., Luo, J.: Location-based influence maximization in social networks. In: Proceedings of the 24th ACM International on Conference on Information and Knowledge Management, pp. 1211–1220. ACM (2015)
7. Kempe, D., Kleinberg, J., Tardos, É.: Maximizing the spread of influence through a social network. In: Proceedings of the Ninth ACM SIGKDD International Conference on Knowledge Discovery and Data Mining, KDD 2003, pp. 137–146. ACM, New York, NY, USA (2003)
8. Li, G., Chen, S., Feng, J., Tan, K.-L., Li, W.-S.: Efficient location-aware influence maximization. In: Proceedings of the 2014 ACM SIGMOD International Conference on Management of data, pp. 87–98. ACM (2014)
9. Leskovec, J., Krause, A., Guestrin, C., Faloutsos, C., VanBriesen, J., Glance, N.: Cost-effective outbreak detection in networks. In: Proceedings of the 13th ACM SIGKDD International Conference on Knowledge Discovery and Data Mining, pp. 420–429. ACM (2007)
10. Gonzalez, M.C., Hidalgo, C.A., Barabasi, A.-L.: Understanding individual human mobility patterns. Nature **453**(7196), 779–782 (2008)

11. Ou, W.: Extracting user interests from graph connections for machine learning in location-based social networks. In: Proceedings of the MLSDA 2014 2nd Workshop on Machine Learning for Sensory Data Analysis, p. 41. ACM (2014)
12. Jiang, J., Lu, H., Yang, B., Cui, B.: Finding top-k local users in geo-tagged social media data. In: 2015 IEEE 31st International Conference on Data Engineering (ICDE), pp. 267–278. IEEE, 2015
13. Cong, G., Jensen, C.S., Wu, D.: Efficient retrieval of the top-k most relevant spatial web objects. Proc. VLDB Endow. 2(1), 337–348 (2009)
14. Hjaltason, G.R., Samet, H.: Distance browsing in spatial databases. ACM Trans. Database Syst. 24(2), 265–318 (1999)
15. Mohaisen, A., Hopper, N., Kim, Y.: Keep your friends close: Incorporating trust into social network-based sybil defenses. In: INFOCOM 2011. 30th IEEE International Conference on Computer Communications, Joint Conference of the IEEE Computer and Communications Societies, 10–15 April 2011, Shanghai, China, pp. 1943–1951 (2011)
16. Goyal, A., Lu, W., Lakshmanan, L.V.S.: Celf++: optimizing the greedy algorithm for influence maximization in social networks. In: Proceedings of the 20th International Conference Companion on World Wide Web, pp. 47–48. ACM (2011)
17. Singer, Y.: How to win friends and influence people, truthfully: Influence maximization mechanisms for social networks. In: Proceedings of the Fifth ACM International Conference on Web Search and Data Mining, WSDM 2012, pp. 733–742. ACM, New York, NY, USA (2012)

Emotion Analysis for the Upcoming Response in Open-Domain Human-Computer Conversation

Xiang Li and Ming Zhang(✉)

School of EECS, Peking University, Beijing, China
{lixiang.eecs,mzhang_cs}@pku.edu.cn

Abstract. Emotion analysis is one of the most active domains, hence attracts lots of attention of researchers in the natural language processing field. However, most of existed works are involved in classification tasks of the current sentence, lack of analysis of upcoming sentences. On the other hand, with the development of automatic human-computer dialogue systems, a response given by the computer side should become increasingly like human beings, for instance, the ability of expressing sentiment or emotion. The challenges lies in how to predict the emotion of a nonexistent sentence currently, which make this problem quite different from traditional sentiment or emotion analysis. In this paper, for the scenarios of open-domain conversation, we propose an architecture based on deep neural networks to predict the emotion before giving the response. In particular, we use a bidirectional recurrent neural network to get the embedding of the current utterance, and joint the representations of its retrieval results, to obtain the best emotion classification of the upcoming response. Experiments based on an annotation dataset demonstrate the effectiveness of our proposed approach better than traditional methods in terms of accuracy, precision, recall, and F-measure evaluation metrics. Then the following is some analysis of the results and future works.

Keywords: Emotion analysis · Deep learning · Neural networks
Open-domain · Human-computer conversation

1 Introduction

Emotion expression is one of the most important activities when human beings communicate with each other. For traditional face-to-face conversation, individuals often take delight in indicating their emotion by facial expression and tone. Recently with the prosperity of social media, emotion could be presented through many ways such as text including emotion words or emoji, and users are gradually accustomed to deliver their current emotion on web platforms like Facebook and Twitter, in order to make their messages more lively.

© Springer Nature Switzerland AG 2018
L. H. U and H. Xie (Eds.): APWeb-WAIM 2018, LNCS 11268, pp. 352–367, 2018.
https://doi.org/10.1007/978-3-030-01298-4_29

Table 1. Two examples of emotion in human-human conversation

Sentences	Emotion
Human1:好久没来看，帖子依旧在	none
(Long time no see, my post is still here)	
Human2:是呀，这感觉还不错吧	happiness
(Yes, it feels pretty good)	
Human1:这么明显的骗你，你都信！	anger
(So obvious a lie to you, unexpectedly you believe it!)	
Human2:他现在也已经不理我了，又消失了	sadness
(He ignores me now and disappears again)	

Hence, a large number of researchers in the natural language processing (NLP) field have paid attention to emotion analysis, which is usually a classification task, aiming to classify a span of text into one of several pre-defined emotion categories [1,17]. There are some different ways to define these categories according to affective science, and one of the most frequent way contains eight classes, i.e., like, surprise, disgust, fear, sadness, happiness, anger and none, especially for the emotion analysis of Chinese short text [1].

On the other hand, as one of the most challenging problems in artificial intelligence (AI), automatic human-computer dialogue systems have developed rapidly this several years. In order to make the computer side capable to communicate at the human level, it is necessary for the dialogue systems to express emotion proactively [2]. Thanks to the studies of controlled response generation, which means adding particular information into the generation process [3], it is possible to obtain an emotional response, if given the emotion category it should express [2]. Therefore, it becomes important to predict the correct emotion for the upcoming response before generation.

Under the scenarios of open-domain human-computer conversation, it is quite different from traditional sentiment or emotion analysis:

- The biggest challenge lies in that the prediction should be completed before generation, so that the result could be regarded as an extra input containing emotional information for the upcoming response, instead of classification on an existed sentence or document;
- Another point is that for open-domain scenarios, the content of conversation could refer to any area, and the utterances are usually with short length and casual expression, which make this task more challengeable [1];
- Last but not least, it is still lack of large-scale annotated dataset for emotion analysis, especially for the scenarios of open-domain human-computer conversation, which make supervised learning more difficult.

For traditional studies on sentiment or emotion analysis, most of them focus on classifying the current sentence. However, emotion of the upcoming response in conversation scenarios is quite different from the current utterance, so it is not suitable to simply propagate the same emotion category. As the examples showed in Table 1, under scenarios of human-human conversation, the second person may

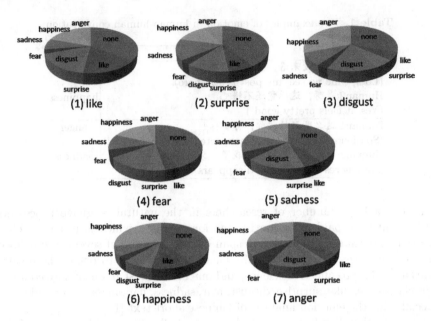

Fig. 1. For one-round conversation manually input by humans, we plot the proportion of emotion expressed in the first utterance, when given a particular emotion of the second person. The emotion consists of seven categories except none, i.e., like, surprise, disgust, fear, sadness, happiness, and anger. We aim to find out the relation of the emotion between the two utterances in one-round conversation. Generally speaking, most of the emotions expressed in the second utterances are different from its context, which make the problem more complicated.

start to present his emotion (as the first example, from none to happiness), or there probably is an emotional transformation with the conversation proceeding (as the second example, from anger to sadness). Moreover, in order to find out this difference quantitatively, we also investigate over one thousand pieces of one-round human-human conversation, which contain emotional expression in the sentence given by the second person according to human annotation. As shown in Fig. 1, our observation is that, only a small percentage of sentences maintain the same emotion category with their previous utterances. Therefore, it is necessary to introduce extra information to predict the upcoming response in conversation scenarios.

Thanks to the maturity of information retrieval (IR), we could collect the requisite information from retrieval-based methods. With the development of social media, accumulate numbers of individuals send posts and reply other people on these platforms like Twitter[1] and Baidu Tieba[2]. These kinds of resource could be regarded as human-human conversation, so that it is possible to construct data collections with large scales for human-computer dialogue systems. Since

---

[1] http://www.twitter.com.

[2] https://tieba.baidu.com.

conversation tasks could be transformed to information retrieval scenarios, one kind of methods to get the upcoming response is selecting an existing sentence from a large scale of corpus as the best reply to the input post [4,5]. The advantages of retrieval-based conversation systems are that the computer side could behave almost like a human and the replies usually have high correlation with the input posts.

Thus, in this paper, we take advantage of the information in retrieval results, to predict the emotion of the upcoming response. Our motivation is that retrieval results could be regarded as candidate responses, which indicate the directions of the final upcoming response, so they probably contain key information of the upcoming emotion that is difficult to be found out only according to the current utterance. Therefore, it is natural to combine the current utterance and its retrieval results. To be specific, we propose a novel architecture based on deep neural networks to deal with this combination and learn emotion classification. Based on the layer of word embeddings, we get the embedding of the current utterance using a bidirectional recurrent neural network (RNN) improved by Long Short-Term Memory (LSTM), and then concatenate it with the sentence embeddings of retrieval results, which are also obtained by a bidirectional LSTM model. Finally, we use full connection layers and a softmax function to complete the emotion classification. Comparing to the traditional methods of emotion analysis, our approach could integrate information from retrieval results and better predict the emotion category.

To sum up, the main contributions of this paper are as follows.

- To the best of our knowledge, we are the first to address the problem of emotion analysis for the upcoming response under open-domain human-computer conversation scenarios, which is quite different from traditional emotion classification.
- We propose a novel classification model using deep neural networks to solve this problem, combining the current utterance and its retrieval results into a hybrid framework, for the purpose of obtaining more information of the upcoming response.
- Empirical experiments demonstrate the effectiveness of our approach, with a better performance competing to traditional methods in terms of different evaluation metrics.

2 Related Work

2.1 Conversation Systems

With the increasing effect to people's everyday life, open-domain human-computer conversation systems have attracted much attention of industrial and academic communities. Many state-of-the-practice systems are developed and have amounts of users, including Siri of Apple, Xiaobing of Microsoft, and Dumi of Baidu. There is also a rising trend for academic studies to tackle the correlation and flexibility challenges, mainly with retrieval-based or generative methods.

Information retrieval is a classical topic which has developed for decade years. The traditional task is getting a ranked list of documents related to the given query, with a process consisting of pre-filter and re-ranking [27]. After retrieving candidates from a large scale of documents, features such as latent structures [6] are calculated to execute a ranking model, like learn-to-rank [7] or semantic matching [8]. When considering scenarios of human-computer conversation as selecting the most appropriate response for the current utterance, it is natural to adapt information retrieval technology for both traditional vertical dialogue systems and open-domain conversation. Leuski and Traum developed a virtual human dialogue system named NPCEditor, using information retrieval techniques [9]. Higashinaka et al. proposed a method to filter noise for candidate sentences in dialogue systems, with techniques of syntactic filtering and content-based retrieval [10]. Retrieval-based approaches could search an enough intelligent response while combining with high-quality and large-scale of conversation data resources [5,11].

In addition to information retrieval, another kind of methods is based on generation. Compare with retrieval-based methods that use instances already existing, generation-based ones are regarded more flexible. Besides filling templates [12] and paraphrase generating [13], statistical machine translation and neural networks are two main kinds of technologies to generate the best response. Based on techniques of Statistical Machine Translation, Langner et al. generate replies by translating the internal dialogue state [14], while Ritter et al. propose a data-driven approach to generate the reply for posts in Twitter [15]. Recently, with neural networks and deep learning being demonstrated useful for many natural language processing tasks, recurrent neural networks are used in conversation systems increasingly, for example, Shang et al. formalize the generation of reply as a decoding process and used recurrent neural networks for both encoding and decoding [16]. The encoder-decoder framework based on neural networks are also adapted to other complex scenarios, such as context-sensitive generation in conversation systems [18].

Since the technology of information retrieval could be used to obtain the response under conversation scenarios and demonstrated effective, it is natural for us to believe that the retrieval results could indicate possible directions of the real upcoming response and exploit key information in them.

2.2 Sentiment Analysis

Traditional sentiment or emotion analysis is a significant research task which has attracted many researchers in the domain of natural language processing. Research work on sentiment analysis often focuses on classifying the polarities of positive and negative, or extends to the third polarity of neutral, or sometimes adds fine-grained classes like a spectrum such as very positive and very negative. Emotion analysis is a kind of classification task aiming to distinguish several predefined emotion categories, such as happy, sad, and so on, while sometimes the emotion classification could be multi-label.

With the development of natural language processing, many theories and technologies have been used to deal with traditional sentiment or emotion analysis. Since some words could provide clear clues of sentiment or emotion classification, an effective series of approaches is lexicon-based models, which rely heavily on dictionaries [19,20]. Another kind of methods with high performance is feature-based models using traditional classifiers, which is called distant supervision by leveraging sentences or documents with human annotation. Support vector machine and conditional random fields based machine learning models are used to implement the emotion classification of web blog corpora [21]. Wen and Wan mined class sequential rules to improve performance of the support vector machine for emotion classification in microblog texts [1]. Other theories like statistical machine translation [22], graph-based approach [24] and topic model [23,25,39] are also used to deal with this kind of tasks.

Recent years, with the development neural networks, research work appears continuously using deep learning approaches to improve the performance of tasks in the natural language processing field, including sentiment or emotion analysis. Since semantic features and latent information could be represented by its embedding, one series of methods is to add specific information indicating sentiment or emotion categories into the word embeddings while training by neural networks [26,28,29]. Proposing novel structures of neural networks is another kind of approaches with high performance [30–33], which means adapting the theory of deep learning to these tasks. Furthermore, context of human interaction are considered to improve sentiment or emotion analysis under some specific scenarios, especially on social networks [34–36].

Under human-computer conversation scenarios, there are also some research works about sentiment or emotion topics. A neural learning approach is proposed to estimate the sentiment of the upcoming response in dialogue systems, while it only distinguish the sentiment polarities of positive and negative, or adding neutral as extension, instead of several categories of emotion, and it only consider about the information in the conversation process, without extra information such as retrieval results [37]. Another series of works is emotional conversation generation, as one kind of controlled response generation, aiming to gift the computer side ability of expressing emotion [2,38,40]. However, this kind of works regard the emotion category as a given input, instead of calculating it.

To the best of our knowledge, under scenarios of open-domain human-computer conversation, it still lacks works of emotion analysis for the upcoming response, and the prediction result could be regarded as the input of emotional conversation generation. Therefore, we propose a novel classification model utilizing retrieval results to solve this problem.

3   Approach

3.1   Task Definition

Given the current utterance $X = \{x_1, x_2, ..., x_T\}$, our aim is to train a classification model which could predict the emotion possibility $P(y|X), y \in Y$ for

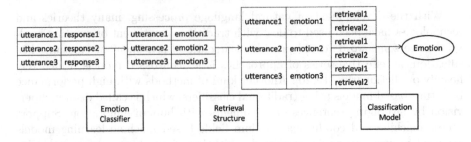

Fig. 2. The whole process to classify the emotion category of the upcoming response.

the upcoming response of the utterance X, under open-domain human-computer conversation scenarios. Y is the pre-defined emotion set mentioned before, e.g., $Y = \{like, surprise, disgust, fear, sadness, happiness, anger, none\}$.

3.2 Overview

The whole process of our proposed approach is shown in Fig. 2, which generally consists of three parts.

- Due to the lack of large-scale annotated dataset for emotion analysis, especially for the scenarios of open-domain human-computer conversation, the first part is a traditional emotion classifier to label the emotion of the response for each data item. The original training data consist of pairs of (utterance, response), and after the classification, they will become pairs like (utterance, emotion), since actually, we only need the emotion label of the response instead of its content. Here we use a bidirectional LSTM model as the classifier, which has better performance than traditional methods [2].
- Then, the second part is a retrieval structure, to get the retrieval results of the current utterance. Hence, after the retrieval process, each data item will be extended to tuples like (utterance, emotion, retrieval results), as the final input of our proposed classification model, noting that the emotion label is needed only in the training process.
- For the final part, it is our proposed classification model based on deep neural networks, to predict the emotion of the upcoming response. The model calculate the sentence embeddings of the current utterance and its retrieval results respectively, and deliver their combination to a softmax layer.

3.3 Bidirectional LSTM

Since natural language sentences could be regarded as sequences, it is natural to use recurrent neural networks to model sentences and get their embeddings. For each hidden layer, the inputs are the current word embedding as well as the last hidden layer, until the end of the sentence, and the final hidden layer is regarded as the embedding of the whole sentence, which could represent all the sequential information. In practice, due to the increasing sparsity with the propagation going on, the Long Short-Term Memory (LSTM) [41] is often used to improve its performance.One LSTM unit could be regarded as a hidden state in the RNN structure, which could remember the information of words far away from the current word in the sentence. The specific calculation is given by

$$f_t = \sigma(W_f[h_{t-1}, x_t] + b_f) \tag{1}$$

$$i_t = \sigma(W_i[h_{t-1}, x_t] + b_i) \tag{2}$$

$$C'_t = tanh(W_C[h_{t-1}, x_t] + b_C) \tag{3}$$

$$C_t = f_t * C_{t-1} + i_t * C'_t \tag{4}$$

$$o_t = \sigma(W_o[h_{h-1}, x_t] + b_o) \tag{5}$$

$$h_t = o_t * tanh(C_t) \tag{6}$$

where W's are weights and b's are bias terms. x_t is the word embedding; h_t is the hidden state at time step t; and the signal "*" denotes element-wise product of two vectors.

Bidirectional LSTM is an important variance of the RNN structure, which has been demonstrated with high performance under lots of scenarios [42].For each time step t, there are two hidden states, both connected with the input layer and the output layer. Thus, it could be regarded as two RNN chains that one propagates from the beginning to the end of the sentence, and another is from the end to the beginning, sharing the same input and deciding the output together. The advantage of this structure is not only to utilize the past information of the current word, but also the future information after it.

Therefore, bidirectional LSTM has high performance on traditional emotion classification task and we directly use it to label the response as mentioned before. Besides, it also is an important part of our proposed classification model to predict the emotion of the upcoming response, which will be introduced later. Instead of giving output at each time step, we only concatenate the final hidden states of the two RNN chains as the sentence embedding in this paper, because our classification scenarios aim at whole sentences.

3.4   Retrieval Structure

To get the retrieval results of the current utterance, we use a retrieval framework based on textual similarity to search out k candidate responses just like retrieval-based dialogue systems, which are semantically related to the utterance and could indicate the directions of the final upcoming response. In this part, we describe in detail the retrieval-based section, which adapts typical frameworks in the domains of search engine or advertisement selection, as shown in Fig. 3, working in a two-step retrieval-and-ranking strategy.

To start, the only input is the current utterance, and we use it to retrieve up to ten hundred candidate replies, from a large scale of conversation campus. The campus consists of post-reply pairs like one-round conversation, so the candidates

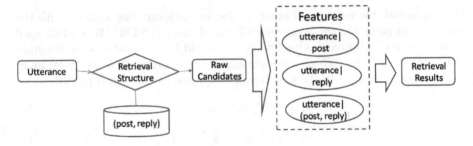

Fig. 3. The retrieval structure using a two-step strategy, the same with typical frameworks in the domain of information retrieval.

picked up should have at least one post textually similar to the utterance. This step is accomplished based on standard keyword retrieval structures, similar to the Lucene[3] and Solr[4] system.

After the coarse-grained retrieval, we have a set of raw candidate replies, and next we need to rerank these candidates in a fine-grained fashion, using much richer information from different perspective. Semantic meanings of a post and its reply are both important, so features should be exacted respectively focusing on three aspects, which means between the utterance and the post, the reply, and the post-reply pair, including textual similarity, measures of word embeddings, and so on. Finally, we get a ranking model and then use it to get top k replies, as the retrieval results of the current utterance.

3.5 Classification Model

Having the retrieval results of the current utterance, naturally the most important part is our proposed classification model to predict the emotion category of the upcoming response. As shown in Fig. 4, the whole structure model the current utterance and its retrieval results respectively and combine them together to make the final classification, based on deep neural networks.

For the left part in Fig. 4, we aim to obtain sentence embedding of the current utterance. A word embedding layer is at the bottom to represent input words in the utterance, and then put into a bidirectional LSTM structure introduced before, given by

$$s_t^{u+} = LSTM(s_{t-1}^{u+}, e(x_t)) \quad and \quad s_t^{u-} = LSTM(s_{t+1}^{u-}, e(x_t)) \qquad (7)$$

where x_t is the input word and $e(x_t)$ indicates its embedding. "u" in the superscript means modeling the utterance; "+" means the direction from the beginning to the end; yet "−" is the opposite. Then, we concatenate the two final hidden states of the two directions and after an extra full-connection layer, we

---

[3] http://lucene.apache.org.
[4] http://lucene.apache.org/solr.

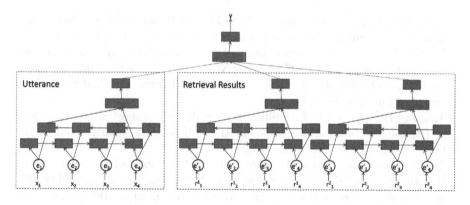

Fig. 4. The structure of our proposed model to predict the emotion category of the upcoming response. Note that there are k retrieval replies on the right part, and we only figure out two for simplicity.

complete the modeling process of the current utterance and get its final sentence embedding s^u, e.g.

$$s^u = \sigma(W_u[s_T^{u+}; s_1^{u-}] + b_u) \qquad (8)$$

where W_u is the weight matrix, b_u is the bias term, and ";" is the concatenation operation of two vectors.

For the modeling process of the retrieval results, as the right part in Fig. 4 shown, the calculation is similar to that for the utterance, which means

$$s_t^{i+} = LSTM(s_{t-1}^{i+}, e'(r_t^i)) \quad and \quad s_t^{i-} = LSTM(s_{t+1}^{i-}, e'(r_t^i)) \qquad (9)$$

$$s^i = \sigma(W_r[s_T^{i+}; s_1^{u-}] + b_r) \qquad (10)$$

where r's are words in retrieval results and "i" in the superscript indicates the i-th retrieval result, with a range of 1 to k.

Then, we combine the two parts through concatenating sentence embeddings of the current utterance and all k retrieval results, and after another full-connection layer, we put it into the final softmax layer to make the emotion classification, e.g.

$$P(y|X) = softmax(W_2\sigma(W_1[s^u; s^1; ...; s^k] + b_1) + b_2) \qquad (11)$$

4   Experiments

4.1   Datasets and Setups

For the whole process of our emotion analysis, there are 4 datasets as follows.

– Conversation Dataset. To construct the index base for our retrieval section, we collected massive resources from Chinese forums, microblog websites, and community QA platforms such as Baidu Zhidao, Baidu Tieba, Douban forum, Sina Weibo[5], and totally extracted nearly 10 million post-reply pairs.

- NLPCC Datasets. These two datasets were used in challenging tasks of emotion classification in NLPCC2013 and NLPCC2014[6]. We use them to train a traditional emotion classification model with bidirectional LSTM to label the replies of our training data.
- Training Dataset. Over 1.3 million post-reply pairs collected the same way as the conversation dataset, with emotion labels of the replies given by the bidirectional LSTM classifier.
- Test Dataset. Also post-reply pairs with emotion labels of the replies. Note that the emotion labels here are annotated by humans as the "ground truth", different from those of training dataset. Each item is annotated by 3 individuals in an independent and blind fashion, and finally we get an annotated dataset consisting of 1996 pairs, with eight categories mentioned before, and its kappa score $\kappa = 0.427$, showing moderate inter-rater agreement.

Metrics. We use several evaluation metrics to demonstrate the effectiveness of our proposed model. The first one is accuracy, which could indicate the correctness directly. Note that the type of none is less meaningful, yet occupies a high proportion. So more important series of metrics is precision, recall and F-measure, which is usually used for emotion analysis [1].

Training Settings. For the training process, we use cross-entropy objective as the loss function and all dimensions of embedding vectors are 128. Different retrieval results should share the same variate to be trained including word embeddings, parameters of bidirectional LSTM and the full connection layer in our model design, yet different from those for modeling the current utterance. However, for the consideration of running time, finally we make k equal to 1.

4.2   Baseline Algorithms

To demonstrate the effectiveness of our proposed model, we include the following methods as baselines. Besides traditional approaches, we also use some basic neural network structures to model the current utterance and make classification. For fairness, we conduct the same data cleaning and layer dimensions in neural networks for all algorithms. Specifically, we filter out utterances containing words with very low frequency, and also utterances containing over 50 words. Basic data pre-processing is also done, including word segmentation and so on.

SVM. SVM is one kind of traditional classification model to construct hyperplanes according to pre-defined features. Here, it is primarily based on filtered word features, and also some secondary features such as the emotion category of the current utterance given by the traditional emotion classifier. Besides, we use its linear version due to the large scale of our training data.

---

[5] http://zhidao.baidu.com,   http://tieba.baidu.com,   http://douban.com,   http://weibo.com.

[6] http://tcci.ccf.org.cn/conference/2013|2014/.

Table 2. Performance of the emotion classification for the upcoming response

Method	Accuracy	Macro average			Micro average		
		Precision	Recall	F-measure	Precision	Recall	F-measure
SVM	0.4118	0.1669	0.0283	0.0484	0.2222	0.0335	0.0581
CNN	**0.4284**	0.1571	0.0021	0.0041	0.1667	0.0026	0.0052
LSTM	0.4238	0.2413	0.0165	0.0309	0.2639	0.0167	0.0314
Bi-LSTM	0.4279	0.3231	0.0183	0.0346	0.2658	0.0185	0.0346
Our method	0.4259	**0.3892**	**0.0786**	**0.1308**	**0.4026**	**0.0819**	**0.1361**

CNN. Convolutional neural networks is a kind of structure good at extracting local features. Instead of the way of full-connection, a neuron in the convolutional layer could only have particular numbers of connections from the last layer. After the convolutional operation, there is often a pooling layer to integrate the information. Specifically, the convolutional layer has 128 filters, with a window size of 3, and here we use max pooling.

LSTM. A recurrent neural network structure improved by Long Short-Term Memory units, which is introduced before.

Bi-LSTM. The variance of LSTM with two recurrent directions, which is also introduced before.

4.3   Performance

In this section, we show the performance of our proposed model against other baselines, and report the performance of emotion classification in all the mentioned evaluation metrics in Table 2. Since retrieval results could give key information for the real upcoming response in some degree, it is obvious that our method utilizing retrieval results perform better than those only with the current utterance, whatever SVM or basic neural network structures, with a comparable accuracy and higher precisions, recalls, and F-measures. However, sometimes the retrieval results are general replies without any emotional information, such as "I think so", so the recall is always lower than the precision, which have large room for improvement.

4.4   Analysis

To analyze the performance of our proposed model specifically, we also investigate the F-measure for each emotion category. Some emotion categories have higher F-measures than the macro-average F-measure, while others not, especially for the "fear" emotion. The reason is that it occupies only a low proportion in practical conversation, so that could not be adequately trained. Thus, our observation is that emotion with higher appearance tends better performance, such as the "like" emotion, with a higher percentage than the other six emotion categories.

Table 3. A case study of the emotion classification for the upcoming response

Utterance	我在等学校装空调 (I am waiting for the air conditioner installed by the school)	None
Response	什么学校哇，这么好(What school? So nice)	Like
Retrieval	好牛(Awesome)	Like

Table 3 shows a case study, in which the current utterance is just a statement without any clear emotion, but the upcoming response comes up with "like" emotion with the expression of "so nice" to the "school". There is no clue to deduce this emotion only from the content of the utterance, yet the retrieval result could indicate that its response will probably be in the "like" emotion category.

5   Conclusion

In this paper, under open-domain human-computer conversation scenarios, in order to solve the problem of emotion analysis for the upcoming response, we propose an approach of jointing representations of the current utterance and its retrieval results using deep neural networks, and deeply analyze its performance through experiments. Empirical results demonstrate our approach better than traditional methods in terms of different metrics. For the future work, one direction is to propose more progressive models to consider contextual information in the conversation process, and another may be a global model to joint the emotion analysis and the controlled response generation, in stead of giving the emotion before the generation process.

Acknowledgements. National Natural Science Foundation of China NSFC Grant (NSFC Grant Nos.61772039, 91646202 and 61472006).

References

1. Wen, S., Wan, X.: Emotion classification in microblog texts using class sequential rules. In: AAAI Conference on Artificial Intelligence, pp. 187–193 (2014)
2. Zhou, H., Huang, M., Zhang, T., Zhu, X., Liu, B.: Emotional chatting machine: emotional conversation generation with internal and external memory. arXiv preprint arXiv:1704.01074 (2017)
3. Mou, L., Song, Y., Yan, R., Li, G., Zhang, L., Jin, Z.: Sequence to backward and forward sequences: a content-introducing approach to generative short-text conversation. In: International Conference on Computational Linguistics, pp. 3349–3358 (2016)
4. Yan, R., Song, Y., Wu, H.: Learning to respond with deep neural networks for retrieval based human-computer conversation system. In: Proceedings of the SIGIR, pp. 55–64 (2016)

5. Ji, Z., Lu, Z., Li, H.: An information retrieval approach to short text conversation. arXiv preprint arXiv:1408.6988 (2014)
6. Wu, W., Lu, Z., Li, H.: Learning bilinear model for matching queries and documents. J. Mach. Learn. Res. **14**(1), 2519–2548 (2014)
7. Liu, T.Y.: Learning to rank for information retrieval. Found. Trends Inf. Retr. **3**(3), 225–331 (2009)
8. Giunchiglia, F., Yatskevich, M., Shvaiko, P.: Semantic matching: algorithms and implementation. J. Data Semant. **IX**, 1–38 (2007). Springer, Berlin, Heidelberg
9. Leuski, A., Traum, D.: Npceditor: creating virtual human dialogue using information retrieval techniques. AI Mag. **32**(2), 42–56 (2011)
10. Higashinaka, R., et al.: Syntactic filtering and content-based retrieval of twitter sentences for the generation of system utterances in dialogue systems. In: Rudnicky, A., Raux, A., Lane, I., Misu, T. (eds.) Situated Dialog in Speech-Based Human-Computer Interaction. SCT, pp. 15–26. Springer, Cham (2016). https://doi.org/10.1007/978-3-319-21834-2_2
11. Wang, H., Lu, Z., Li, H., Chen, E.: A dataset for research on short-text conversations. In: Proceedings of the conference on human language technology and empirical methods in natural language processing, pp. 935–945 (2013)
12. Sugiyama, H., Meguro, T., Higashinaka, R., Minami, Y.: Open-domain utterance generation for conversational dialogue systems using web-scale dependency structures. In: Proceedings of the SIGDIAL, pp. 334–338 (2013)
13. Mairesse, F., Young, S.: Stochastic language generation in dialogue using factored language models. Comput. Linguist. **40**(4), 763–799 (2014)
14. Langner, B., Vogel, S., Black, A. W.: Evaluating a dialog language generation system: comparing the mountain system to other NLG approaches. In: Annual Conference of the International Speech Communication Association (INTERSPEECH), pp. 1109–1112 (2010)
15. Ritter, A., Cherry, C., Dolan, W. B.: Data-driven response generation in social media. In: Proceedings of the Conference on Human Language Technology and Empirical Methods in Natural Language Processing, pp. 583–593 (2011)
16. Shang, L., Lu, Z., Li, H.: Neural responding machine for short-text conversation. In: Annual Meeting of the Association for Computational Linguistics, pp. 1577–1586 (2015)
17. Li, X., Yan, R., Zhang, M.: Joint emoji classification and embedding learning. In: Chen, L., Jensen, C.S., Shahabi, C., Yang, X., Lian, X. (eds.) APWeb-WAIM 2017. LNCS, vol. 10367, pp. 48–63. Springer, Cham (2017). https://doi.org/10.1007/978-3-319-63564-4_4
18. Sordoni, A., et al.: A neural network approach to context-sensitive generation of conversational responses. In: The Annual Conference of the North American Chapter of the Association for Computational Linguistics, pp. 196–205 (2015)
19. Turney, P. D.: Thumbs up or thumbs down?: semantic orientation applied to unsupervised classification of reviews. In: Annual Meeting of the Association for Computational Linguistics, pp. 417–424 (2002)
20. Taboada, M., Brooke, J., Tofiloski, M., Voll, K., Stede, M.: Lexicon-based methods for sentiment analysis. Comput. Linguist. **37**(2), 267–307 (2011)
21. Yang, C., Lin, K. H. Y., Chen, H. H.: Emotion classification using web blog corpora. In: IEEE/WIC/ACM International Conference on Web Intelligence, pp. 275–278 (2007)
22. Lambert, P.: Aspect-level cross-lingual sentiment classification with constrained SMT. In: Annual Meeting of the Association for Computational Linguistics, pp. 781–787 (2015)

23. Hai, Z., Cong, G., Chang, K., Liu, W., Cheng, P.: Coarse-to-fine review selection via supervised joint aspect and sentiment model. In: Proceedings of the SIGIR, pp. 617–626 (2014)

24. Wang, X., Wei, F., Liu, X., Zhou, M., Zhang, M.: Topic sentiment analysis in twitter: a graph-based hashtag sentiment classification approach. In: Proceedings of the CIKM, pp. 1031–1040 (2011)

25. Yang, M., Peng, B., Chen, Z., Zhu, D., Chow, K.P.: A topic model for building fine-grained domain-specific emotion lexicon. In: Annual Meeting of the Association for Computational Linguistics, pp. 421–426 (2014)

26. Tang, D., Wei, F., Yang, N., Zhou, M., Liu, T., Qin, B.: Learning sentiment-specific word embedding for twitter sentiment classification. In: Annual Meeting of the Association for Computational Linguistics, pp. 1555–1565 (2014)

27. Li, X., Mou, L., Yan, R., Zhang, M.: Stalematebreaker: a proactive content-introducing approach to automatic human-computer conversation. In: International Joint Conference on Artificial Intelligence, pp. 2845–2851 (2016)

28. Zhou, H., Chen, L., Shi, F., Huang, D.: Learning bilingual sentiment word embeddings for cross-language sentiment classification. In: Annual Meeting of the Association for Computational Linguistics, pp. 430–440 (2015)

29. Ren, Y., Zhang, Y., Zhang, M., Ji, D.: Improving twitter sentiment classification using topic-enriched multi-prototype word embeddings. In: AAAI Conference on Artificial Intelligence, pp. 3038–3044 (2016)

30. Tang, D., Wei, F., Qin, B., Zhou, M., Liu, T.: Building large-scale twitter-specific sentiment lexicon: a representation learning approach. In: International Conference on Computational Linguistics, pp. 172–182 (2014)

31. Dong, L., Wei, F., Zhou, M., Xu, K.: Adaptive multi-compositionality for recursive neural models with applications to sentiment analysis. In: AAAI Conference on Artificial Intelligence, pp. 1537–1543 (2014)

32. Zhang, M., Zhang, Y., Vo, D. T.: Gated neural networks for targeted sentiment analysis. In: AAAI Conference on Artificial Intelligence, pp. 3087–3093 (2016)

33. Zhao, Z., Liu, T., Hou, X., Li, B., Du, X.: Distributed text representation with weighting scheme guidance for sentiment analysis. In: Li, F., Shim, K., Zheng, K., Liu, G. (eds.) APWeb 2016. LNCS, vol. 9931, pp. 41–52. Springer, Cham (2016). https://doi.org/10.1007/978-3-319-45814-4_4

34. Vanzo, A., Croce, D., Basili, R.: A context-based model for sentiment analysis in twitter. In: International Conference on Computational Linguistics, pp. 2345–2354 (2014)

35. Ren, Y., Zhang, Y., Zhang, M., Ji, D.: Context-sensitive twitter sentiment classification using neural network. In: AAAI Conference on Artificial Intelligence, pp. 215–221 (2016)

36. Li, S., Huang, L., Wang, R., Zhou, G.: Sentence-level emotion classification with label and context dependence. In: Annual Meeting of the Association for Computational Linguistics, pp. 1045–1053 (2015)

37. Bothe, C., Magg, S., Weber, C., Wermter, S.: Dialogue-based neural learning to estimate the sentiment of a next upcoming utterance. In: Lintas, A., Rovetta, S., Verschure, P.F.M.J., Villa, A.E.P. (eds.) ICANN 2017. LNCS, vol. 10614, pp. 477–485. Springer, Cham (2017). https://doi.org/10.1007/978-3-319-68612-7_54

38. Zhang, R., Wang, Z., Mai, D.: Building emotional conversation systems using multi-task Seq2Seq learning. In: Huang, X., Jiang, J., Zhao, D., Feng, Y., Hong, Y. (eds.) NLPCC 2017. LNCS (LNAI), vol. 10619, pp. 612–621. Springer, Cham (2018). https://doi.org/10.1007/978-3-319-73618-1_51

39. Hai, Z., Cong, G., Chang, K., Cheng, P., Miao, C.: Analyzing sentiments in one go: a supervised joint topic modeling approach. IEEE Trans. Knowl. Data Eng. **29**(6), 1172–1185 (2017)
40. Yuan, J., Zhao, H., Zhao, Y., Cong, D., Qin, B., Liu, T.: Babbling - the HIT-SCIR system for emotional conversation generation. In: Huang, X., Jiang, J., Zhao, D., Feng, Y., Hong, Y. (eds.) NLPCC 2017. LNCS (LNAI), vol. 10619, pp. 632–641. Springer, Cham (2018). https://doi.org/10.1007/978-3-319-73618-1_53
41. Hochreiter, S., Schmidhuber, J.: Long short-term memory. Neural Comput. **9**(8), 1735–1780 (1997)
42. Graves, A., Schmidhuber, J.: Framewise phoneme classification with bidirectional LSTM and other neural network architectures. Neural Netw. **18**(5–6), 602–610 (2005)

A Recurrent Neural Network Language Model Based on Word Embedding

Shuaimin Li[(✉)] and Jungang Xu

University of Chinese Academy of Sciences, Beijing, China
lishuaimin17@mails.ucas.ac.cn, xujg@ucas.ac.cn

Abstract. Language model is one of the basic research issues of natural language processing, and which is the premise for realizing more complicated tasks such as speech recognition, machine translation and question answering system. In recent years, neural network language model has become a research hotspot, which greatly enhances the application effect of language model. In this paper, a recurrent neural network language model (RNNLM) based on word embedding is proposed, and the word embedding of each word is generated by pre-training the text data with skip-gram model. The n-gram language model, RNNLM based on one-hot and RNNLM based on word embedding are evaluated on three different public datasets. The experimental results show that the RNNLM based on word embedding performs best, and which can reduce the perplexity of language model significantly.

Keywords: Recurrent neural network · Language model · Word embedding
Perplexity · One-hot

1 Introduction

As an important subarea of artificial intelligence, natural language processing has been a hot topic among many researchers. There are many challenging tasks in natural language processing, such as question answering system, machine translation, text generation, etc. The modeling method of language model has always been the basic task of natural language processing, which is the premise to realize other more complicated tasks.

The origin of the language model was to solve the problem of speech recognition. In speech recognition task, the computer needs to know whether a word sequence can make a meaningful sentence for human beings [1].

Early researchers established the language model based on artificial rules, they judged whether a sentence is meaningful by judging whether the sequence of text conforms to the grammar rules. However, because of the complexity of grammar rules, this method cannot cover all linguistic phenomena and is difficult to be used to deal with large-scale real texts. At present, the statistical language model has become the main method, which can summarize the language knowledge, obtain the connection probability between words, and determine whether the sequence is reasonable according to the joint probability of word sequences.

L. H. U and H. Xie (Eds.): APWeb-WAIM 2018, LNCS 11268, pp. 368–377, 2018.
https://doi.org/10.1007/978-3-030-01298-4_30

N-gram language model is a typical generated model, which assumes that the current word probability is only related to n words before, and the probability of word sequence is obtained by the chain rule [2]. However, since many n-tuples are not observed in the training corpus, that is, data sparseness is serious, which lead to zero probability phenomenon.

In recent years, with the development of deep learning technology, researchers have tried to apply artificial neural network technology to language model. In 2003, Yoshua Bengio first used the neural network into the learning of language model [3] and trained a distributed word representation. In 2010, Tomas Mikolov first used recurrent neural network to train language model [4]. The deep neural network language model solves the problem of zero probability caused by sparse data, and reduces the difficulties caused by a large number of parameters. At present, deep neural network is widely used in language model training, word vector representation and other natural language processing tasks.

In this paper, based on Tomas Mikolov's work, a recurrent neural network language model based on word embedding is proposed. Furthermore, the validity of RNNLM based on word embedding is verified on three different public datasets.

The rest of the paper is organized as follows. A brief overview of recurrent neural network language model is described in Sect. 2. The recurrent neural network language model based on word embedding is proposed in Sect. 3. Experimental methodology and experimental result analysis are given in Sect. 4. Finally, the conclusions and future work are summarized in Sect. 5.

2   Recurrent Neural Network Language Model

In 2010, Mikolov first proposed a method of training language model with recurrent neural network, i.e., recurrent neural network language model [4]. The structure of recurrent neural network language model is shown in Fig. 1.

In Fig. 1, input(t) denotes the current input of recurrent neural network, context(t) denotes the history information, and output(t) denotes the output of the network. Recurrent neural network uses the loop of the hidden layer to enhance the use of historical information. Therefore, time stamp t is introduced for the input layer, hidden layer and output layer in the network. The calculation of hidden layer $h(i)$ in the recurrent neural network is shown in Eq. 1.

$$h(i) = sigmoid(e(w_i) + Wh(i-1)) \tag{1}$$

where w_i denotes the i-th word in the sequence, $e(w_i)$ denotes the word vector of the i-th word, and $h(i)$ denotes the hidden layer corresponding to the i-th word in the corpus, which is obtained by the combination of the word embedding of the target word and the hidden layer $h(i-1)$ corresponding to the previous word. What's special about this is that instead of using the n-gram approximation, recurrent neural network language model uses the iterative method to model all the target words above.

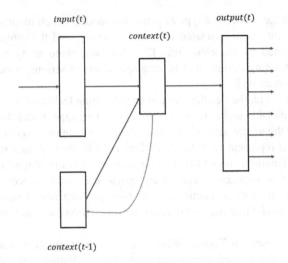

Fig. 1. Structure of recurrent neural network language model

3 A Recurrent Neural Network Language Model Based on Word Embedding

In this paper, a word embedding representation method is used to improve the recurrent neural network language model based on one-hot representation. The unfolding of the recurrent neural network language model based on word embedding is shown in Fig. 2.

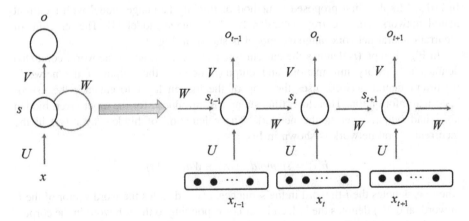

Fig. 2. Unfold of recurrent neural network language model

In Fig. 2, x is the input word sequence, x_t is a formal representation for a word in the word sequence, s_t is the hidden layer state which represents the historical information of the sentence, o_t is output state and it indicates that a vector whose size is equal to the size of vocabulary size, U, V, W are the parameter matrices of recurrent neural network. The modeling process of RNNLM based on word embedding has four main steps as follows.

3.1 Data Processing

Segment training text.

The datasets used for training language model usually consists of raw texts. In order to convert plain text into the input of the network, the input text data needs to be broken up first. In this part, NLTK [5] is used to realize sentence segmentation and word segmentation.

Merge 'sparse words'.

Generally, some words in training text appear rarely which are called 'sparse words', these 'sparse words' in training text are merged and marked as 'UNKNOWN_TOKEN' in this paper. The reasons for doing this are as follows: (a) Large vocabulary leads to a great reduction in the speed of training; (b) Sparse words have little information about the context and history so that they cannot get good training results. This is just like the process of learning a language for human, the mastery of a new word must be acquired in the process of learning different contexts. Here, the vocabulary size is set to 8000 for merging the sparse words, in other words, the words appeared no more than 8000 times in the text will be marked as 'UNKNOWN_TOKEN'.

Add sentence start and sentence end token.

In order to process the text by sentence, the corresponding token at the start and the end of the sentences are added.

3.2 Word Embedding Representation

In 1986, Hinton first proposed the idea of distributed word representation [6]. In 2013, Mikolov proposed CBOW (Continuous Bag-of-Words) model and Skip-gram model to obtain the distributed word representation and the method is verified to close the distance of the context related words. This distributed word representation method is usually called word embedding [7]. In this paper, word embedding is applied to the construction of RNNLM, the dataset is pre-trained with the skip-gram model to obtain the word embedding of each word in the training set. Here, the size of word vector is set to 200.

3.3 Construction of Recurrent Neural Language Model Based on Theano

Construction of recurrent neural network layers.

The model in this paper is constructed by Theano [8]. The structure of the network is built by setting up the hidden and output layer of RNN. As the memory link of the network, the hidden layer is represented by Eq. 2.

$$s_t = f(Ux_t + Ws_{t-1}) \tag{2}$$

where f is the activation function of the hidden layer, and hyperbolic tangent function is generally used as the activation function of the hidden layer, which is defined as Eq. 3.

$$f(z) = tanh(z) = \frac{1}{1+e^{-z}} \tag{3}$$

The output layer is represented by Eq. 4.

$$o_t = softmax(Vs_t) \tag{4}$$

Parameter Matrix and Vector Dimension Setting of RNN.

In this paper, the size of vocabulary is set to 8000. Considering the complexity of the operation, the size of the hidden layer is set to 100. Therefore, according to the Eqs. 2 and 3, the dimension of the various parameters matrix and vectors in RNN are listed as follows.

$$x_t \in R^{8000}$$

$$o_t \in R^{8000}$$

$$s_t \in R^{100}$$

$$U \in R^{100 \times 8000}$$

$$V \in R^{8000 \times 100}$$

$$W \in R^{100 \times 100}$$

Word Predictions.

The RNN returns the computational output from the Softmax function, here each o_t represents the probability distribution vector of the words in the vocabulary.

3.4 Supervised Training Based on Recurrent Neural Network

The goal of network training is to obtain the network parameters U, V and W of RNNLM by inputting a number of text datasets, in this way, a recurrent neural network language model based on word embedding will be obtained. Here, U is the current input parameter matrix, V is the parameter matrix of the history state, and W is the parameter matrix of the current state. In each round of the training, the error vector is calculated according to cross-entropy criterion as is shown in Eq. 5, in this equation, o represents the predictions of the model and y represents the real label. The parameters of the network are updated through the stochastic gradient descent algorithm and back propagation through time algorithm.

$$L(y, o) = -\frac{1}{N} \sum_{n \in N} y_n log o_n \qquad (5)$$

4 Experiments

4.1 Metric

The major evaluation metric of the language model is perplexity [9]. This section mainly introduces the calculation method of perplexity of language model on test text.

If the test text T is composed of l_T sentences, m denotes the number of words in the sentence, w_i denotes the word to be predicted, w_{i-n+1}^{i-1} denotes the first n words before w_i, then the joint probability of the sentence is defined as Eq. 6.

$$p(s) = \prod_{i=1}^{m} p(w_i | w_{i-n+1}^{i-1}) \qquad (6)$$

The joint probability of the whole test dataset is defined as Eq. 7.

$$p(T) = \prod_{j=1}^{l_T} p(s_j) \qquad (7)$$

Before calculating the perplexity of the language model, it is necessary to calculate the cross-entropy of the model on the given test data set. The cross entropy is used to measure the difference between the probability distributions of language model and natural language. The smaller the perplexity is, the model behaves better and it will be closer to the real probability distribution. For the test corpus which consists of W_T words, the cross entropy of the model on the whole test data set is defined as Eq. 8.

$$
\begin{aligned}
H_p(T) &= -\frac{1}{W_T} log_2 p(T) \\
&= -\frac{1}{W_T} \sum_{j=1}^{l_T} (log_2 p(s_j)) \\
&= -\frac{1}{W_T} \sum_{j=1}^{l_T} \sum_{i=1}^{m_{s_j}} (log_2(p(w_i | w_{i-n+1}^{i-1})))
\end{aligned} \qquad (8)
$$

Based on cross entropy of the model on the whole test data set, the perplexity is defined as Eq. 9.

$$ppl(T) = 2^{H_p(T)}$$

$$= 2^{-\frac{1}{W_T}\sum_{j=1}^{l_T}\sum_{i=1}^{ms_j}(log_2(p(w_i|w_{i-n+1}^{i-1})))} \tag{9}$$

The meaning of the perplexity is the geometric mean of the candidate words after each word when the language model is used to predict a linguistic phenomenon. The lower the perplexity is, the stronger the constraint ability to the context and description ability of the language model are, and the better the performance of the language model will be.

4.2 Datasets

Three different plain text data sets are adopted to evaluate the language model, which includes reddit comments Reddit_2015_08, Reddit_2015_09 from Google's big query [10] and public dataset Penn Treebank [11], which is downloaded from Mikolov's webpage. The details of the data sets are listed in Table 1.

4.3 Experimental Design

In the experiments, SRILM [12] is used to implement the n-gram language model. Similarly, Theano framework is used to implement RNNLM based on one-hot and RNNLM based on word embedding. Then n-gram language model, RNNLM based on one-hot and RNNLM based on word embedding are trained and tested on three different datasets listed above. In the experiment, learning rate is set to 0.001 and it is adjusted after every round, hidden layer size is set to 100.

4.4 Experimental Results and Analysis

For evaluating the performance of RNNLM based on word embedding, we choose n-gram language model and RNNLM based on one-hot as baseline model. Firstly, the three models are trained on three different data sets until the model converges, after the training, the corresponding parameters of the model will be obtained. Then, the obtained models are tested on the different test data sets. The perplexities of n-gram language model, recurrent neural network language model based on one-hot and recurrent neural network language model based on word embedding on three datasets are listed in Table 2.

Table 1. The details of the data sets

Datasets	Reddit_2015_08	Reddit_2015_09	Penn Treebank
Size	5.12 M	5.10 M	4.77 M
Number of sentences	56042	57729	42067

Table 2. The perplexities of three models on different datasets

Datasets	n-gram	RNN (one hot)	RNN (word embedding)
Reddit_2015-08	508	272	**227**
Reddit_2015-09	553	252	**224**
Penn Treebank	870	418	**318**

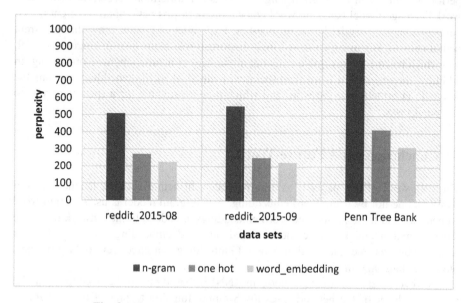

Fig. 3. The perplexity of each model on different data sets

From Table 2, the perplexity of the recurrent neural network language model based on word embedding is 227, 224 and 318 on the three different data sets. From Fig. 3, it is obviously that the perplexity of RNNLM based on word embedding is the least on all the three datasets among these models. With further analysis, as is shown in Table 3, perplexity decrease percentages of RNNLM based on word embedding against other two models can be obtained.

Table 3. Perplexity decrease percentages of RNNLM based on word embedding against other two models

Datasets	n-gram	RNN (one hot)
Reddit_2015-08	55.3%	16.5%
Reddit_2015-09	59.5%	11.1%
Penn Treebank	63.4%	23.9%

In Table 3, the perplexity decreases percentages of recurrent neural network language model based on word embedding on three datasets against n-gram model are 55.3%, 59.5% and 63.4% respectively. The perplexity decreases percentages of recurrent neural network language model based on word embedding on three datasets against recurrent neural network language model based one-hot are 16.5%, 11.1% and 23.9% respectively. It can be found that recurrent neural network language model performs better than the n-gram language model. Furthermore, recurrent neural network language model based on word embedding performs better than recurrent neural network language model based on one-hot. Experimental results show that recurrent neural network is playing an increasingly important role in the field of language model, and which has the no negligible potential in the field of natural language processing. In addition, the distributed word representation method of word embedding contains the semantic information of the word and which is more suitable for a specific corpus to carry out tasks.

5   Conclusions and Future Work

The core work of this paper is to improve recurrent neural network language model based on one-hot by using word embedding. A skip-gram model is used to train word embedding and these pre-trained word embeddings are added to the network to realize the recurrent neural language model based on word embedding. The experimental results show that the introduction of word embedding can effectively reduce the perplexity of language model.

However, there are some problems in RNNLM based on word embedding. Firstly, the output layer of the network uses the Softmax function to predict the probability distribution of the next word, which may result in high complexity of the model training and the difficulty of network parameter adjustment; Secondly, a sample neural network is used in this paper, which is too easy compared with the complicated neural network such as LSTM (Long Short-Term Memory) network. Thirdly, the datasets used in this paper are small and untagged, which may result in the training of the language model without any grammatical supervision.

Considering about these problems, our future work will focus on the following two points: (1) Try to modify the output layer by classifying the original vocabulary, therefore, the word character can be predicted first and then the next word is predicted in the corresponding word character part, this method may reduce the complexity of the model prediction; (2) Try a more complicated neural network, such as LSTM; (3) Tagged data sets can be used to combine the traditional grammar knowledge with the neural network training, which means that we can combine the traditional language model based on rules with the neural network language model.

Acknowledgements. This work is supported by the Beijing Natural Science Foundation under Grant No. 4162067.

References

1. Jun, W.: The Beauty of Mathematics. Posts and Telecommunications Press, Beijing (2012)
2. Brown, P.F., Desouza, P.V., Mercer, R.L., Pietra, V.J.D., Lai, J.C.: Class-based n-gram models of natural language. Comput. Linguist. **18**(4), 467–479 (1992)
3. Bengio, Y., Ducharme, R., Vincent, P., Jauvin, C.: A neural probabilistic language model. J. Mach. Learn. Res. **3**(2), 1137–1155 (2003)
4. Mikolov, T., Karafiát, M., Burget, L., Černocký, J., Khudanpur, S.: Recurrent neural network based language model. In: Proceedings of the Eleventh Annual Conference of the International Speech Communication Association, pp. 1045–1048. International Speech Communication Association, Makuhari, Chiba (2010)
5. Bird, S., Loper, E.: NLTK: the natural language toolkit. In: Proceedings of the ACL 2004 on Interactive Poster and Demonstration Sessions, p. 31. Association for Computational Linguistics (2004)
6. Hinton, G.E.: Learning distributed representations of concepts. In: Proceedings of the Eighth Annual Conference of the Cognitive Science Society, vol. 1, p. 12. Oxford University Press, Oxford, UK (1986)
7. Mikolov, T., Chen, K., Corrado, G., Dean, J.: Efficient estimation of word representations in vector space. ArXiv Preprint arXiv:1301.3781 (2013)
8. Theano 0.9.0 documentation. http://deeplearning.net/software/theano/. Accessed 10 Dec 2017
9. Jian, Z.: Research on recurrent neural network language model in continuous speech recognition. The PLA Information Engineering University, Zhengzhou, pp. 19–20 (2014)
10. Recurrent neural networks tutorial. http://www.wildml.com/2015/09/recurrent-neural-networks-tutorial-part-2-implementing-a-language-model-rnn-with-python-numpy-and-theano/. Accessed 12 Jan 2018
11. Marcus, M.P., Marcinkiewicz, M.A., Santorini, B.: Building a large annotated corpus of English: the Penn Treebank. Comput. Linguist. **19**(2), 313–330 (1993)
12. Stolcke, A.: SRILM-an extensible language modeling toolkit. In: Proceedings of the Seventh International Conference on Spoken Language Processing, pp. 901–904 (2002)

Inferring Social Ties from Multi-view Spatiotemporal Co-occurrence

Caixu Xu[✉] and Ruirui Bai

Soochow University, Suzhou, China
csxucaixu@gmail.com

Abstract. Recently, social ties inferring in spatiotemporal data has attracted widespread attentions. Previous studies, which focused on either co-occurrence or context, do not fully exploit the information of spatiotemporal data. In order to better use the spatiotemporal information, in this paper we introduce two novel co-occurrence feature, namely, topic co-occurrence feature and context co-occurrence feature. The former feature is extracted by the topic model on carefully constructed bag-of-words. The latter feature is extracted by natural language processing tools on carefully constructed context sequence, which considers context, co-occurrence and mobility periodicity simultaneously. These two novel co-occurrence feature are both based on time and space perspectives. Then we infer social ties from these multi-view co-occurrence feature (including baseline co-occurrence, topic and context co-occurrence). The experiments demonstrate that the two novel co-occurrence feature contribute to the social tie inferring significantly.

Keywords: Social ties · Spatiotemporal co-occurrence · Topic co-occurrence Context co-occurrence

1 Introduction

In recent years, spatiotemporal data has attracted interest from more and more people. Spatiotemporal data usually include time and space information, where time dimension information is represented by check-in time and space dimension represented by check-in location. Specifically, because of the fashionable usage of mobile devices, users can easily share spatiotemporal information with their friends. This phenomenon inspires companies to use spatiotemporal data to mine user behavior patterns and offer customized services to them. Therefore, it is meaningful to mine the social tie between people hidden in this spatiotemporal data. These social tie offer an opportunity to understand users' requirements, such as friend recommendations or targeted advertisements for Internet companies [1].

Intuitively, users with higher social tie would have a greater chance to appear together at the same location, such as colleagues meeting in workdays or friends spending time together at a coffee shop. The methods inferring relationship have been widely studied [2–6] through co-occurrence feature and current context feature. Different from these works, we infer social tie from *Multi-View Co-occurrence* (*MVC*). As shown in Fig. 1, we apply the strong explanatory co-occurrence feature as baseline. At

© Springer Nature Switzerland AG 2018
L. H. U and H. Xie (Eds.): APWeb-WAIM 2018, LNCS 11268, pp. 378–392, 2018.
https://doi.org/10.1007/978-3-030-01298-4_31

the same time, we introduce one novel feature named topic co-occurrence feature. Additionally, we further combine context and co-occurrence information as context co-occurrence feature. The two novel features can both capture people's periodic mobility.

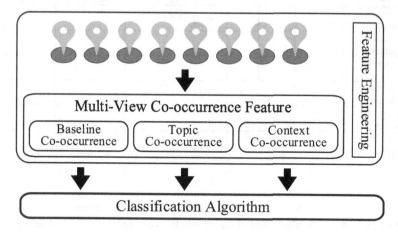

Fig. 1. The overview of Multi-view Co-occurrence

In summary, the main contributions to this paper are as follows:

(1) We carefully build spatiotemporal bag-of-words from both temporal and spatial aspect, and the user is regarded as word. Then we use Latent Dirichlet Allocation algorithm to extract topic feature representing user co-occurrence.

(2) We carefully construct context sequence from two different aspects. The context sequence includes co-occurrence, context and time periodicity information simultaneously. Then we present a novel method to extract context co-occurrence feature based on the context sequence. Our method carefully transfers the user-pair relationship in spatiotemporal data to word-pair relationship in sequence.

(3) The two novel co-occurrence feature contribute to the social ties inferring significantly. In the subset of Brightkite, the topic feature leads 9.1% improvement than the baseline in AUC indicator, context co-occurrence feature leads 9.3% improvement.

The remainders of this paper are organized as follows. Section 2 discusses related work. Section 3 describes our methodology in detail. Section 4 reports our experiments. Finally, we make conclusions in Sect. 5.

2 Related Work

We categorize the related works into three groups based on their focus: trajectory based methods [7–9], context based method [6] and co-occurrence based methods [2–5]. We compare our method with prior works in Table 1.

Table 1. Comparison between MVC and prior works

Characteristics	RWCFR [6]	EBM [3]	TAI [4]	SCI [5]	Multi-View Co-occurrence		
					Baseline Co-occurrence	Topic Co-occurrence	Context Co-occurrence
Location Diversity		√		√	√		
location context	√						√
temporal context							√
location co-occurrence		√	√	√	√	√	√
temporal co-occurrence		√	√	√	√	√	√
mobility periodicity				√	√	√	√

Trajectory based methods relaxed the concept of co-occurrence and use similarity in trajectory to measure likelihood of friendship between two people [4]. Chen et al. [7] applied frequent sequential pattern mining technology to extract the sequence of places that a user frequently visits, then use them to model his mobility profile combined with semantics of spatiotemporal information. [8, 9] focused on measuring user similarity using trajectory patterns, and [8] provide a tool named MinUS which integrates the technologies of trajectory pattern mining on discovering user similarity.

In context based method, the context includes social context, personal preferences context, location context and temporal context. Bagci et al. [6] proposed a random walk based context-aware friend recommendation algorithm (RWCFR). Depending on the location-based social network, they build a graph according to the current context (i.e. social relations, personal preference and location) for user. The method demonstrate that the context can describe the users' social tie. However, spatiotemporal data usually includes co-occurrence information, making full the use of spatiotemporal co-occurrence can further enhance the prediction accuracy.

Co-occurrence based methods had been shown to improve accuracy of social relationship estimation than trajectory based methods because of the co-occurrence feature [4]. Grandall et al. [2] demonstrated that the co-occurrence feature contributes to inferring social ties based on the experiments with a dataset of 38 million geo-tagged photos from Flickr. They also had shown that the probability of a social tie increases as the number of co-occurrence times increases and the temporal range decreases. Pham et al. [3] proposed an entropy-based model (EBM) that estimates the strength of social connections by analyzing people's co-occurrences in space and time through diversity and weighted frequency. Zhou et al. [4] proposed a Theme-Aware social strength Inference (TAI) approach that mines theme (also called the unit for co-occurrence) from co-occurrence behaviors, and then leverages the theme to measure the social strength of two persons. Njoo et al. [5] proposed a unified framework called SCI framework (Social Connection Inference framework). The SCI framework quantified three key co-occurrence features (i.e. diversity, stability and duration), and then aggregate co-occurrence features using machine learning algorithms to predict the social ties.

In summary, [2–5] illustrated the importance of co-occurrence feature which is also considered in our proposed two novel co-occurrence features. [6] shows that location context contribute to social tie prediction, therefore the context co-occurrence introduce location context. The check-in time sequence can reflect social tie between users to some context, and the characteristic is included in the context co-occurrence feature which is not involved in [6]. Different from [3–6], the context co-occurrence has novelty, which is not a traditional fusion. Peoples' mobility periodicity is also an import characteristic [10], and our proposed two novel features both take it into account. Generally, the characteristic of each view in MVC is shown in Table 1.

3   Methodology

In this section, we first describe how to generate baseline co-occurrence feature from co-occurrence times and location diversity. Moreover, we present the method to generate topic co-occurrence feature from location and time aspect. Finally, we describe how to generate context co-occurrence feature based on two carefully constructed context sequence.

3.1   Baseline Co-occurrence Feature

Times Co-occurrence Feature. The number of co-occurrence is powerful signal to infer social tie, which lead us to choose it as one of baseline feature. Intuitively, the more the times of co-occurrence between two users, the stronger the strength of social tie. More formally, co-occurrence set $\psi_{x,y}^z \in \psi_{x,y}$ quantifies the meeting frequency between users u_x and u_y in the location l_z during time threshold Δt. The parameter Δt can be set to different granularity (1 h, 2 h or 24 h). The $\psi_{x,y} = \{\psi_{x,y}^{z_1}, \psi_{x,y}^{z_2}, \ldots, \psi_{x,y}^{z_m}\}$ is the meeting frequency set for all meeting locations between users u_x and u_y. The $|\psi_{x,y}|$ is co-occurrence times, which is the number of two users appearing together.

Diversity Co-occurrence Feature. We also consider location diversity as the baseline feature, which is considered by [3, 5]. Variation in the meeting places between users is useful for reducing the possibilities of coincidences. For example, the co-occurrence number of user u_1 and u_2 is equal that of user u_1 and u_3, however u_1 meets u_2 several times in the same location, u_1 meeting u_3 a few times in several locations. The meeting occasions in the former are more likely to happen by chance than those in the latter. The reason is that the possibility of meeting in more diversified locations is lower than the possibly of meeting in the same location. Therefore, the location diversity feature is determined by Eq. 1.

$$Diversity(u_x, u_y) = -\sum \psi_{x,y}^z \log(\psi_{x,y}^z) \tag{1}$$

3.2 Topic Co-occurrence Feature

The topic feature was used in the paper [11, 12] in other domain, and we transfer it to apply in spatiotemporal data domain. The topic feature is mainly from two aspects.

Location Topic Co-occurrence Feature. There are lots of check-in location information in spatiotemporal data. In a certain location (e.g., a specific longitude and latitude), all users in the same location can be represented as a document and each user as a word. After removing less frequent location, we form vocabulary words from location-based spatiotemporal data. In order to mine co-occurrence feature between two users, we choose the Latent Dirichlet Allocation algorithm [13] to mine topic co-occurrence feature. We use a sparse matrix $x_{W \times M}$ to represent the bag-of-word representation of all locations, where there are $1 \leq m \leq M$ locations and $1 \leq w \leq W$ user. LDA allocates a set of thematic topic labels, $z = \{z_{w,m}^k\}$, to explain non-zero elements in the location-user co-occurrence matrix $x_{W \times M} = \{x_{w,m}\}$, where $1 \leq w \leq W$ denotes the word index in the vocabulary, $1 \leq m \leq M$ denotes the document index, and $1 \leq k \leq K$ denotes the topic index. Usually, the number of topics K is provided by us. The nonzero element $x_{w,m} \neq 0$ denotes the number of user check-in mth location. The objective of LDA inference algorithms is to infer posterior probability from the full joint probability $p(x, z, \theta, \phi)$, where z is the topic labeling configuration, $\theta_{K \times M}$ and $\phi_{K \times W}$ are two non-negative matrices of multinomial parameters for document-topic and topic-word distributions, satisfying $\sum_k \theta_m(k) = 1$ and $\sum_w \phi_w(k) = 1$. Both multinomial matrices are generated by two Dirichlet distributions with hyperparameters α and β. For simplicity, we consider the smoothed LDA with fixed symmetric hyperparameters. We use a coordinate descent (CD) method called belief propagation (BP) [14] to maximize the posterior probability of LDA,

$$p(\theta, \phi | x, \alpha, \beta) = \frac{p(x, \theta, \phi | \alpha, \beta)}{p(x | \alpha, \beta)} \propto p(x, \theta, \phi | \alpha, \beta). \qquad (2)$$

The output of LDA contains two matrices $\{\theta, \phi\}$. The $\phi_{K \times M}$ can is the location topic co-occurrence feature of each user, which is useful for us.

Time Topic Co-occurrence Feature. The process of time topic co-occurrence feature is similar to that of location topic co-occurrence feature. In this situation, we see all user that check in the same day as a document, and see each user as a word. Then we use LDA to generate two matrices $\{\theta, \phi\}$. The $\phi_{K \times M}$ can is time topic co-occurrence feature of each user. The time granularity is set as day because people check-in usually present the characteristic of periodicity [10]. For example, the middle class check-in every morning and night in the company.

3.3 Context Co-occurrence Feature

In this section, we first give some important notations definition. Then, we carefully construct context sequence from two aspects (location-time and time-location) to represent spatiotemporal co-occurrence and context information. Specifically, we propose a new method to extract context co-occurrence feature based on two context sequence respectively.

Notation Definition. In Table 2 we list the notations of parameters that we use. We denote $u \in U$ as the user, and $c \in C$ as the check-in data. Each c reflects the appearance of a user u at a specific location l at a specific time t with the form of $\{u, t, l\}$. Each user has many check-ins c information. The $C_t = \{c_1, c_2, ..., c_n\}$ represent all check-ins during the same time period. $C_{t\&l}$ is a sequence of elements in C_t that ranked according location shortest distance principle. $C_t^{Sequence} = \{C_t^1, C_t^2, ..., C_t^N\}$ is a sequence that elements ranked according time order. $C_l = \{c_1, c_2, ..., c_m\}$ represents all check-ins in the same location. $C_{l\&t}$ is a sequence of elements in C_l that ranked according time order. $C_l^{Sequence} = \{C_l^1, C_l^2, ..., C_l^M\}$ is a sequence that elements rank according location shortest distance principle. The θ is a context sequence which only consists of user id. The parameters relationship is that $\sum_{i=1}^{M} |C_l^i| = \sum_{j=1}^{N} |C_t^j| = |C| = |\theta|$ and the θ usually includes many repetitive user id.

Table 2. Notation of parameters

Variable	Notation
$u \in U$	u is a user id; U is all different user id set $\{u_1, u_2, u_3, ...\}$
c	$\{u, t, l\}$, a user check in at specific location l at specific time t
C_t	$\{c_1, c_2, ..., c_n\}$, all c at the same time period
$C_{t\&l}$	Elements in C_t rank according shortest distance principle
$C_t^{Sequence}$	$\{C_t^1, C_t^2, ..., C_t^N\}$, the elements C_t rank according time order
C_l	$\{c_1, c_2, ..., c_m\}$, all c at the same location l
$C_{l\&t}$	Elements in C_l rank according time order
$C_l^{Sequence}$	$\{C_l^1, C_l^2, ..., C_l^M\}$, the elements C_l rank according shortest distance principle
θ	A sequence that capture spatiotemporal context co-occurrence information

Location-Time Context Co-occurrence Feature. We first generate context co-occurrence sequence, and the generation process of location-time context sequence is given in Algorithm 1. *SortLocationByDistance* function produce sequence $C_l^{Sequence}$. The elements in $C_l^{Sequence}$ are ranked according distance shortest principle: if there is no location before, the first location is chosen randomly; otherwise, the location closest to the former location is assigned as the current location; the location closest to the $(M-1)th$ location is assigned as the Mth location; and so on. *SortTime* function uses quick sort algorithm to rank according time order because time is one-dimensional information. The returned value of the Algorithm 1 is the location-time context sequence θ consisted of user id, shown in Fig. 2 (B). Note that the same ellipse color represents the same location in Fig. 2 (B) and this context sequence capture strong location co-occurrence, meanwhile including shortest location context and time context.

The context co-occurrence feature is not simple fusion between context and co-occurrence, different from traditional approaches. We artfully use the toolkit word2vec to extract context co-occurrence feature through context sequence. The tool takes as its

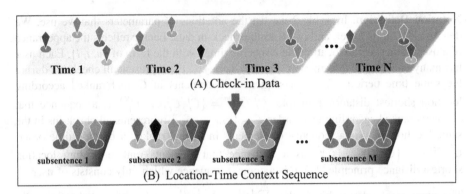

Fig. 2. The rhombuses with different colors denote different users, the ellipses with different colors denote different locations, and the time change is denoted by the shade of the background. (A) shows that at a fixed time period, different users check-in at different locations. (B) shows location-time context sequence. Note that the color of rhombuses, the color of ellipses and the change of background shade.

input a large corpus (corpus also can be seen as a sequence consist of words) and produces a dimensional space, with each unique word being assigned a vector in the space [15]. These word vectors are positioned in the vector space that words that share common contexts in the corpus are located in close proximity to one another in the space [15, 16]. The word vectors is context co-occurrence feature that we need. The feature includes spatiotemporal co-occurrence and context information.

More formally, given a context sequence $\theta = \{u_1, u_2, u_3, u_4, \ldots, u_T\}$ with spatiotemporal semantics representation, our objective is to maximize the average log probability

$$\frac{1}{T} \sum_{t=1}^{T} \left[\sum_{j=-k}^{k} log p(u_{t+j} | u_t) \right], \tag{3}$$

where T is number of elements in θ and k is the size of the window. The inner summation goes from $-k$ to k to compute the log probability of correctly predicting the user u_{t+j} given the user in the middle u_t. The outer summation goes over all users in the context sequence. The values of the two ends of the window are filled by the boundary value. Every user u is associated with two learnable parameter vectors, w_u and v_u. They are the "input" and "output" vectors of u respectively which can be learned [16]. The probability of predicting the user u_i given the user u_j is defined as

$$p(u_i | u_j) = \frac{\exp(w_{u_i}^T v_{u_j})}{\sum_{l=1}^{U} \exp(w_l^T v_{u_j})}, \tag{4}$$

where U is different users in the context sequence θ. The optimization approach is using stochastic gradient descent and the gradient is computed using backpropagation rule [16]. Each user's context semantic feature v_u (also called word vector in the

Natural Language Processing domain) can be learned. The word vector v_u captures word context and word co-occurrence information, which is extremely useful for us.

The toolkit word2vec is usually used to find synonyms in document (a sequence consist of words), and we innovatively apply it to finding user-pair relationship. Context feature captures sequence representation that context sequence has. Therefore, context co-occurrence feature also includes spatiotemporal co-occurrence and context. The feature with co-occurrence and context is different from above literatures [2–6], and not simply merges context and co-occurrence together.

Day-Location Context Co-occurrence Feature. The generation process of time-location context sequence is given in Algorithm 2. *SortTimeByGranularity* function can produce $C_t^{Sequence}$ quickly. The time parameter τ can be accurate to different value. The elements in C_t rank according distance shortest principle. The returned value of the Algorithm 2 is the time-location context sequence θ, shown in Fig. 3 (B). Note the distance between different ellipses in Fig. 3 (B). The time-location context sequence capture time co-occurrence, meanwhile including time order context and shortest location context.

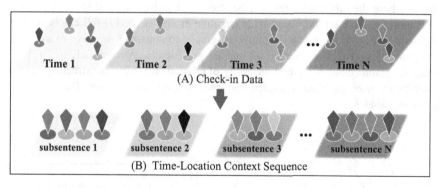

Fig. 3. (A) shows that at a fixed time period, different users check-in at different locations. (B) shows time-location context sequence. Note that the color of rhombuses, the color of ellipses and the change of background shade.

The time parameter τ is set to day. As shown in Fig. 3 (A), every chunk is one day i.e. time 1 is the first day; time 2 is the second day and so on. The elements in the time chunk C_t rank according distance shortest principle. Day-location is special time-location context sequence, because people's periodic movement is based on day [10] such as people commute on working day, people check in home at night. The periodicity of people's movement is a unit of day. Therefore, this context sequence captures people's mobility periodicity, including day co-occurrence, time order context and shortest location context. We do not adopt other parameter τ because the check-in data in location-based social networks is very sparse in time. Of course, if the strong time co-occurrence is needed, and the smaller parameter τ can be assigned.

After achieving the time-location context sequence $\theta = \{u_1, u_2, u_3, u_4, \ldots, u_T\}$ through above process, we use word2vec toolkit to extract context co-occurrence feature. The generation of feature is similar to that of the location-time context feature.

4 Experiments

In this section, we first describe datasets. Second, we describe how to generate three types of co-occurrence feature in detail. Moreover, we describe classifier algorithm learning. Finally, we evaluate our performance.

4.1 Datasets

Our experiments are based on a subset of the real dataset, Brightkite and Gowalla [17]. The original dataset is a global dataset, and the people in the global dataset have a heterogeneous nature. For the isomorphism of the data, we choose the data in the eastern US. The check-in data is handled to a triplet $< u, t, l >$, where the l is represented by longitude and latitude. The user-pair data is handled to a triplet $< u_1, u_2, label >$ where the *label* indicates whether two users exist relationship.

The original dataset did not provide negative examples [5, 17] (all labels are true). As shown in Fig. 4, we use non-connected graphs to construct negative examples. In the figure, u1 and u2 are friends, u2 and u3 are friends. We see {u1, u2, u3} as a connected graph considering the transitivity of relationship. After constructing negative examples, the user-pair data is divided into training and testing dataset. The overview is shown in Table 3.

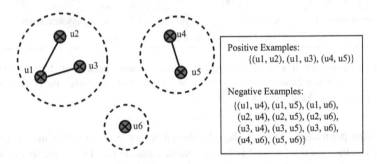

Positive Examples:
    {(u1, u2), (u1, u3), (u4, u5)}

Negative Examples:
    {(u1, u4), (u1, u5), (u1, u6),
    (u2, u4), (u2, u5), (u2, u6),
    (u3, u4), (u3, u5), (u3, u6),
    (u4, u6), (u5, u6)}

Fig. 4. Negative sample construction

4.2 Multi-view Co-occurrence Feature Generation

Baseline Co-occurrence Feature. There are mainly two types of features, which are times co-occurrence and diversity co-occurrence. In our experiment, the parameter Δt of the feature is set to 1 h (i.e. 3600 s), 2 h (i.e. 7200 s) and 24 h (i.e. one day) respectively. According to these three types of granularity, three types of times co-occurrence sets in different location are generated respectively. Then, we can achieve

Table 3. Experiment datasets

Datasets	Brightkite	Gowalla
Checkins	837,161	732,205
Nodes (Users ID)	5,966	10,585
Train Data (user-pair)	291,543	87,567
Test Data (user-pair)	26,674	5,155

times co-occurrence and diversity co-occurrence based on specific granularity. The combination of the two type of features serves as input to the classifier (i.e. baseline-3600, baseline-7200, baseline-day). Based on the combination of the above three granularities, we call it baseline-merge feature that is used as the object of comparison with the other two novel features.

Topic Co-occurrence Feature. There are mainly two types of features, which are location and time topic co-occurrence feature. For both two type of features, we assigned $K = 100$ dimensional topic features for each location and each time period respectively. The time period is set as day (i.e. 24 h) to the time topic feature. We call these two types of features Topic-Day and Topic-Location respectively.

Context Co-occurrence Feature. Location-time sequence with spatiotemporal co-occurrence and context information can be generated through Algorithm 1. To achieve the time-location sequences through Algorithm 2, the parameter τ is assigned to day (i.e. 24 h). Then, two types of context sequences θ with co-occurrence and context information can be achieved. The word2vec provides an implementation of the skip-gram architecture which is in accord with our objective function [16], so we choose the skip-gram architecture. The context co-occurrence feature size is set as 200 and the window of max skip length between users is set as 10 (the parameter k in formula (3) is 10). The learning rate is set as 0.01 and other parameters are default. After the toolkit learning from context sequences, the context co-occurrence features with spatiotemporal information (the vector parameter v_u called word vector in NLP domain) can be learned. Each user is mapped to two types of 200 dimensional context co-occurrence features. The two context co-occurrence features represent co-occurrence and context information in spatiotemporal data. In our experiment, we call these two types of feature context-location-time and context-day-location.

4.3   Classification Algorithm

The multiple classifiers can be trained through three different views with different degree of co-occurrence information. In our experiment, we choose the XGBoost classifier to make prediction. It is a supervised learning method that uses a tree boosting technique. For a given datasets with n examples and m features $D = \{(x_i, y_i)\}(|D| = n, x_i \in R^m, y_i \in [0, 1])$, a tree ensemble model uses K additive functions to predict the output, as follows:

$$\bar{y}_i = \sum_{k=1}^{K} f_k(x_i), f_k \in F, \tag{5}$$

where F is the space of the regression trees [18]. The output $\bar{y}_i$ is the relative probability of a user-pair relationship strength.

The user is represented with the co-occurrence feature and then the user-pair features combination as the input. Considering relationship is bidirectional in each view feature, a relatively larger vector position value is placed ahead of the corresponding position value. There are some primary parameters: the booster parameter is set as gbtree; the max depth is 3 which avoid overfitting; the boosting learning rate is 0.1; the objective function is binary logistic; the early stopping is 10; and other parameters are set as default. Multiple classifiers are used to predict social ties in the test data, and multiple types of result can be achieved. Generally, the higher the prediction result, the higher strength the two users' social ties.

4.4 Performance Evaluation

The classifier will output a list of top U user-pair that have the higher social tie. We use recall and prediction metrics on top U user-pair to evaluate the prediction results. Generally, increasing U will increase recall but decrease precision. Fix a certain U, the higher recall and precision correspond to the better prediction performance. The definition of recall@U is

$$R@U = \frac{\text{The number of true user-pair in top U}}{\text{The total number of true user-pair}}. \tag{6}$$

Similarly, the definition of precision@U is

$$P@U = \frac{\text{The number of true user-pair in top U}}{U} \tag{7}$$

We also use the area under the ROC curve (AUC) [19] evaluated on the test data, which is the standard scientific accuracy indicator. Generally, we use AUC, R@U and P@U to evaluate the overall predictive performance.

Baseline Co-occurrence Feature. We compare different granularities on baseline co-occurrence in Tables 4 and 5. Coarse-grained time co-occurrence can achieve better result because of the sparseness of the dataset in the time dimension. The baseline-day achieve the best result compared to the baseline-3600 and baseline-7200, which largely depend on user movement periodicity. For example, people commute on working day, and people check in home at night [10]. Meanwhile, the baseline-day is also coarse-grained time granularity. The baseline-day contribute to the baseline-merge significantly. In all baseline features, the baseline-merge achieve the best precision and recall in top 500, because baseline-3600 and baseline-7200 contribute to short time co-occurrence and baseline-day contribute to periodic co-occurrence, which satisfy complementary principle. However, the baseline-merge does not increase in U at

13000, mainly because the sparsity of the dataset in the time dimension makes fine-grained time co-occurrence without any effect and only the baseline-day is in effect.

Table 4. Performance on Brightkite subset

	AUC	Precision@U 1000 (Top 3.8%)	Recall@U 1000 (Top 3.8%)	Precision@U 13000 (Top 50%)	Recall@U 13000 (Top 50%)
Baseline-3600	0.552	0.789	0.131	0.526	0.540
Baseline-7200	0.566	0.828	0.164	0.533	0.554
Baseline-day	0.670	0.907	0.300	0.583	0.676
Baseline-merge	0.671	0.913	0.317	0.583	0.676
Topic-day	0.762	0.849	0.189	**0.633**	**0.825**
Topic-location	0.631	0.826	0.163	0.57	0.641
Context-day-location	**0.764**	**0.925**	**0.359**	0.622	0.789
Context-location-time	0.653	0.859	0.202	0.581	0.668

Topic Co-occurrence Feature. In the subset of Brightkite dataset, the topic-day features are better than topic-location, because the sparseness of the location is stronger than that of the time, and more difficult to infer the social ties. However, the conclusion is exactly the opposite in Gowalla because sparseness of the time is slightly stronger than that of the location. The sparseness of the time in Gowalla subset leads that topic co-occurrence feature is not good as the baseline-merge. The baseline-merge consider time co-occurrence from different granularities, which is more overall than the topic-day in the Gowalla subset. In the Brightkite subset, the topic-day better portray coarse-grained co-occurrence and takes second place in the AUC indicator. As shown in Figs. 5(a) and 6(a), with the U increasing, the topic-day achieved the best performance on the precision and recall compared to other views.

Context Co-occurrence Feature. In two datasets, the context co-occurrence feature achieve the best performance on AUC because it capture both context and co-occurrence information, other views only capturing co-occurrence information. The context-day-location feature achieve the better performance on AUC than context-location-time in Brightkite subset, because of the sparseness of location dimension information. Due to the sparseness of time in Gowalla subset, the context-location-time feature achieve the better performance on AUC than context-day-location. The context co-occurrence feature usually works better than corresponding topic co-occurrence feature. As shown in Figs. 5 and 6, the solid line of the corresponding color is above the dotted line in most of the time, because the context co-occurrence feature captures both co-occurrence and context information compared to the topic co-occurrence feature. From Figs. 5 and 6, we can also conclude that the context co-occurrence feature overall exceeds the baseline.

In summary, the two novel feature we proposed for extracting co-occurrence have their own advantages over the baseline. We emphasize the context co-occurrence feature because it captures the spatiotemporal context, co-occurrence and periodic mobility simultaneously. In general, it is better than the baseline and topic feature in AUC on the current two dataset subset.

Table 5. Performance on Gowalla subset

	AUC	Precision@U 500 (Top 10%)	Recall@U 500 (Top 10%)	Precision@U 2500 (Top 50%)	Recall@U 2500 (Top 50%)
Baseline-3600	0.562	0.719	0.198	0.528	0.532
Baseline-7200	0.573	0.748	0.218	0.535	0.545
Baseline-day	0.736	0.939	0.445	0.621	0.719
Baseline-Merge	0.738	**0.944**	**0.454**	0.621	0.719
Topic-day	0.705	0.811	0.27	0.616	0.709
Topic-location	0.727	0.858	0.321	0.628	0.737
Context-day-location	0.761	0.827	0.287	0.649	0.788
Context-location-time	**0.782**	0.879	0.348	**0.656**	**0.805**

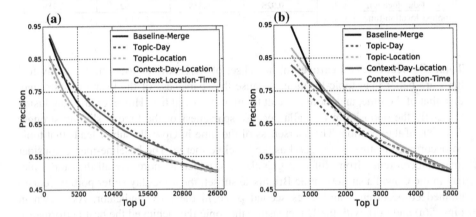

Fig. 5. (a) Precision on Brightkite subset. (b) Precision on Gowalla subset

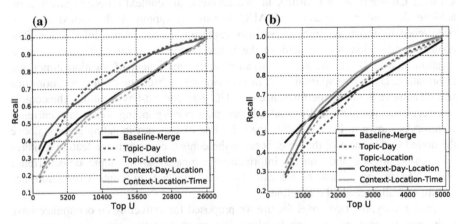

Fig. 6. (a) Recall on Brightkite subset. (b) Recall on Gowalla subset

5 Conclusion

In this paper, we infer social ties from multiple spatiotemporal co-occurrence, where the topic and the context co-occurrence features are presented. The two proposed co-occurrence feature are both from space and time aspects. The latter represents spatiotemporal context, co-occurrence and peoples' periodicity mobility simultaneously. The experiment results demonstrate that our two novel feature contribute to social ties inferring significantly.

References

1. Machanavajjhala, A., Korolova, A., Sarma, A.: Personalized social recommendations: accurate or private. Proc. VLDB Endow. **4**, 440–450 (2011)
2. Crandall, D., Backstrom, L., Cosley, D., Suri, S., Huttenlocher, D., Kleinberg, J.: Inferring social ties from geographic coincidences. In: Proceedings of the National Academy of Sciences of America, pp. 22436–22441 (2010)
3. Pham, H., Shahabi, C., Liu, Y.: EBM: An entropy-based model to infer social strength from spatiotemporal data. In: Proceedings of the 2013 ACM SIGMOD International Conference on Management of Data, pp. 265–276 (2013)
4. Zhou, N., Zhang, X., Wang, S.: Theme-aware social strength inference from spatiotemporal data. In: Li, F., Li, G., Hwang, S.-w, Yao, B., Zhang, Z. (eds.) WAIM 2014. LNCS, vol. 8485, pp. 498–509. Springer, Cham (2014). https://doi.org/10.1007/978-3-319-08010-9_56
5. Njoo, G.S., Kao, M.-C., Hsu, K.-W., Peng, W.-C.: Exploring check-in data to infer social ties in location based social networks. In: Kim, J., Shim, K., Cao, L., Lee, J.-G., Lin, X., Moon, Y.-S. (eds.) PAKDD 2017. LNCS (LNAI), vol. 10234, pp. 460–471. Springer, Cham (2017). https://doi.org/10.1007/978-3-319-57454-7_36
6. Bagci, H., Karagoz, P.: Context-aware friend recomendation for location based social networks using random walk. In: Proceedings of the 25th International Conference Companion on World Wide Web, pp. 531–536 (2016)
7. Chen, X., Pang J. and Xue R.: Constructing and comparing user mobility profiles for location-based services. In: Proceedings of the 28th Annual ACM Symposium on Applied Computing, pp. 261–266 (2013)
8. Chen, X., Kordy, P., Lu, R., Pang, J.: MinUS: mining user similarity with trajectory patterns. In: Calders, T., Esposito, F., Hüllermeier, E., Meo, R. (eds.) ECML PKDD 2014. LNCS (LNAI), vol. 8726, pp. 436–439. Springer, Heidelberg (2014). https://doi.org/10.1007/978-3-662-44845-8_29
9. Chen, X., Lu, R., Ma, X., Pang, J.: Measuring user similarity with trajectory patterns: principles and new metrics. In: Chen, L., Jia, Y., Sellis, T., Liu, G. (eds.) APWeb 2014. LNCS, vol. 8709, pp. 437–448. Springer, Cham (2014). https://doi.org/10.1007/978-3-319-11116-2_38
10. Cho, E., Myers, S.A., Leskovec, J.: Friendship and mobility: user movement in location-based social networks. In: Proceedings of the 17th ACM SIGKDD international conference on Knowledge discovery and data mining, pp. 1082–1090 (2011)
11. Huang, Y., Zhu, F., Yuan, M., et al.: Telco churn prediction with big data [C]. In: Proceedings of the 2015 ACM SIGMOD International Conference on Management of Data, pp. 607–618. ACM (2015)

12. Liu, G., Nguyen, T.T., Zhao, G., et al.: Repeat buyer prediction for e-commerce [C]. In: Proceedings of the 22nd ACM SIGKDD International Conference on Knowledge Discovery and Data Mining, pp. 155–164. ACM (2016)

13. Blei, D.M., Ng, A.Y., Jordan, M.I.: Latent dirichlet allocation [J]. J. Mach. Learn. Res., pp. 993–1022 (2003)

14. Zeng, J., Cheung, W.K., Liu, J.: Learning topic models by belief propagation[J]. IEEE Trans. Pattern Anal. Mach. Intell., pp. 121–1134 (2013)

15. Mikolov, T., Chen, K., Corrado, G., Dean, J.: Efficient Estimation of Word Representations in Vector Space (2013)

16. Mikolov, T., Le, Q.V., Sutskever, I.: Exploiting Similarities among Languages for Machine Translation (2013)

17. Leskovec, J., Krevl, A.: SNAP Datasets: Stanford Large Network Dataset Collection (2014)

18. Chen, T., Guestrin, C.: XGBoost: A Scalable Tree Boosting System (2016)

19. Bradley, A.: The use of the area under the ROC curve in the evaluation of machine learning algorithms. In: Pattern Recognition. (1997)

Author Index

Printed in the United States
By Bookmasters